Michael Kuhn, Doris Weidemann (eds.)
Internationalization of the Social Sciences

MICHAEL KUHN, DORIS WEIDEMANN (EDS.)
Internationalization
of the Social Sciences
Asia – Latin America – Middle East – Africa – Eurasia

Bibliographic information published
by the Deutsche Nationalbibliothek
The Deutsche Nationalbibliothek lists this publication in the Deutsche Nationalbibliografie; detailed bibliographic data are available in the Internet at http://dnb.d-nb.de

© 2010 transcript Verlag, Bielefeld

All rights reserved. No part of this book may be reprinted or reproduced or utilized in any form or by any electronic, mechanical, or other means, now known or hereafter invented, including photocopying and recording, or in any information storage or retrieval system, without permission in writing from the publisher.

Cover layout by: Kordula Röckenhaus, Bielefeld
Proofread and Typeset by: Doris Weidemann
Printed by: Majuskel Medienproduktion GmbH, Wetzlar
ISBN 978-3-8376-1307-0

Distributed in North America by

Transaction Publishers
New Brunswick (U.S.A.) and London (U.K.)

Transaction Publishers   Tel.: (732) 445-2280
Rutgers University   Fax: (732) 445-3138
35 Berrue Circle   for orders (U.S. only):
Piscataway, NJ 08854   toll free 888-999-6778

Contents

Internationalization of the Social Sciences: Introduction   11
MICHAEL KUHN, DORIS WEIDEMANN

Asia

China's Historical Encounter with Western Sciences
and Humanities   21
HE HUANG

Internationalization of Japanese Social Sciences:
Importing and Exporting Social Science Knowledge   45
KAZUMI OKAMOTO

Internationalization of Social Science in South Korea:
The Current Status and Challenges   67
KWANG-YEONG SHIN, SANG-JIN HAN

Internationalizing Education and the Social Sciences:
Reflections on the Indian Context   87
PRADEEP CHAKKARATH

Indonesian Experiences: Research Policies and the
Internationalization of the Social Sciences   115
I KETUT ARDHANA, YEKTI MAUNATI

Latin America

The Current Internationalization of the Social Sciences
in Latin America: Old Wine in New Barrels?   135
HEBE VESSURI

The Americanization of Argentine and Latin American
Social Sciences — 159
TOMÁS VÁRNAGY

Rethinking International Cooperation in the Human
Sciences of Brazil — 175
RENATO JANINE RIBEIRO

MIDDLE EAST

Internationalization of the Humanities and Social Sciences:
Realities and Challenges in Jordan — 191
ABDEL HAKIM K. AL HUSBAN, MAHMOUD NA'AMNEH

Internationalization of Social Sciences: The Lebanese
Experience in Higher Education and Research — 213
JACQUES E. KABBANJI

AFRICA

The Internationalization of South African Social Science — 237
JOHANN MOUTON

EURASIA

Internationalization of Social Sciences and Humanities
in Turkey — 265
SENCER AYATA, AYKAN ERDEMIR

Internationalization of the Social Sciences and Humanities
in Russia — 285
IRINA SOSUNOVA, LARISSA TITARENKO, OLGA MAMONOVA

Social Sciences and Humanities in Ukrainian Society:
The Difficulty of Integration into International Structures — 307
IGOR YEGOROV

Challenges for Research and Research Policies in Belarus
as a Post-Communist »Developing Country« — 333
LARISSA TITARENKO

Challenges of International Collaboration in the
Social Sciences   353
DORIS WEIDEMANN

Facing a Scientific Multiversalism—Dynamics of
International Social Science Knowledge Accumulations
in the Era of Globalization   379
MICHAEL KUHN

Authors   411

Acknowledgments

Collecting and editing contributions by colleagues from sixteen different countries and of nearly as many scientific disciplines was a highly exciting endeavour. It included all the difficulties that different working and communication styles entail and confronted authors with various degrees of annoying editorial interference. This book is indeed the result of the patience, intercultural tolerance, and scientific professionalism of all contributors to this volume. It is the product of an international collaborative knowledge production process and of the manifold abilities international collaboration requires from everybody. In fact, the history of this joint venture would easily provide scientific material for another book. After all is now done we would like express our gratitude to all authors who so dedicatedly supported our joint project. Our thanks also go to Olivia Berger, Vasco da Silva, Yvonne Franke, Hanka Jattke, Anne Winkler, and Gina Zimmermann who helped with the preparation of the manuscript.

As English is a second or foreign language to nearly all contributors we would also like to express our admiration and gratefulness for the efforts authors took to make their thoughts accessible to an international (English-speaking) audience. Jack Rummel provided sensitive and highly professional language editing support. He meticulously checked and thoughtfully amended all our chapters. Needless to say, all remaining mistakes are our own responsibility.

Publishing a book is not for free. Since we discreetly used the money we received from the European Commission, we should mention this here. However, though the publication rules of the Directorate General for Research demand the formulation of »policy recommendations« we should admit here that this book does not contain any. Still, we can recommend European and other »policy makers and stakeholders«, as they like to be called in European jargon, to read this book. It contains many policy recommendations because there are none.

Michael Kuhn and *Doris Weidemann*

Internationalization of the Social Sciences: Introduction

MICHAEL KUHN, DORIS WEIDEMANN

Approaching Internationalization

In a world of economic, technical, and financial globalization any question about the internationalization of the social sciences might easily appear as outdated and superfluous. It is obvious that since their inception in nineteenth-century Europe, social sciences have spread internationally and are now firmly established in scientific institutions, university curricula, and disciplinary organizations around the world. Social scientists today are linked through global networks and associations, international journals, and travel. If we take internationalization to refer to scientific activities (research, publication, travel, communication, funding) that transcend national borders, there is no doubt that the internationalization of the social sciences is a firmly established reality. Still, internationalization refers not to a state or final stage of development but signifies an ongoing process. It is the nature of this process, its preconditions, deviations, and possible future that this book is interested in.

The contributions by scholars from different parts of the world that are collected in this volume demonstrate that this process neither rests on universally shared goals nor on uniform rules and point out the multiple paths, contradictions, and conflicts internationalization implies. They also raise questions concerning academic power structures and claims to universal knowledge that touch on the epistemological foundations of social sciences in general—questions that, we believe, are central to future debates on international collaboration in the social sciences.

Internationalization is usually considered a desirable development. It has therefore in recent years been heavily promoted in the social sciences. The European Union, for example, provides huge amounts of research funds for international social research. National governments and funding organizations likewise stress the »international dimension« as an indicator of high-quality research and encourage international cooperation. There are several reasons for the increased attention to international research:

- In an era of globalization knowledge about foreign societies and policies has gained importance, especially since the anticipated arrival of a »multipolar« world makes knowledge about different regions indispensible. Sometimes national institutions simply wish to extend their knowledge about neighboring countries or look for »best practice« examples to solve local problems. Social sciences contribute to an understanding of foreign social realities; at the same time they also provide the analytical categories—such as »modernization«, »globalization«, or »multipolarity«—that allow conceptualizing international change and its effects in scientific discourse and beyond.
- As social phenomena—for instance, the production and distribution of wealth, migration, or identity formation—increasingly develop a global dimension, research needs to »go international« to sufficiently understand them. In this respect, the internationalization of research is a by-product of the globalization of social realities.
- Political aims of internationalizing research are less often addressed as they interfere with the ideology of »disinterestedness« of science (Merton). Yet, as the case of the former Soviet Union shows, developing international research ties is also a means to strengthen political alliances—a motive that also underlies current research funding policies, especially the inception of international mobility programs.
- Internationalization is intrinsically linked to scientific goals: taking research findings and theories across national borders tests their robustness and validity. The investigation of cross-national variation helps to increase knowledge and identify moderating factors and limits of general theories. The ultimate aim is to arrive at universally valid knowledge and theories, and it is this goal that has implemented an international orientation of the sciences from their beginning.
- Finally, another motive arises at the interface of political and scientific goals of scientific excellence: internationalization makes a broader talent pool available, and the hope to attract the best scholars worldwide to join their expertise in a certain project promises to en-

hance the quality of scientific outcomes and to strengthen local competitiveness on global (scientific and economic) markets.

Despite these factors, the internationalization of the social sciences is lagging behind developments in the natural sciences, which is indicated by the fact that most research is carried out within the confines of national borders and much of its results are published in national languages (other than English). This is especially true for large countries, such as Germany, France, Russia, or Brazil, that show a considerable degree of self-sufficiency (Crawford/Shinn/Sörlin 1993: 4; Hakala 1998). In general, the close entanglement of social sciences and national states that Wallerstein et al. (1996) observe (and criticize) has counteracted internationalization trends. As a result, career paths, funding, and education in the social sciences are still predominantly national (Crawford et al. 1993).

Because internationalization is linked to scientific as well as to political concerns and has recently been propelled by the growing influence of non-Western science communities, the internationalization of the social sciences has received considerable research attention. Below, we briefly review some characteristic aspects of these debates before sketching the central aims of the present publication.

Structures of the Scientific Field

A rough typology of the research field finds two general strands of argumentation that in a first attempt may be labeled »descriptive« versus »normative«. While the goal to internationalize introduces an overarching normative component into the entire discussion, research on the internationalization of social sciences still largely relies on what we call a *descriptive approach*: internationalization is studied with an interest in the quantity of international activity while its aims and contents are not generally critically discussed. Research thus focuses on international ties of researchers and research communities; on global disciplinary structures, a worldwide system of journals, conferences, and associations; or on the number of established multinational research teams. Based on this approach, bibliometric studies investigate the frequency of international coauthorship, or contributions to international journals and statistics take stock of international travel, university partnerships, amount of funding for international research, and other transnational science-related activities. This approach—that is also characteristic of current research policies and university statistics—usually takes the existing mode of universality of social sciences for granted.

A second strand of argumentation becomes visible in contributions by authors who critically discuss the fact that the social sciences that were so successfully spread around the globe are institutionally, theoretically, and methodically closely oriented at the Euro-American model and thus not readily applicable to non-Western contexts. As it focuses on the question of how to arrive at »better« or »true« international collaboration, this approach may be labeled *normative*. It rests on insights of postcolonial studies, feminist and indigenization movements that have criticized social categories, theories, and methods as essentially drenched in European thought. Debates on orientalism and gender have revealed Eurocentric and male biases of assumptions and theories. Consequently, it was questioned whether the spread of Western-dominated social science provided a sound basis for international science, and calls for »true internationalization« that would include previously excluded perspectives were voiced. According to this view, true internationalization rests on equal participation of scientists of different provenience and focuses on qualitative aspects—a new orientation of the social sciences—rather than on the quantity of traditional international ties that would only strengthen the dominance of Euro-American social sciences. Because the claim to universal validity of findings and theories cannot be upheld in a world that systematically excludes the less privileged, the inclusion of multiple experiences and viewpoints becomes mandatory. Insofar as »internationalization« can be considered a prerequisite of valid scientific theories, it gains the status of an epistemological imperative.

Despite efforts to overcome Eurocentric biases and exclusion mechanisms, many of the inherited fault-lines are still in place: this not only holds true for female and minority representation in research, but also for international structures that preconfigure and guide international collaboration. Reviewing discussions on the internationalization of the social sciences, it seems to us that Eurocentric traditions are not absent from discussions of internationalization itself. Two observations may serve as indicators:
- Calls for internationalization by researchers from the United States and Europe draw on observations that »traditional« science has exhausted its potential to explain current social and transnational phenomena. They do not usually take up postcolonial criticism of ethnocentric parochialism but diagnose insufficient knowledge about the »rest of the world« or the need to understand global phenomena that

have started to affect the »First World« (such as migration or economic global dependency). This perspective is even more pronounced in research programs, such as the European Commission's Seventh Framework Program that invites project proposals on »Europe in the world«.
- Discussions on internationalization usually ignore the fact that in many places, social sciences have never been »national« in the same sense that they have been in Europe but have possessed a strong international orientation from their beginning.

It is apparent from this short overview that in discussing internationalization a broader perspective is called for that includes the viewpoints and experiences of researchers from different research communities and countries. If much of the discussion reflects the needs and experiences of »Western« scholars, researchers from »peripheral« science communities have much to offer for a better understanding of the many facets and challenges of internationalizing social research.

Applying a popular model of global academic power structures (Alatas 2003) to the issue of »internationalization« makes systematic differences visible.[1] It is therefore worthwhile to shortly dwell on Alatas's definitions of different players in the social sciences before returning to the issue of internationalization: S. F. Alatas differentiates between *social science powers* (United States, United Kingdom, France) that have a global reach of theories and ideas and *peripheral* (academically dependent, usually Third World) social science communities that »borrow« research agendas, theories, and methods from the social science powers. A third category is made up by *semiperipheral social science powers* (e.g., Australia, Japan, Germany, the Netherlands) that hold an intermediate position: while they are dependent on the social science powers they also exert considerable influence on the peripheral science communities.

Challenges of internationalization come in a different guise to each of these communities; staying with the roughly drawn picture of the three different types of research communities, we may envision the following viewpoints:

---

1 As all simple typologies Alatas's model blurs the many distinctions that exist between world regions, disciplines, and local science communities. It also does not grasp dynamic developments and interrelations between communities. We believe, however, that with respect to the issue of internationalization this model may serve an important heuristic function.

To the *social science powers*, internationalization is not generally a disturbing affair. Their reach is, by definition, global in nature, and they have the means and prestige to attract scholars globally. If they integrate international dimensions, aspects, and perspectives into their research agendas they do so because it seems academically or politically rewarding to do so. They are heavily engaged in international cooperation, but it is cooperation on their own terms, in their own languages, based on their own theories and agendas. According to their understanding there is no need to internationalize since they are international already, and if they feel a need to internationalize, international cooperation will be extended. Globalization is perceived as a chance to extend international dialogue and move closer to the aim of creating universal science for global humanity. Much of the above reviewed research positions relate to the outlook of social science powers.

Because academic social sciences came as a Western transplant, *peripheral science communities* have usually seen quite a strong international (center-to-periphery) engagement in their social sciences. Since they are, using Alatas's definition, dependent on means, theories, institutions, and scholars of the social science powers, their concern is not to become international but to find ways to adjust social science agendas to local (national) concerns. Because of dependency structures, international cooperation is usually automatically understood as cooperation with social science powers. Even where there are initiatives to develop regional international networks (as within Latin America or across Asia) they are—as the chapters of this volume point out—rarely referred to as internationalization.

The *semiperipheral science communities* (especially those on the European continent) find themselves in yet another position: Endowed with strong national social science communities, solid national science traditions and national scientific languages, they face a unique dilemma: if they want to exert international influence they have to adopt language, media, theories, and agendas of the social science powers. Yet, because of their strong national traditions and local infrastructures there is a strong resistance to this way of internationalization, which is held to imply a »loss of culture« and »loss of voice«.

Even if we do not wish to follow the above line of argumentation, the center-periphery-model underlines that internationalization may hold vastly different connotations in different social science communities that are not usually addressed in internationalization discourses. Yet, it also points to the need of a more dynamic interpretation of academic relationships in an era of globalization that witnesses the emergence of new economic and scientific powers as well as the marginalization of established communities. Some of the most significant changes were experienced by those science communities that have seen a severe degrading following the breakdown of the socialist East-bloc states. After a short-lived interest of the social science powers in research that was carried out behind the »iron curtain« (and that went along with a considerable amount of systematic exploitation), social science communities have slid toward peripheral positions and are still struggling with identity crises that (as the chapters on Ukraine, Belarus, and Russia illustrate) influence international engagement until today. At the same time, following economic development, formerly peripheral social science societies move toward the semiperiphery. Clearly, this situation holds potential for new alliances, agendas, and initiatives that have as yet escaped attention of social science communities that define themselves only with respect to the social science »center«. One of the results of this ongoing repositioning is that the position of social science powers is questioned by former (semi-) peripheral science communities. The scientific alliances of European science communities, for example, do not only change their position within the European Union but aim to challenge the United States as the leading social science power.

It is the aim of this book to capture perspectives from different (semi-) peripheral science communities that illustrate current positions and concerns and that hint at the multiple interrelations and dynamics that accompany the repositioning of actors in an era of globalization.

Background and Agenda of This Book

The conceptualization and history of this book are closely linked to two EU-funded projects that both addressed the issue of internationalization of the social sciences.[2] In most cases, earlier versions of the contributions to this book were presented and intensively discussed by an inter-

---

[2] In particular, the book is the outcome of work-package 3 of the »GLOBAL SSH« project (www.globalsocialscience.org) and work-package 7 of the »ESSHRA« project (http://esshra.tubitak.gov.tr) that were both funded under the Framework Programme 6 of the European Commission. The GLOBAL SSH project was coordinated by the Swedish Collegium for Advanced Study (SCAS), and the ESHHRA project by the national science council of Turkey (TUBITAK). Some chapters of this book incorporate data collected by questionnaire surveys and focus group interviews about the social science communities in Russia, China, Japan, and Turkey in collaboration with local cooperation partners. The reports about the work-packages of both projects can be found at www.knowwhy.net.

national group of scholars on three workshops that were held in Beijing in 2007 and in Ankara and Paris in 2008. The majority of chapters were thus subjected to multilateral perspectives and have benefited from the critique, comments and ideas of all workshop participants.

Because »internationalization« contains the idea of the »national«, it is sometimes considered a phenomenon of the past that has been superseded by an era of the truly »global«. By using the term, we wish to stress our belief that national boundaries are still a valid category for much of the current education, research, and funding in the social sciences. In fact, the incessant call for carrying the endeavor of internationalization further gives proof to the diagnosis that it has not been generally achieved. The fact that book chapters largely address national science communities mirrors the national context and constitution of social science communities. With respect to the topic of internationalization this reproduces a contradiction that permeates the (national) world of social sciences and that cannot easily be overcome in academic discussions.

Still, we endorse the idea of a global approach to our subject that allows for different scientific standpoints and approaches. We have tried to address the subject not in a linear manner that would expose the idea and agenda of internationalization from a single disciplinary or geographical standpoint, but to present and interlink different standpoints without imposing an overall analytical grid or theoretical point of orientation. In a first exercise, this global approach highlights the fact that internationalization means different things from different angles and local perspectives. Yet it also helps to identify developmental patterns and shared concerns that might otherwise escape our attention. In a longer perspective, it might turn out that a global approach will establish (global) topics and working modes that are different from currently known scientific collaboration practices and that might be accompanied by new forms of organizational structures or (international) schools of thought.

It might seem ironic and even unforgivable that a book about the internationalization of the social sciences does not contain chapters on those countries that have played the most important role in internationalization efforts: the United States and Europe. Globalization and shifting hegemonic structures have doubtlessly affected science policies, research topics and higher education in these areas, as well, and the resulting changes would be an interesting topic in its own right—it has been omitted here for the purpose of giving space to less-frequently discussed research communities. Yet it follows from the current academic power structures that discussing the latter cannot ignore the former: U.S. and European interests, theories, and funding are factors that affect social science communities all over the world, and discussing internationalization in Korea, Lebanon, or Ukraine inevitably also sheds light on »Western« internationalization practices. In a way, each chapter therefore also—often inadvertently—contributes to an understanding of the internationalization of North American and European social sciences.

This book wants to expand the scope of discussion by inviting readers to redirect their view toward the (semi-) peripheral science communities and to discover shared experiences and concerns. We thus follow Oommen (1991: 82) who recommends that »internationalization to be authentic and fruitful should consciously design for a multidirectional flow of sociology, particularly strengthening the flow from the weak to strong centres. The project should not simply aim at ›educating‹ the non-Westerners but *learning* from them.«

Intellectual learning is never a simple affair: in our case it requires openness to controversial standpoints and tolerance for varying text formats that we have not (entirely) trimmed to fit »Western« standards. In many cases, texts show traces of emotional involvement of the authors that we consider an important part of authentic testimony and that were therefore not »cleared away«. We certainly also hope that readers will find the intellectual journey enjoyable.

References

Alatas, S.F. (2003): Academic Dependency and the Global Division of Labour in the Social Sciences. Current Sociology, 51 (6), 599-613.

Crawford, E./Shinn, T./Sörlin, S. (1993): The Nationalization and Denationalization of the Sciences: an Introductory Essay. In: E. Crawford (Ed.): Denationalizing Science. The Contexts of International Scientific Practice (pp. 1-42). Dordrecht: Kluwer

Hakala, J. (1998): Internationalisation of Science. Views of the Scientific Elite in Finland. Science Studies, 1, 52-74.

Oommen, T. K. (1991): Internationalization of Sociology. A View from Developing Countries. Current Sociology, 39 (1), 67-84.

Wallerstein, I./Juma, C./Fox Keller, E.F./Kocka, J./Lecourt, D./Mudimbe, V.Y./Mushakoji, K./Prigogine, I./Taylor, P./Trouillot, M.-R. (1996): Open the Social Sciences. Report of the Gulbenkian Commission on the Restructuring of the Social Sciences. Stanford: Stanford University Press.

China's Historical Encounter with Western Sciences and Humanities

HE HUANG

Introduction

In the Middle Ages, China, as well as the Muslim world, was one of the most important powerhouses of science and technological innovations, while Europeans' efforts to explore the scientific realm were drowned by the Catholic Church's draconian theocracy. For several hundred years, China was accustomed to enjoying its superior status on science and technology. However, from the sixteenth century onward, European avidity and energy were released from struggles between monarchs and the pope, between two social powers, one being the Catholic Church, the other being the church's opposition consisting of the Protestant Reformation and the Humanism. In consequence, Europe caught up in rapid paces, finally surpassed China, and dominated in science and technology. In 1668, a debate on astronomy was held at the court of Imperial China, which demonstrated the superiority of European scientific methodology. Had the Chinese scholars and the imperial court been penetrated by the perception of this shocking event, a scientific revolution could have occurred in China, and China could have possibly caught up with Europe in science while maintaining the vital force of traditional Chinese thoughts. Unfortunately, Chinese recognition of this European superiority had to wait another 172 years. Thence China lost an early chance of proactively embracing and assimilating the elements of scientific internationalization.

It was in 1840–42 that the British Royal Navy, equipped with far more advanced warships and sophisticated guns, defeated China's imperial army and imposed the humiliating Nanking Treaty on China. Though completely aware of European superiority in science and technology, Chinese authorities nonetheless still deemed their social system and lifestyle better than Westerners' until at the turn of the twentieth century when Chinese intellectuals began to acknowledge China's also lagging behind in social sciences and humanities. In 1895, the Chinese imperial army was unexpectedly defeated by Japan, a tiny country for millennia regarded as inferior to the Chinese Empire. By dint of this very humiliation, more and more Chinese were recruited in the ideological army that advocated uprooting »corrupt and moribund« Chinese tradition and exploring Western science and policy concepts. In the meantime, the imperial court began to institute a modern education system. In 1898, the Imperial University of Peking, which later became the prestigious Peking University, was set up. This can be considered as the beginning of the institutionalization of Western sciences and humanities, consisting of both natural and social sciences. In later decades, numerous modern universities, spates of high schools, and primary schools cropped up and supplanted private or public schools at which Chinese classics, mainly Confucian works, had been taught and studied for centuries. Finally Chinese science and humanity were internationalized but unfortunately not in a very constructive way. Chinese traditional thinking used its capacities to absorb the new impact from the West of internationalization. This situation is the basic background and starting point for observing the development of social sciences and humanities in modern China.

After the fall of the Qing Dynasty in 1912 and an ensuing period of warlord rivalries, the Nationalist Party and the Communist Party began to vie for the ultimate political power in China. Under the Nationalists' reign in 1927–49, mimicking the Anglo-American model, China built up a modern comprehensive and all-round institutional system of science. However, unlike the later Soviet institutions, this system fell short of deeply influencing Chinese daily life. In 1949 the Nationalists were militarily defeated and retreated into Taiwan, an island off the southeast coast of China. The communists thus created in mainland China a gigantic social and economic revolution that began to penetrate every aspect and level of the socioeconomic system. Accordingly, the academia built after the then dominating Anglo-American model was reformed again, adjusting it entirely to the Soviet model. Thus China not only mistimed the installation of modern scientific institutions, but also chose with the Soviet academic system the wrong model, which had at least two major differences from the Anglo-American model.

Firstly, even though the Soviet state of China claimed to be atheist, science in the Soviet ideology, unlike in West Europe and North America, is not only science per se but is like a religion. Accordingly, the Soviet state believed philosophically in Marxism and politically in scientific socialism, which thus functions as a kind of systemic religion, which is, however, by no means a religion in a traditional sense. While traditional religions believe in otherworldly Gods, which do not need to verify their very existence, Soviet ideology based on scientific socialism regards science as its »god« and the scientific nature of this »god« requires its scientific verification, both concerning the substance of the doctrines' correctness and even its very existence. With this in mind, even the political legitimacy of the Soviet regime was based not just on politics but on the theory of scientific socialism, just as mediaeval monarchs claimed that their legitimacy originated from the pope's anointment. Thus the Soviet government attached much more importance to science as its ideological basis than its Western counterparts ever did and do.

However, this very Soviet cognizance of science yields thoroughly sour outcomes. The belief, which desperately needs eternity, perfection, and sereneness, unnecessarily suffers from often violent disturbances triggered by the progress of daily scientific study, which usually crop out in West Europe and North America, while the scientific studies under the regime of scientific socialism are inhibited by ubiquitous out-of-date scientific principles or doctrines. There is a tragic story illustrating the contradictory concept of science as the semireligious basis of the Soviet system. In early Soviet times, Nikolai Vavilov repeatedly criticized the non-Mendelian concepts of Trofim Lysenko, who was skilled of manipulating political power to silence opposition and eliminate his opponents within the scientific community. Then Vavilov was arrested in August 1943 and died of malnutrition in a prison. As a result, the science of biology in the Soviet Union was virtually demolished. Thus the Soviet preparation for a technological revolution based on Mendelian genetics and molecular biology deteriorated to the level of Zaire. It's noteworthy that China's biology also suffered from Lysenko's doctrines, because China blindly accepted Soviet scientific ideas, though one could assume that biology is a relatively ideology-neutral science. In the social sciences and humanities, which are much more ideology relevant, the ideologization of knowledge penetrated the generation of knowledge much more. Lysenko needed Stalin's personal support to force other scientists to scrap Mendel-Morgan ideas. However, in the realm of social sciences and humanities, scholars were not supposed to doubt Marxist economical assumptions or its epistemological basis, dialectical materialism, though these theories could not correctly grapple with state-of-art socioeconomic changes since the turn of the twentieth century. This is, at least partly, the reason why countries hosting an academic institutional system following the Soviet model only make very few scientific or technological innovations despite their enormous investments of capital and human resources in science and technology studies and in research. In the West, Lysenko cases could not happen because the Western science system is institutionally separated from religious preoccupations and from any political regimes' legitimacy. All these Soviet liabilities were implanted into the People's Republic of China. Now, after the Soviet model has fallen in its mother country, its basic ideological frame still dominates China's academic system.

Second, in the Soviet model, the private sector of science and humanities was virtually eliminated. Any institution of science and humanities was »public«, or to be precise, was an offshoot of the Soviet bureaucracy. Lysenko's career success was a success of a political power struggle in the bureaucracy, not of scientific progress as such. While the spirit of challenging and doubting is the driving force of scientific improvement, the subordination of science under the bureaucratic Soviet model urges obedience, a lethal impediment for scientific and technological innovations. If a scholar in the United States feels oppressed in public universities or research institutions, he or she can continue his or her career in private institutions, whose freedom of publishing is guaranteed by the country's constitution. Personal persecutions as daily happening in socialist countries are unthinkable for Western scholars. In contrast, if a scholar in the Soviet Union published views at odds with official ideology, he or she would quickly become scientifically homeless. There was no private institution to which they could escape. In consequence, scholars are discouraged from initiating innovative research projects. In the West, a scholar carrying out research innovations is also confronted with the risk generated from the gap between any new ideas and its feasibility. But besides such a »natural« risk, in the Soviet counterpart any scientist has to weigh the risk of sabotaging the doctrines of scientific socialism. From the perspective of bureaucracy, any innovative idea implies a systemic risk, which thus becomes the hotbed of innovation. Thus, when dealing with the natural risk of any innovative ideas, the Western system is more innovation friendly, although the West is not at all a paradise for scientists, who also are confronted with all kinds of nuisances, such as pressure of marketization and bureaucratization. However, due to its rigorous private sector, the Western system, especially the American system, allows more flexibility in the organization of research and is therefore concomitantly more efficient, while individualism, diversity, and freedom, all those beneficial factors for innovations, were strangled by the tight grip of the Soviet type of bureaucracy.

From 1952 into the 1980s, besides the two Soviet traits just mentioned, China had its unique characteristics. First, China went further with the Soviet cognizance of blurring the border between science and humanities. In China's organization of the academy, the disciplines of the humanities, consisting of literature, philosophy, and history are pigeonholed as a branch of science. In modern Chinese the term *humanity* is often phrased as »humanistic science«. Scientific standards and methodology are considered the sole correct approach in China's humanities studies. This tendency of scientism was prevalent in all communist countries but its intensity and extent varied along the geographical line of West-East. In general, scientism prevailed less in the western half of the Soviet bloc but more in the East.

In East Europe and the Soviet Union, there was the religion of Christianity. Although religious ideas were disparaged in communist ideology and their traditions and organizations persecuted by the communist movement, their influence on social life still existed and penetrated into every social scale, including the social sciences and even more the humanities. As the ultimate stronghold of humanities, religion played the role of checking scientism from rampant excess.

However, in China the influence of religious ideas on science is a very different story. China had had no dominant religion for millennia. Confucianism functioned as a religion in many respects but it is not a true religion in that it does not believe in an eternal and absolute God. Thus the existence of Confucianism seems much more fragile than that of Christianity. The core of the Confucian values was the faith to the monarch. And this faith was sustained by the imperial examination system by which the emperor built and maintained a meritocracy in which capable officials were selected from common people, based on ability or merit. Unfortunately, the imperial examination system was repealed by radical reformers in 1905, and the monarchy was prematurely overthrown in 1912. Had history given China enough time to modify Confucianism and adapt it to the modern world, the core values of Confucianism could have been precious assets on the basis of which a modern Chinese society could have been built. Without the counterbalancing and moderating influence of religion on the sciences, the claws of scientistic thinking were consequently more deeply put into the body of China's science and humanities than it was put into these institutions in East Europe or Soviet Union. This belief in science was vividly heralded by Chen Duxiu, one of the most prominent precursors of the Chinese Communist Party, who wrote that science could solve every problem of every aspect of human life, including social and moral life. He announced that this idea is a maxim which needs no proof and people should believe it with no doubt.

Second, China encountered the Great Proletarian Cultural Revolution, which virtually succeeded in uprooting the Chinese cultural traditions with a continual history of thousands of years. Every revolutionary movement, which lauds the »brand new«, and thus more advanced and better, tends to deprecate traditional values and cultures. However, once the revolutionary regime's grip of political power is secured, it rarely fails to revive these traditions, which usually contribute much to social stability. As a corollary, the loss of traditional values caused by revolutions is thus normally diminished.

Mao's China was, however, a conspicuous exception. Since 1953, when the cease-fire of the Korean War took effect, communist rule was firmly solidified. But the condemnation and destruction of Chinese traditions were not ordered to a halt. On the contrary, they were greatly deepened and expanded. In the first half of the twentieth century, Chinese elite scholars as a rule held very negative views on Chinese traditional culture but their influence could not encompass most social scales nor affect the Chinese humanities. Not only common people were beyond the reach of these culturally self-denying ideas; but the Chinese traditional culture, including academic studies on traditional Chinese classic works, in other words, studies on Chinese humanities, were maintained. In fact, once could even say that elite scholars were merely theoretically hostile to Chinese traditions in the humanities.

After the Communist Party came to power, the country began to be solidly organized by an unprecedented bureaucracy that could through modern media deliver official ideological ideas to every ear of the gigantic society of China. As a consequence, if the top leader of the regime decided to devalue traditional Chinese culture, this could be effectively done by the mighty and ruthless bureaucratic machine, which was unimaginable before the Communist era. This was, in fact, what Mao's China did between 1949 and 1979, culminating in the Cultural Revolution (1966–76). During this time, studies of traditional Chinese humanities were virtually deracinated. Social sciences thus lost a valuable native source.

They could have had other two foreign sources: the West and the Soviet bloc. Due to the Cold War, however, all scientific communication with the West got entirely cut off. In the early 1960s a Sino-Soviet ideological conflict broke out, which isolated Chinese sciences and humanities from the last foreign source and communication. With all three sources of scientific communications gone, Chinese science and humanities were finally pushed to the brink of an all-out collapse in 1979.

How is it that there was a more powerful force in China than in other socialist countries pressing its leaders to entirely uproot the cultural tradition? Two points could serve as the explanation.

First, since the removal of the imperial examination system and the monarchy, respectively in 1905 and 1912, which pulverized the self-respect and self-confidence of the traditional culture with a core of Confucian values, both Chinese authorities and common people have despised Chinese traditions. Second, the blood-shedding upheavals and anarchy between 1912 and 1949, which lasted for a period of nearly two generations and cost tens of millions of lives, made people callous to human life and humanities. Civility got trampled, and the law of the jungle and cynicism prevailed. Consequently few people cared about the elimination of the trappings of civilization, or traditional humanities.

Contemporary China's Social Sciences and Humanities in the Context of Globalization

On August 28, 2008, the *Southern Weekend*, one of the most influential newspapers in China, published a comment on the Beijing Olympics. The reporter mentioned that a foreign counterpart had asked him an interesting question on August 9: »The opening ceremony of the Olympics is very good. But it has no elements of Mao or revolutionary things. How come?« The astonished Chinese reporter replied: »If you did not ask me this question, I could have never thought of it.« This response speaks from most Chinese hearts. Yes, Mao and revolutions are no more the emblems of China. Chinese people are desperately drawing a line from the revolutionary past just as they were eager to say goodbye to their ancient traditions in the first three quarters of the twentieth century. As the programs of the 2008 Beijing Olympics opening ceremony demonstrated, now the Chinese are willing to be more regarded as inheritors of the brilliant ancient Chinese civilization symbolized by Confucianism, classic arts, and technological inventions of the compass, paper making, gunpowder, and movable type printing than deemed fans of communist and revolutionary achievements. It is a stark change, in contrast to the situation in 1979 when Deng Xiaoping, then the Chinese paramount leader, kicked off an epic policy of »Reform and Open up«, which managed to dismantle the barrier isolating China from the outside world. So after early internationalization of Chinese sciences and humanities and later virtual isolation from the West and Soviet academic circles, in China a new wave of internationalization has been set in motion. This wave is not merely a repeat or a simple revival of the previous one. In the previous wave, national borders were clearly distinct, although international communication of sciences and humanities grew intensive. At least in China, science and humanity studies mainly took place within the borders and international communication was far from daily life of scholars.

In today's wave of internationalization of academic activities, both the intensity and the depth have skyrocketed. This very change owes much to two facts. First, the end of the Cold War in 1991 dismantled the ideological barrier that had impeded international academic communication. Thus all views could be placed and debated on the same platform. Second, continual economic development all over the world and new technological advancements such as the Internet blur national boundaries and bind together the whole planet, making the world a global village. As a corollary, a new type of internationalization takes shape: globalization. This trend coincides with China's internal policy of »Reform and Open up«, which accordingly exerts great influence on China's sciences and humanities. History repeats itself: in 1840, China was opened up by Western impact, and in 1979 China opened up again under the pressure of Western-dominated internationalization, often referred to as globalization.

Deng intended to pave the way for realizing »four modernizations« (modernization of industry, agriculture, military defense, and science and technology), China's national strategic objectives initiated in Mao's era. Furthermore, in Deng's design, China would rise to the level of an average developed country by the middle of the twenty-first century. These objectives included the refurbishings of the rigid and dilapidated science and humanity institutions where scientific and technological innovations might be hatched and in more cases through which Western advanced scientific achievements and technology could be imported, digested, and put into daily application. On March 30, 1979, Deng said at a work conference:

»We did recognize that our natural sciences are backward, compared to foreign countries. Now it's time for us to recognize that so are our social sciences (in respects that could be compared). We are at a very low level. Statistics needed for social science research have been unavailable for many years. . . . It is noteworthy that the reason of the backwardness lies in the flaws of the directions of the Central Committee of the Party and its offshoots at each level. There are too many taboos and not enough aid and support.« (Deng 1994: 180-181)

Two months before the speech Deng had visited the United States. It was the first time a paramount Chinese leader visited the United States since the birth of the People's Republic of China in 1949. Apparently he was impressed by the prosperity of U.S. economy and the advancement of U.S. science and technology. Later he asserted that science and technology constitute the primary production force. In the official scientific socialist ideology based on Marxism, »production force« is a magic term. In Marx's theory, production force is the root factor and the most revolutionary element of the social development, which is regarded by communist theories as the heart of human civilization. From then on, Chinese natural scientists did not need to worry about persecution due to innovative ideas, or in most cases, ideas imported from the West that were at odds with official scientific socialist ideology. This ideology had been modified by Deng as a fledging theory under the condition of »the primitive phase of socialism« instead of an accomplished and unquestionable cluster of doctrines. In other words, issues of natural science were opened to debate and virtually all the ideological taboos about it were removed. Lysenko affairs were eliminated. International collaborations on science and technology began to be encouraged and facilitated.

From this liberation of thoughts the social sciences and humanities (SSH) also benefited, though to a lesser extent and depth. The more social science disciplines were relevant to contemporary political ebbs and flows, the less liberty scholars enjoyed. As a result, economics and sociology gained more in prosperity than political science. Economics developed so that the market economy, the core concept of West economics, was included in the official ideology, which had been unthinkable in the Soviet model. Sociologists also exerted their influence on the government and the society by making acceptable the idea that civil society, a typical Western concept, ought to be erected as the cornerstone of the Chinese society. Scholars of the humanities could take the full liberty of advocating ancient traditional Chinese culture and values, which in Mao's era were deemed reactionary feudal poisons. Western humanity, once considered the symbol of corrupt and depraved capitalism and forbidden, again took its step into China.

However, political scientists have been muted on political reform because leaders of the regime correctly recognized that political reform is a juggling act requiring both Herculean strength and delicate operative adroitness. A small bungling can topple the dominoes and lead to a terrible anarchy and blood-shedding chaos, as occurred in the wake of the fall of the Qing Dynasty. In general, studies on sciences and humanities were restored in the light of Deng's policy. It's not an exaggeration to say that Deng played the role of messiah, saving the life of China's sciences and humanities from death.

Unfortunately, the saved life was left with a string of ills. All the ills could be imputed to the unhealthy body that is overwhelmed by an enormous intake of external nutrients. In other words, globalization helps reinstall China's sciences and humanities but tramples their Chineseness and thus undermines China's ability to contribute meaningful science and humanities research to the globe.

Marginalization of Humanities

In Deng's important speech of March 30, 1979, there is a conspicuous absence of humanities. Probably he shared the opinion on science and humanity with Chen Duxiu. Since science could grapple with every issue of human life, study of the humanities is not necessary or it is a branch of social science. But there is another possibility, that Deng feels that humanities of different nations, unlike natural and social sciences, are not comparable. It is not Deng's fault but it really reflects the situation of the humanities in China. Studies on humanities were given great importance before 1979 but were distorted by blind adherence to Marxist and Maoist doctrines. Since 1979, humanities have been rapidly marginalized both in academics and in mainstream social life. Within the academic circle, more and more proportions of financial and human resources are being diverted into natural sciences and some social sciences such as economics. In the late 1970s and early 1980s, humanity scholars were significantly more influential than scientists and the most excellent high school students preferred humanity studies, such as linguistics and literature, history and philosophy. Later, especially after 1992, when the market economy was legalized and facilitated by the communist government, studies in the humanities have been more and more regarded as useless. In the imperial era, learning Chinese classic humanities was a prerequisite for the selection of individuals as officials, the most admirable profession at that time. In Mao's era, learning and researching humanities (or the »humanistic sciences«) dominated by Marxism could strongly bolster one's pursuit of positions as party or government cadres, also the most admirable profession at that time. Traditional Chinese humanities, as mentioned above, have been virtually wiped out by the Cultural Revolution. And from 1992 onward, Marxist humanities have been losing ground. Western humanities have poured into China but have been heavily chafed by the still extant communist ideology. Additionally, humanity studies could not be directly used as tools for personal prosperity nor for GDP growth. So nowadays research and study of the humanities in China are in great trouble. The humanities are the source, soul, vital force, and soil for the growth of sciences. Ku Hung-ming, one of the most prominent essayists in English during the Victorian and Edwardian eras, commented:

»Today, those so called progressive Chinese are pursuing science for building railways and aeroplanes. They could never understand the essence of science. In European history, those who devoted themselves to science and endeavoured to push science progress, those who made possible building railways and aeroplanes, at the beginning had no idea of railways and aeroplanes. They devoted themselves to science and contributed great efforts to improve science, just because their hearts and souls were eager to explore the awe-striking mystery of this limitless universe.« (Huang 1996: 38)

He was right. Such hearts and souls could not be cultivated by science education but by humanity education, which needs state-of-art refurbishments by humanity studies and research. This will be discussed in detail later.

Bureaucratization, Industrialization, Commercialization

The marginalization of the humanities is one of the aftershocks of the radical movement culminating in the Cultural Revolution. As mentioned above, the revolution smashed the Chinese cultural tradition, which could be dated back to ancient times. In the mean time, SSH institutions and traditions erected in modern times were also demolished. The chaos led to an irrevocable loss for Chinese SSH: peer review, the basis of excellence evaluation and the key engine of the mechanics of academic institutions, had its spine broken. It is relatively easy to restore the peer-review system for natural sciences in China by using the peer review system from other countries, in particular, from West Europe or North America, because natural sciences have a cosmopolitan nature. However, SSH studies are innately parochial. According to an article written by Peter Weingart and Holger Schwechheimer,

»In a recent study of international social science research commissioned by the UK's ESRC the authors state: ›Most social science research is done within national and local boundaries, and, most often, by individual scholars rather than the research teams which populate medical and natural science research‹ (Forbes/Abrams 2004). […] This is probably even more true for the humanities.« (Weingart/Schwechheimer 2007)

As a consequence, once the SSH peer review system in China was deracinated, it was far from easy to restore it by setting up a foreign model. The vital force of the shattered system could not recover within several generations. Thus China's SSH, compared to its natural sciences, is more inclined to be brought down as a prey by three modern »monsters« that are nemeses of creativity and innovations: bureaucratization, industrialization, and commercialization.

Bureaucratization

As mentioned above, imitation of the Soviet model brought China a bureaucratic leviathan for its academic system. In China no private institution could be permitted. Every legal academic institution is public so it does not need to deal with the competition which is a daily routine for contemporary private enterprises. So a benign stimulant for innovations is absent. Financial and human resources are not dynamic and thus not efficiently utilized. China has no liberal intellectuals but bureaucratic intellectuals instead. Liberal and independent ways of thinking are always scuttled. Many problems crop up in this thread of bureaucratic thinking. For example, many scholars and officials are considering abolishing traditional Chinese medical science, merely because its scientific nature is in doubt. This very science, due to its absolute parochial nature and intransigent resistance to quantification, should be regarded as one of the humanities rather than science. If this destructive initiative takes effect, the world will suffer from a great loss of a precious cultural heritage, which can relieve pain caused by diseases or even save many lives, whether it is a science or not. Were traditional Chinese medical science situated in the West, as it was in China before 1949, its fate would be decided by the market, not by a bureaucratic writ. The allocation of resources complies with the logic of the planned economy. Ideas, which lead to academic projects, as a rule are not spontaneously inspired by academic practices or contemplations but by whims in a bureaucrat's mind. Scholars have to submit themselves to the erratic, nonprofessional management of government funds, which are virtually the sole source of funding. What's more, government funds lopsidedly go to projects involving issues related to official ideology. For example, the Party Central Propaganda Department built up a prize of »Five ones«, exclusively in favor of Marxist theories and works. In contrast, natural science in China is far luckier. It's irrelevant to ideology so a major hitch is removed. And funds for natural science are far more ample. There are for natural science at least three national great prizes, without commensurate counterparts in SSH. Poor organizing is dealing another blow to

Chinese SSH. The traditional mentor-disciple model imbedded in convoluted bureaucracy seems an anachronism. So is the notion of being »the Emperor's mentor«, which was prevalent in imperial eras but has formally vanished. Nonetheless this very notion was actually ingrained consciously or unconsciously in the minds of Chinese scholars. In the West, the SSH circle is an existence basically independent from the government bureaucracy. SSH evolves independently while the government could occasionally glean some fruits for policy use. In China, scholars have to guess about senior officials' political ends and then give up their own ideas and serve these political ends.

With the native peer review system dismantled by the Cultural Revolution, a bureaucratic evaluation system fills in the vacuum. This system, unlike the deceased one, could run without belief, heart, and soul. Although there are no more bloodthirsty predators such as Lysenko, the bureaucratic machine is able enough to sap creative vitality and extinguish virtually every spark of innovation. In the academic bureaucracy, the power of policymaking and allotting financial and human resources are not in the hands of scholars but in the grip of administrative staffs who probably have not enough expertise on academics. Professionals thus are unwilling to devote themselves to pure academic careers, which require brilliant talent and great passion but fall short of personal prosperity and thus of social respect. An ambitious person in an academic institution would rather invest his time and energy to get an administrative position with which he or she could spend much less effort but economically and professionally benefit from the bureaucratic power. A typical successful Chinese academic career requires early attainment (in one's thirties) of the title of professor or senior research fellow. With this starting point, the young title holder then takes the »genuine« first step of his career by vying for bureaucratic positions. For a bureaucracy, standardized management is a favorable convenience. Organic, vital peer review, which could handle issues with rich subtleties that are as a rule the hotbed for innovations, is shoved aside by administrative and political pragmatism, radically and effectively reducing academic diversity, vitality, and resilience to very few, easily manageable indicators, such as quantities of works published or numbers of citations in local or nationwide core academic journals. Some key universities and research institutions, in pursuit of so called internationalization, which is per se an indicator for evaluating institutions, use the Social Sciences Citation Index (SSCI) or Art and Humanities Citations Index (A&HCI) journals. The implementing of quantitative measure of evaluation instead of a qualitative one paves the way for another leviathan's intrusion: industrialization.

Industrialization of Scientific Knowledge Productions

In pursuit of career promotion, most SSH scholars in China have to publish a sufficient quantity of works in local or nationwide core academic journals. Furthermore, quite a few universities require their master's and Ph.D. candidates to publish a fixed minimum number (say, two to five pieces) of work in local or nationwide core journals to attain their degree. As a result, with a huge amount of work awaiting publication every year, an industry of paper and dissertation production takes shape. Papers and dissertations are produced in such large numbers that their quality is diluted. Maybe it is not that much as »garbage in, garbage out«, but it is not far from it. Consequently, Chinese academics seem to prosper. However, with froth skimmed and bubbles broken, the true growth or progress of SSH is marginal. What's more, the story hasn't come to an end yet. The volume of these journals is so limited that a large portion of scholars and students cannot not find enough space in the journal to publish their work. Thus many journal bosses have decided to make profits from this insatiable demand.

Commercialization

Here comes the third problem: commercialization. Many journals created additional pages to accommodate those unqualified but required works. The authors of these papers pay for the utility of these spaces. There are three winners and one loser of this game. Students could get the degree diplomas; scholars could get career promotions; journal bosses could become millionaires. There is a loser: the very enterprise of Chinese academics. Money, instead of anything else, holds the key of academic excellence evaluation. Instead of local or nationwide journals, using SSCI and A&HCI journals whose bosses do not dare to trade their prestige with lucre, some institutions manage to effectively mitigate the foul aftermaths of academic commercialization. However, due to the innate parochial nature of SSH, works not written in the native language and not reviewed by native peers find great difficulty in receiving relevant and beneficial feedback and evaluations.

Interactions between China's Higher Education and the Pressure of Globalization

Since Deng's »Reform and Open up« policy took effect in 1979, China's higher education institutions have been overhauled. Apparently impressed by the American model, the reformers have taken measures to imitate it. In the name of catching up with the advanced American system, Communist Party bureaucracy retreated from universities just as they scaled back from enterprises and governments. However, unlike enterprises that continued to be free of the party bureaucracy, universities saw a reversal of the trend of noninterference from 1989 onward. The socialist organizing frame was restored while the outside environment grew more and more capitalist. With a gutted tradition and weak humanity education, the senile Soviet system of China's higher education struggles for life by growing more and more bureaucratized, industrialized, and commercialized, in response to the globalization dominated by America.

Like research institutions, China's primary schools, high schools, and colleges are actually offshoots of an education department of the government at a certain administrative level. Unlike private enterprises that were extinct before 1979 but are thriving nowadays, each school does not answer to its private or public shareholders, but to its higher-level bureaucratic officials. In other words, China's schools are not autonomous or self-sufficient in financial and human resources. This was not in the case before 1949. As mentioned previously, this educational structure is a legacy of the Soviet model. The economic segment of the Soviet bureaucratic model has long been repealed but the academic segment has been kept roughly intact. Bureaucratization, kicked off in Mao's era, has intensified and expanded during the post-Mao economic prosperity. The other two problems mentioned previously, commercialization and industrialization, also predate China's higher education model.

Universities undertake a double task: research and education. The three problems mentioned above stifle education as well as research.

First, bureaucratization slashes general education (in the West it is called liberal learning), putting each broken piece into the pigeonholes along the disciplinary lines. For bureaucratic convenience, every Chinese student must choose a discipline as his or her specialty before he or she enters a college, so that he or she can be pigeonholed into a certain department and be more easily manageable. And it is extremely difficult for a student to transfer from a discipline to another. This practice makes no sense. Often a student will not know much about a discipline unless he or she has spent a couple of years in studying and thus comparing a variety of courses from different disciplines. As a sour result, a student is forced to start a career that he or she knows little about it. It is reasonable to say that this is detrimental to this student's career. Given that most students could not be lucky enough to happen to choose a favorite discipline, the quality of education is apt to be severely hamstrung. We know that in the United States, college students do not need to choose an unfamiliar specialty before they enter the universities. They receive two years of general education encompassing courses about various disciplines and then choose a specialty which hits their interest. There are quite a few cases in which a student registered at the Department of History in his undergraduate era ends up being admitted to a graduate program at the Department of Physics. Such cases are unthinkable in China. Today many universities recognize this liability and take some measures to temper it. However, they yield few sweet fruits because they are subject to the cold and hard bureaucracy. A typical measure is to teach courses of a discipline to students specializing in another discipline. It is a misunderstanding about general education. General education requires a small bunch of courses but with high intensity and depth. The measure above exponentially stretches the length of the course list, making students dabble in every discipline. And the students could not take these shallow courses seriously. But as every such course is added, the bureaucratic evaluation system could give a point to the university by reporting that this university has gained an advancement in pushing an indicator of the »general education«.

Second, bureaucratization cripples diversity, which is a stimulant for innovations. For the departments of education at various administrative levels, the greater the diversity the universities which are obliged to answer to the education departments have, the more difficult they are to manage. After all, bureaucrats prefer the most reduced and simplified indicators. A salient example is the mechanism by which the universities admit and recruit freshmen. Before 1949 the educational bureaucracy was so weak that the minister of education had less influence on higher education issues than the president of Peking University. At that time different universities recruited freshmen in different seasons, based on different evaluation systems on merits. Different universities had different sets of questions in their entrance examinations. After 1949, the immense education bureaucracy could not bear this anarchical situation. A uniform set of recruitment measures consisting of an evaluation system and entrance examination were implemented for each university, which was uniformly organized as an offshoot of a department of education in local or central government. Before 1977, high school teachers or officials' recommendations were taken seriously as a merit evaluation with a weight comparable to the score points of the College Entrance Examination (CEE), which is uniformly organized and has a uniform set of test questions. Unfortunately, these reasonable practices were to be removed later. Since 1977, evaluation on the merits of a prospective freshman for admission to colleges has depended solely on the score points of the CEE. Even at this stage at which virtually every »odd« ingredient of the evaluation system has been shaved off, the bureaucracy's inborn momentum for quantification, reduction, and simplification has not been exhausted. Owing to some effects of educational globalization, this determined bureaucratization continued to advance. Before the late 1980s the test questions of the CEE were far from quantified or standardized. A considerable amount of qualitative questions could perceive some unique talents of a certain test-taker. Since the late 1980s the notion of standardized tests has been internationalized from the United States to China. The CEE, as the most important test in China, consequently was unlucky enough to accept and practice such a so-called advanced measure imported from advanced countries. Questions that could not be answered by merely ticking a choice from »ABCDE« have been trimmed off and minimized. As a result, more and more freshmen are skilled at playing useless and stupid »ABCDE« games but incapable of doing basic reading and writing. However, in the United States, this quantified, standardized, machinelike measurement test, the Standardized Aptitude Test (SAT), has not caused as many troubles, because U.S. universities do not take SAT score points as the sole source of evaluation. Far less bureaucratized, U.S. universities take teachers' recommendations and students' leadership performance in voluntary activities or in sports, as measures in weight at least equivalent to SAT score points. Due to its natural tendency to simplification, the bureaucracy is not able to install the whole U.S. system in China but was ready to slice the nonorganic segment off and attach it to China. In the Chinese environment this kind of internationalization invariably produces unpleasant outcomes. The admission process to Chinese graduate schools is similar.

Third, bureaucratization, in some cases combined with industrialization and commercialization, is marring China's higher education system. Before the start of »Reform and Open up«, there were no such results. The typical Soviet model, unlike the higher education system nowadays, requires no payments for tuition or room and board from students. On the contrary, every student was considered a worker or cadre and paid a salary. This system had a variety of shortcomings but to some extent and degree it guaranteed equal opportunity access to higher education. As long as a student deserved an offer of university admission, he or she did not need to worry about the financing of schooling. People in poverty and those well-to-do were equal in pursuing the education opportunities. After about 1990, China's universities began to unravel the Soviet tradition and tax tuition and other fees on students. Nearly two decades have passed and the industrialization and commercialization of China's education have developed to maturity. Now if you get admitted to a university but cannot afford the tuition and other fees, you must forfeit the opportunity of higher education. In the United States, the problem of inequality between the poor and the well-to-do is dealt with somewhat by ubiquitous financial aid for students. As a result, in the U.S. higher education system youth from poor and wealthy families begin at the same starting point. Thus the loser has the hope to be the winner. Class conflicts could be pacified and social stability could be maintained. In China, the fees are internationalized but the financial aids are not. As a result, China's higher education system widens the gap between the well-to-do and the poor, worsens the relations among different classes, and impairs social stability.

Some harmless design in the Soviet model ends in poisons for today's higher education, as it is catalyzed by commercialization. For instance, in the Soviet model of bureaucratic arrangement, there are undergraduate professional training specialties such as law, management, and medicine. These professional trainings are not true university education. They are just training workshops targeting know-how required by prospective jobs. They are interested in »how«, instead of »why«, which is the essence of higher education. In line with Robert M. Hutchins, »Ideal education is one that develops intellectual power. It is not one that is directed to immediate needs; it is not a specialized education, or a pre-professional education; it is not a utilitarian education. It is an education calculated to develop the mind« (Hutchins 2001). In the planned economy of Soviet system, general education was given greater importance than professional training. So the former attracted more talented students. However, unlike in the planned economy, in the market economy the most talented and ambitious Chinese youths of seventeen or eighteen years old are pushed or attracted to the professional training undergraduate programs and are pigeonholed into a niche of the education bureaucracy, just because these students have the upper hand in the job market when they graduate. Enterprises are apt to hire students well prepared in job training. As a consequence, the best prospective citizens of the nation will not receive a true university education. Partly because of these false exemplars, students in other specialties also have job-hunting paranoia full in their minds during the four years of university life. In corollary, general education gets virtually annihilated and universities become more and more like job-training workshops. In contrast, usually U.S. undergraduate specialties do not include professional trainings. No undergraduate students could benefit from being registered in a job training program. This policy guarantees that most U.S. university students can invest their money, energy, and time on true general education for four years, without the noisome diversion of wasting their most precious time in life on job-hunting paranoia.

Today's students will be tomorrow's teachers, researchers, or scientists. Consequently what is taught in higher education, especially the undergraduate education, holds the key of understanding a nation's present and future academics.

A typical American undergraduate curriculum, particularly at the elite universities such as the Ivy League, has at least three characteristics different from its Chinese counterpart. First, its courses signal the notion of a true general education, which is centered with social sciences and humanities. Second, these SSH core courses require intensive reading on Western classic cultures, ideas and values (CIVs). Third, these courses are arranged with not only professors' teaching but also discussion seminars that usually consist of a small group of a dozen students or so.

Although Chinese higher education bureaucracy takes great lengths to imitate American universities, it falls short of implanting the three characteristics mentioned above, which are the essence of the American model because bureaucratic management is innately incompatible with these organic, humanistic measures. First, as enunciated previously, bureaucratization pulverizes general education. And due to their nature of being more compatible with quantified management favored by bureaucrats, natural sciences are centered on the platform of academics. SSH is regarded as »soft sciences« and thus unimportant. A saying from Mao Zedong effectively indicates the view of the society on SSH: »Universities must be maintained. I mean mainly the universities of natural science and technology.« Second, at China's universities, except in departments specializing in traditional Chinese CIVs, undergraduate curriculum requires virtually zero Chinese classic readings, let alone intensive reading in depth. If you read an undergraduate curriculum in China, you have every reason to say that is not a Chinese curriculum. The curriculum is full of so-called advanced Western CIVs and unfortunately includes few courses studying the most important classic books of the Western civilization but consists of a myriad of introductions, summaries of stylish academic publications. Even reading on these lousy publications is not intensive. And more and more pressure is put on university professors to teach in English. In bureaucrats' eyes, English is an academically advanced language and teaching in Chinese is an inferior practice because English symbolizes the advanced wave of internationalization. Third, China's universities have almost no student seminars. They are literally teaching machines that rarely respond to students' feedback. Actually little feedback from students, which are an important pool of fresh ideas and inspiration, could be delivered to the professors and integrated into the courses. And another bad habit is noteworthy. In China, the more prominent the professor, the less likely it is that he or she will teach freshmen and sophomores who are at the most suitable age of contacting inspiring ideas from distinguished scholars. Maybe bureaucrats think that because undergraduate students are at too low a level in study, good professors should not deign to teach them. In contrast, U.S. star professors often teach undergraduate students.

Depletion of the Spirit of Initiative and the Concomitant Passive Role in Academic Internationalization

There is a perception that non-Western countries are academically colonized due to their lack of funds. Thus Western countries pour money in and then automatically dominate the academic agenda of non-Western academics. This scenario is true in some places but in my opinion, in many other cases, it misses the point. Actually, at least in China, academic funding comes mainly from an indigenous base. Nonetheless the Chinese academy is grievously colonized, not by its Western counterparts but by itself. This is because Chinese authorities chose a wrong path in response to the impact of the West since the middle of the nineteenth century.

During China's long history many outside influences have been absorbed into the country. For example, from the first to the eighth centuries, Indian CIVs, especially Buddhism, managed to penetrate every social scale in China. In spite of this, Indian CIVs never dominated any agenda of China's CIVs. Chinese scholars chewed the imported culture and ideas, digested their ingredients adaptive to Chinese indigenous values, and assimilated the nutrients of this international cultural communication. In doing so, Chinese CIVs improved itself without losing its self-confidence or self-respect.

But the internationalization from the nineteenth century onward tells a totally different story. As mentioned previously, Chinese ruling elites chose to repudiate their own valuable CIVs in pursuit of so-called advanced and cosmopolitan (actually parochially Western) systems mimicking those of the West. Some Chinese writers offered a metaphor to contend with this notion: with time, a previously valued prominent drawing grows to be perceived as ugly and outmoded. If you wipe out the deposited ink or paint, you could have a blank sheet of paper to draw the brand new and the most beautiful picture. Apparently these writers forget that the »brand new« and »the most beautiful« will grow dilapidated. Additionally, after you spend too much time (say, several generations) in wiping out the deposited ink or paint and attain a blank sheet of paper, you usually find that you have forgot how to draw even a mediocre picture, probably far worse in quality than what you wiped out. Obviously there is a reasonable alternative. It makes sense if these writers kept the old picture and sought or manufactured a new sheet of paper. In doing so time could not be wasted in destroying, and the traditional drawing skills could be passed on to the next generation so that they have the base for improving. It's a pity that Chinese authorities did not choose this right route. Traditional Chinese CIVs, a parallel to its Western counterpart, have been highly respected for millennia. Today China's economy might be awesome, but no one would take modern China's CIVs seriously.

Social sciences and humanities, especially the latter, are innately parochial and greatly depend on native cultural tradition. If a people hastily remove their native CIV traditions, they will be condemned to be a backward nation for a very long era. We know that if you want to make the world different, you must have the spirit of initiative. This very spirit consists in two factors: emotional or rational sensitivity, deep and abiding interest. The two could not be sustainable unless they are steeped in an intense passion. And the passion has to rest on a belief. The belief and the passion could not be solid or sustainable unless the person receives a coherent education in the humanities, which could not be rigorous without national or local traditions. With indigenous culture and its tradition of humanities uprooted in Mao's era, the heart and soul of Chinese academics withered and bureaucratization became rampant. As a result, Chinese scholars lost the spirit of initiative, and this tragic situation has at least two consequences on the nature of China's academic internationalization.

First, China's academic circle has little true consciousness of Chinese issues. Chinese scholars tend to attach more importance to what their Western counterparts say about Chinese issues than to their own contemplation on their national or native Chinese issues. And for Western scholars, non-Western issues, including China issues, are in the periphery of the academic realm, so Western concern about China is usually off the mark. Chinese scholars, although naturally with greater access to indigenous knowledge, give up the initiative of setting the agenda even on Chinese issues, let alone those on other nations' and those cosmopolitan ones. This leads to a never-ending journey for Chinese scholars: they are always waiting for Western, especially American scholars' initiating academic issues and then catching up with the pace of these issues imbedded in the agenda set by American scholars. Bureaucracy, the natural enemy for any innovation, is very glad to push this trend. Since scholars are accustomed to not initiating, follow-up prevails, which is the favorite of the bureaucracy. For instance, China's international studies are reduced to news follow-ups. Scholars are losing their grip of the agenda and surrendering it to CCTV (China Central Television) or *Global Times* (a newspaper), which surely has no agenda of its own but merely fits itself as a bolt into the mass media machine dominated by CNN, ABC, BBC, etc. At least in international studies, Chinese scholars have little understanding of truly substantive issues.

Second, in international collaboration, Chinese academics are rarely able to contribute true substance to the collaborated programs. Their role in international collaboration thus is supplying data and exotic cases for Western ideas or theories. As for ideas and methodology, these are tasks for the advanced Western scholars or institutions. International collaborations in academic activities have been formulated into a bizarre pattern in which Western scholars or institutions offer funds, ideas, methodology, and an agenda while other people in non-Western countries are expected to collect data about their own countries to be integrated into theories in so-called universal, but actually Western, contexts. And Western scholars are expected to help backward nations to catch up in a certain area of science. However, the international academic community could benefit much more if non-Westerners employ their own global perspective, based on their indigenous knowledge, which could be a really substantive contribution to international research collaborations. Moreover, why must non-Western nations render »internationalization« as catchup to advanced industrial countries such as United States, Great Britain, or Germany? Why must scholars from so-called »developing countries« go to London or Boston to meet each other and attain »internationalization«? They could be internationalized by going to Beijing, New Delhi, Istanbul, or Rio de Janeiro, discussing issues capturing our own but maybe not Western attention.

Internationalization, which nowadays could more suitably be phrased as »globalization«, seems a great help for upgrading education of an academically backward nation like China. Unfortunately at least in China reality goes diametrically opposite to this perception. Delicate and tasty food is good for a healthy body but can harm a stomach suffering from indigestion. With weak hearts and souls caused by lousy education in the humanities, China's SSH suffers as much as it benefits from globalization, which could surely be of great help for China's SSH if China had healthy and strong humanities education and research programs. China is ill-prepared for academic globalization unless it restores the traditional Chinese CIVs and thus has a solid base to proactively participate in globalization and make meaningful contributions to the international academic community.

References

Deng, X. (1994): Insist on the Four Basic Principles. In: Deng Xiaoping's Selected Works Guangdong: The People's Publishing House, 180-181.

Huang, X. (1996): Ku Hung-ming's Essays, Volume 2. Haikou: Hainan Publishing House.

Hutchins, R.M. (2001): The Educational Theory of Robert M. Hutchins (Version I). In: http://www.newfoundations.com/GALLERY/Hutchins. html [Date of last access: 23.02.2009].

Weingart, P./Schwechheimer, H. (2007): Conceptualizing and Measuring Excellence in the Social Sciences and Humanities. Available at: http://www.globalsocialscience.org/uploads/Weingart%20paper%20 GlobalSSH%20May%202007.pdf [Date of last access: 23.02.2009].

Internationalization of Japanese Social Sciences: Importing and Exporting Social Science Knowledge

KAZUMI OKAMOTO

Introduction

The nature of Japanese social sciences has historically been international, as exchanges, or rather importing, of knowledge and scholars from Western countries were repeated from the earliest stage in their development in the nineteenth century. Nowadays, their current aim set by governmental organizations is exporting knowledge that Japanese social scientists generate to the other parts of the world. In this sense, their scientific activities seem to have been international for about one and a half centuries due to the intentions of importing/exporting knowledge. It is the government that has always taken initiative to lead the Japanese social sciences to the status of being international. Although the Japanese social sciences seem very international in a certain sense, interestingly it seems rather difficult to grasp what individual scholars do to further the internationalization that the government aims for. This chapter, therefore, tries to explore the same topic but with less policy-based aspects, which could influence the internationalization of Japanese social sciences. The chapter is constructed as following: how and why the Japanese social sciences have been international in respect of their history from the late nineteenth century to the present would be firstly examined. Secondly, the Higher Education system in Japan as well as its contribution to the internationalization of Japanese academia would be described and thirdly, some aspects of Japanese communication styles underlying difficulties in communicating in English language would be discussed. Finally, the recent science policies on internationalization of Japanese social sciences would be presented.

History of Japanese Social Sciences

Emergence of the Social Sciences in Japan

The emergence of the social sciences took place accompanied with the opening of the country to the world. Under the slogan »fukoku kyouhei«, the Meiji government tried to build a centralized nation-state through a policy of policy of strengthening military power and building economic prosperity to keep its independence from the Western countries. To face facing Western military power as epitomized by so-called »Kurofune« (black ships), which signified Western scientific superiority, Japanese leaders urgently sought the import of social and economic mechanisms and thoughts from the Western countries. This was the starting point when also the Western social sciences became attractive for Japan for the first time. Confronted with Western ambitions to open Japan for their hegemonic and business interests, the government was motivated to learn especially about international law, dealing with international relations between countries. Because Japanese leaders sought to establish a society and economy similar to the Europeans, they made it a priority to learn about the social knowledge needed to establish such a society and economy. To do so they found that they would need the knowledge of Western economics to catch up with the obvious power of Western economies, which were invading the world with their international business activities.

Prior to the Meiji period, the Edo shogunate had already sent two people to the Netherlands, which was nearly the only European country that Japan had a connection with by that time, also to study international law (Ishida 1984: 25). Numerous books were translated into Japanese due to the interest in the knowledge of international law. However, the original books were only partly translated and often their basic theories were skipped from translations. It is not because they were unable to be translated into Japanese language, but due to the purpose of the government that only those parts of the books were extracted that seemed useful and profitable in practice for building the new Japan. Thus the social sciences borrowed from the Western countries at that time could be characterized as very politically directed toward exploiting the Western sciences in a timely manner to make the new country as powerful as the Western countries.

Inviting foreign scholars to Japan was also important for the Japanese government, and the invited foreign scholars taught economics as well as international law. With this very political mission initiated and steered by the Japanese government, social sciences, or better, fragments of social sciences, emerged in Japan for the first time.

German Influence on the Japanese Social Sciences in the Nineteenth Century

After some twists and turns, the Japanese social sciences started to establish strong relations with the German social sciences. It is worthwhile to understand why Germany exerted such influence on Japan in the social sciences, because this politically motivated focus on learning from German social sciences to create the new powerful Japan contributed to an historical accident, encountering and importing thoughts that resulted in opposition against the rationale of the Japanese political elite.

The transition of the University of Tokyo could be taken as the best example illustrating how and why the German social sciences were prioritized and why and what caused the historical accident of the Japanese social science mission—and its very Japanese solution.

The University of Tokyo was founded in 1877, and the department of law was only a part of the department of literature at that time. The law taught at the beginning in the university was based on English and French laws rather than the German one. The first turning point occurred in 1882 when the university invited a German economist, Karl Rathgen as a teacher of political science. In 1887, Udo Eggert was also invited from Germany and began teaching finance; thereafter, the tendency of German dominance at the university became stronger and stronger (Ishida 1984: 28). In addition, when a politician names Shigenobu Ohkuma tried to introduce the English style of government in 1881, his group was purged from the government, which strengthened German dominance in the Japanese academe as well as in the Japanese political scene. As a result, the Japanese government began to introduce the German style of government into Japan, which led to German dominance in the whole of Japanese academe and the disappearance of any influences of English and French laws by 1885. Although the genealogy of English and French laws remained mainly at the private universities such as the precursors of Waseda, of which Ohkuma was the founder, and Keiogijuku, because of government suppression, they did not become mainstream concepts in Japan.

Political decisions rather than academic considerations resulted in Japan's choice of Germany as a model nation-state to emulate in its race to become competitive with Western countries. Seeking political models they could import, the Japanese elite found that the German, strictly speaking Prussian, society was very similar to its own, in which the emperor was considered as the center and the most powerful person in every respect. It is said that it was Maurice Block (Moritz Block), a German-French statistician and economist, who suggested that the Iwakura Shisetsudan (Iwakura mission) learn from Prussia. As the result, Iwakura Shisetsudan later met Wilhelm I, Bismarck, and Moltke and became motivated to learn from the country (Ishida 1984: 30). Orientating the creation of Japanese social sciences along German social sciences can thus only be considered as a by-product of this political decision.

Despite that Japan had possibilities to take, for instance, English or French social sciences, it was the German style of government, even more the German constitution and, as a consequence, that the German sciences were the most favoured by the majority of Japanese politically influential people at the time, apparently assuming that the German social sciences were a natural element of a German elite ruling the country. Very much in line with the tradition of feudalistic thinking, the possibility that social sciences would oppose the German government system was obviously not an option for the Japanese knowledge importers. There also seems to be little indication among Japanese scientists that they wanted to learn from the Western countries, or in particular from Germany, genuinely for their intellectual curiosity. In reality most of the previously mentioned import activities were initiated by the government or politicians, and there are no cases where one could trace that Japanese academics took the initiative. They were involved, but only as a part of the Japanese government's missions.

As a consequence of the prioritisation of the German political system and the German social sciences, this resulted in the facts that English and French social sciences could only survive in private educational organizations such as Waseda and Keiogijuku where the governmental policies on science were of less influential to construct the universities' academic frameworks. Considering that the University of Tokyo then was the »imperial« university, the influence of government was naturally very strong and it is not hard to imagine that the policy directed the university to choose the German social sciences not only for the university but also for the country.

Social Transformation of the Japanese Society and Encountering Marxism

While the German dominance of the social sciences import influenced the Japanese academe, the societal situation in Japan changed dramatically. Around 1883, the translation of the words *society* and *sociology* were introduced to Japan. However, the Japanese meaning of »society« (*shakai*) had a totally different meaning from the Western meaning of society, which meant »civil society« in English and »buergerliche Gesellschaft« in German and represents a societal unit incorported in the political system of a country (Ishida 1984: 47). *Shakai*, however, described people engaging in the process of rapid industrialization who were considered as dropouts of the existing unified societal order. In other words, *shakai* meant something negative and problematic that tended to do harm to the political authorities. *Shakai* was a rather odd thing unlike politics, literature, and commerce. In opposition to this mainstream thinking reflected in the negative connotation of the term *society,* some Japanese social scientists founded, »the society of social politics« (*shakai seisaku gakkai*). This organization tried to imitate another German idea, the »Verein fuer Sozialpolitik« in Germany (Ishida 1984: 56), which then widely influenced the social scientists, intellectuals, industrialists, and other professionals who played significant roles in building the new Japan responding to the emergence of new classes in the Japanese society. The emergence of this group had a great impact on the development of economics in Japan. The independence of the Department of Economics from the Department of Law at the University of Tokyo in 1919 implies the beginning of a new era in the Japanese social sciences. The same group of social scientists played another important role by nurturing successive generations of influential Japanese social scientists and developing another academic realm for social sciences in Japan. Due to their activities, Japanese social sciences acquired more independency from politics and started developing out of one discipline toward other disciplines. Thus gradually, the development of the social sciences in Japan was formed by other people other than the politicians and the government. The movement of the society of social politics (*shakai seisaku gakkai*) was notable in the history of Japanese social sciences because nonpolitical people took a lead in developing their own social sciences.

Marxism and Societal Change

Before the Japanese social sciences met Marxism, they were strongly influenced by the new Kantian school in Japan. As the background of this, the fact that many of Japanese scholars studied in Germany, which was considered in Japan as the most advanced country in philosophy at that time, should be noted. They studied the theories of new Kantian school as expounded by Stammler, Lask, and Rickert among others, returned to Japan, and spread the theories. The introduction of the new Kantian school theory was the turning point in the Japanese social sciences because they became differentiated from only one discipline focusing on the analysis of nation-state toward a multiplicity of disciplines, supplied with a differentiated methodological basis by the new Kantian schools theory.

However, the theories of the new Kantian school tended to be considered as too philosophical, abstract, and—more importantly—lacking in any analysis of the more practical living conditions of the Japanese society. Therefore, many scholars gradually shifted their interests to Marxism to compensate for these negative aspects of new Kantian school (Ishida 1984: 100).

Thus Marxism became more appealing to Japanese scholars and gained a lot of popularity in Japan, not only for epistemological reason, but due to the fact that Marxist theories addressed the social phenomena occurring in the developing Japanese society. In fact, at that same time in the period of industrialization, the »working class«, or so-called »the proletariat«, emerged in the Japanese society. With the emergence of this class of people, »*shakai*« (society) gradually lost its negative meaning that was attached in the early Meiji period. The Japanese society started to learn how to deal with »societal problems« it was facing and began to acknowledge them as core problems, thanks to the interventions of Marxism of Japanese scholars influenced by Marxism. This seems to indicate the origin of that Marxism was from then attributed being the generic approach of any Japanese social sciences. Marxism in this context meant reflections on social phenomena with economics being the theoretical basis of all other fields of sciences (Ishida 1984: 106).

Student groups in some universities were founded, which also became more and more interested in Marxist theory. The main activity of these student groups had nothing to do with studying the theory itself but was the organizational basis for their political activities by using the name »social sciences study group« (*shakai kagaku kenkyuusho*). To avoid any confrontation with the authorities, they pretended to be only study groups (Ishida 1984: 105) interested in theories. The »social sciences study group« at the University of Tokyo was established in 1923 and the same movement spread to the other universities in the following year.

In 1924, the nationwide organization of these groups was established and thus »social sciences study groups« became a major trend of radical political activists. Hiding themselves under the umbrella of academics, did not, however, help them remain invisible for long. The government considered these groups as politically active people engaged in opposition against the Japanese policy agenda, with the result that the »social sciences study groups« were banned at the high schools and universities by 1926.

This series of events concerning the abolishment of Marxists could be caricatured as an irony in that it was the government that was very keen to introduce any fashionable and influential sciences in Germany to Japan, which resulted in introducing something unfavorable for the government. Despite that, the trend of Marxism was not at all unimportant in the history of Japanese social sciences development as it guided the intellectuals about theories of knowledge and supplied a common intellectual base to the Japanese social sciences.

However, Marxists (social sciences study groups) were officially banned in the Japanese society, and it took about two decades to see Marxism again in the Japanese social sciences scene after the end of the Second World War.

U.S. Influence on the Japanese Social Sciences

As happened to many countries during that period, Japanese social scientists were utilized as the consultants of the policymakers during the war. After the suppression of Marxism, which the Japanese authorities used as an excuse to suppress any critical thinking, there seems to have been no remarkable scientific criticism on any matters. The authorities instead made it clear that social scientists would be expected to support the move of the country toward the war.

After Japan surrendered in 1945, the United States occupied Japan, and the country was entirely under U.S control. All the prewar societal and political systems were abolished, and the constitution of the country was revised.

As a consequence, U.S social sciences were introduced to Japan and influenced the Japanese social sciences massively from this period. In 1946, Japan started to »import« social psychology and American philosophy such as pragmatism (Ishida 1984: 176). What is most significant here is the fact that the Japanese social sciences started to experience diverse social sciences not only from Germany but from the other Western countries this time without any steering or restriction by the government. This was the beginning of new era for the Japanese social sciences, as now they were going to be dominated by U.S. models and develop stronger relations with the U.S. academe.

Higher Education in Japan

Undergraduates Studies Focusing on Practical Skills

It is widely said that any eighteen-year-old student in Japan could enter a university if they do not mind the level of the university because there are a large number of universities serving a dwindling population of college-aged students (Amano 2006: 186). As there are not such school systems in Japan as the »A-level« in the United Kingdom or »Abitur« in Germany, all high school students are allowed to take any university entrance examinations. This has caused a severe competition for entrance into the better universities or high schools, or even for entrance into the well-known private elementary schools that are affiliated with the well-known private universities to avoid any further severe entrance exam competition at the later stage. Although going to university is not the only choice, the percentage of more than 50 percent of secondary-level students advance to higher education in Japan (Amano 2006: 46; Ministry of Education, Culture, Sports, Science and Technology (MEXT) 2008a), and this might be because there is no selective school system, which decides who could go on to the higher education, and also because of the competitive university market in which each university tries to get a sufficient number of students as the source of revenue. Undergraduate studies are for many students places to move when they finish high school so that they could get better jobs in the future. It normally takes four years to obtain a degree in a Japanese university, and a bachelor degree would be rewarded after successfully writing the graduation thesis.

Accordingly, university curricula focus on practical skills such as information technology and foreign language—usually English—with an emphasis on achieving a high score on the Test of English for International Communication (TOEIC) (Yamada 2005: 181), which many Japanese companies highly value. Regarding foreign language education in the Japanese higher education system, McVeigh (2004) states that »acquiring English skills is vital to climbing up Japan's education-examination ladder toward better employment« (McVeigh 2004: 217).

Thus curricula at the undergraduate level in Japanese universities generally are not so much connected to postgraduate studies but to careers as a company employee. Most of the Japanese undergraduate students even start to find a job before they get degrees, taking for granted that they would get a degree with the four-year-study in university, although there is no guarantee that every student will complete their studies in four years. This is very normal, and the more students of certain universities get jobs with well-known companies, the more people apply to those universities in the future. As the majority of Japanese universities are private, it is also crucial for universities to attract more applicants, or even better applicants, for their own reputation, which leads to a better financial situation and future prosperity. After all, the undergraduate studies in Japan are closely related to whether the students are able to get good jobs after four years and are less aimed at higher academic degrees and academic careers.

Graduate Schools between Research and Labor Market

Japanese graduate schools normally take around five years if one would like to acquire a doctorate, and the majority of students who go onto the graduate schools stop studying when they acquire master degrees. It is quite common that the engineering students tend to go onto the graduate schools to acquire master degrees, which would lead them to better work positions. On the other hand, a smaller number of students in social sciences go onto graduate schools. As a result, students in social sciences are also a minority in Japanese graduate schools even though more than 50 percent of the undergraduates in Japan study social and human sciences (MEXT 2007). There are fewer social sciences graduate students because advanced degrees in the social sciences do not seem to translate into better job opportunities. Those who acquire degrees beyond the bachelor in social sciences tend to be considered as too old and more difficult as employees in the labor market since they do not have technical or practical knowledge and skills which, for example, people who have acquired a master degree in engineering have.

Thus even the postgraduate studies in Japan are partly geared towards the labor market, and it is no wonder that few people choose to study social sciences for higher degrees than bachelors. This tendency for higher education to be considered as the preparatory stage for the Japanese labor market could be seen the current movement to create »professional schools« similar to the ones in the United States, and some universities have already founded law schools or business schools to teach not only theories but more practical aspects so that the students could contribute to the future judicial circle or world of business in Japan (Amano 2006). Thus the Japanese graduate schools are divided into two: one consists of graduate schools that focus on research activities to create new knowledge, and another is made up of professional schools that lead students to the practical fields such as law, business, medical science, and education.

The table below shows the number of postgraduate students in some countries regardless of their major subjects per population of 1,000 people. The number of graduates per 1,000 people in Japan is low. Although a high figure does not guarantee the high quality of scholars, the low Japanese figure might imply that the Japanese academe could be, in number, a real minority in the world.

Table 1: International comparison of education index 2008: Population of graduate school students per 1000 people

Country	Number of postgraduates per 1,000 people
Japan	1.91
U.S	4.51
U.K	3.79
France	8.62
Russia	0.99
China	0.63
Korea	5.75

(MEXT 2008b)

Higher Education Preparing for International Academic Activities

Unlike in the Western academe, a Ph.D. is normally not necessary to become a professor in Japan (Amano 2006). In fact, many Japanese professors with only a master degree could be found in any Japanese university. This might be one of the reasons why also a small number of postgraduates obtain doctoral degrees. Given that a doctoral degree is not a precondition to enter the world of academe, anyone would wonder why they should struggle for some more years toward uncertainty that one might not get a job beyond the academe even with doctoral degrees (Mizuki 2007). Therefore, when it comes to the level of academic degrees obtained by professors in the world, the Japanese professors look academically rather poorly qualified.

The language ability is an essential issue for Japanese people in terms of internationalization, as Japan is a monolingual country and the education system so far has not been very successful in providing the pupils and students with sufficient quality of education in English, despite the fact that they spend so much time learning English from the secondary to the higher education (Yamada 2005). If we refer only to social sciences, although it is not obligatory for students to write their thesis in English, which is seen in some fields of natural sciences in Japan (Nitta 2006), this is gradually changing as many universities set up the curricula to teach more communicative and academic English to match the competence requirements for international activities at the undergraduate level. It is not clear whether this change would also be reflected in the language ability of postgraduates, considering that most of the undergraduates choose to work as company employees.

Finally, there are few opportunities to have any international experiences for postgraduates. They might come across some international projects within the country; however, such projects normally would be different from international projects that include academics coming from different cultural and academic backgrounds, including different languages. Needless to say, few universities teach how to prepare for international collaboration activities. Thus it seems too challenging for young academics to collect any experience and consequently makes it difficult for them to work in any international context. Opportunities for the young researchers to experience any international collaborative research activities only depend on chance or the coincidences that they happen to come across. In other words, there seems little constructive and strategic way to develop young researchers in Japan for future international activities. Although there are some programs that are for the young researchers (Japan Society for the Promotion of Science 2009a, 2009b), these are limited opportunities and very selective. Even though some exchange programs and curriculum reforms were implemented at institutional level, Huang (2009), referring to two national surveys undertaken in 1992 and 2007 on internationalization of Japanese academic profession, concludes:

»fewer numbers of faculty members in 2007 thought that it was no longer necessary that further efforts should be made in undertaking exchange activities with scholars, even for their professional activities, or that their universities should undertake exchange activities with foreign scholars or students. More important, a vast majority of faculty members maintained a rather passive attitude toward a further internationalization of their curriculum.« (Huang 2009: 157)

This discrepancy between the policy and the academic reality is noteworthy in thinking of the perspectives toward international academic activities in Japan. It indicates what different viewpoints Japanese scholars take from ones of the government and their institutions.

Studying Abroad—a Different Type of Academic

It should be mentioned that there are many young academics who go to universities abroad and get degrees, including Ph.D.s (Huang 2009). Since many are going abroad privately without any grant, the exact figures about how many young academics go to foreign universities and to which countries are not documented. However, there are certainly young Japanese academics who make their academic careers in other countries. They are the exception to the aforementioned mechanism, as they are obviously not educated in the Japanese higher education system. This does not imply that they are better than the students in Japanese universities; however, one can assume that are more experienced in terms of learning in an international context, as they must accommodate themselves to different cultural, linguistic, and academic backgrounds. They would have to master one or more foreign languages to achieve degrees in a foreign country and also would learn different ways of thinking from the Japanese ways. An interesting topic for a research study would be to compare the preparedness for international collaboration activities of these young academics who studied abroad and the graduates coming from the Japanese higher education system.

Culture and Foreign Language Communication

Japan is a monolingual country and has never been widely influenced by Western languages such as English, German, and French in terms of the language development. English is taught at the secondary level and is an obligatory subject as English is necessary for university entrance examinations in many cases. Pupils in Japan study English grammar, writing, and reading quite intensively at school for this particular purpose. Therefore, university students must be proficient at a certain level in English to pass the entrance examinations, and needless to say, the postgraduates who are likely to be future scholars cannot be illiterate in English. They, of course, study how to read foreign literature for their studies and also to write academic papers at the postgraduate level.

However, much less communicative English seems to have been taught until quite recently even at the postgraduate level. This situation is gradually changing through the introduction of English language instruction at the primary level (Yamada 2005), but this could not influence the current English level of Japanese scholars.

Although there are no exact studies available, it is commonly agreed among the Japanese academe that most Japanese social science scholars do not seem much appealed to working in any international context. As the report by the Science Council of Japan phrases it, Japanese scholars tend to carry out their discourses in » semai mura shakai« (small village society) (Science Council of Japan 2002). Japanese academics rather seem to prefer to stay in their local universities and to undertake research in specialized topics that are incomprehensible to other academic researchers.

If this rather locally orientated and very specialized academic mentality prevails in the Japanese academe, it seems obvious that such a learning environment might not be conducive for young academics who want to work in any international contexts.

As an interviewed Japanese scholar phrased it, »The professors who don't have the ability of speaking good English don't encourage the students to go out [of Japan]« (individual interviews carried out with Japanese scholars in 2008). More than this, in Japanese academe, professors are very influential for the careers of young academics, and if students aim to become a scholar in the future, students are more likely to focus their academic life on the very local academic networks. International activities, or even any work in a foreign language, especially in English, is not advantageous for young academics who want to foster these local networks for a career.

English language abilities are, however, just one precondition for the participation in international scholarly activities. Japanese scholars seem to be quite unwilling to express themselves in English not only because of their language abilities but also because of the embarrassment or the uncomfortableness when they try communicating in English.

As Japanese scholars expressed in interviews of a small-scale study carried out in the FP 6 Global SSH project: »Japanese communication style is the disadvantage [...] We don't have the culture of enjoying communication or discussion. Most brilliant psychologists finally decided to move to the U.S. because they miss those kinds of communication with colleagues. They miss the discussion« (From individual interviews carried out with Japanese scholars in 2008). First of all, Japanese scholars do not dare to make any mistakes when they speak English. It is a common feeling among Japanese, who have been taught throughout their education that making mistakes is shameful. McVeigh (2004) illustrates reasons why many Japanese people do not dare to speak English as follows: »They are too concerned with their improper pronunciation; they will not speak unless they can produce perfect grammatical sentences; and, they are afraid of making mistakes« (McVeigh 2004: 218).

Hence, it is not hard to imagine that those elites being academically top people in Japan would not like to reveal their awkwardness in front of anyone, especially of other academics from other countries. This mindset casts a long shadow when learning foreign languages. It might be difficult for those who speak English to understand but this mindset seems to make the cultural barrier higher than the mere difficulty of acquiring the language itself.

Second, and possibly even more important, Japanese scholars are not really trained in academics discourses. Strictly speaking, it is not even a matter of training but a part of Japanese communication style that most Western people might not be aware of. In Japanese communications direct and open statements could be considered as being impolite. While the effects of such intercultural communication phenomena are widely researched and discussed also concerning the particular Japanese communication styles, how this particular Japanese communication style effects international academic discourses seems to be widely unresearched.

However, it is most obvious that a communication style in which saying »No« openly to other people is considered as an offence against the communication partner must have major implications for scientific discourses of Japanese scholars with academics from a more direct communication style.

Adding to this type of indirect communication, another factor certainly affects the communication abilities needed for international academic activities. Self-assertion is not considered a virtue in Japan. In Japanese communications it is always better to listen to other people and contemplate what they would like to hear than telling others what he/she think. Therefore, Japanese are not accustomed to expressing directly what they would like to say. Needless to say, such communication style is incorporated in the Japanese language and—vice versa—other languages do not provide the linguistics means for such communication. This creates obviously some particular difficulties for Japanese communicating in English. Hosack's (2005) study describing the difficulties Japanese university students have in carrying out peer review illustrates this challenge originating from a Japanese specific communication style. His study shows that students were very much worried about writing negative reviews for their classmates unless the review was undertaken anonymously. This stems from the students' uneasiness about making negative comments on others' compositions as they might »hurt« the classmates by the comments. Together with the fact that they cannot express their disagreements, this Japanese character that people respect the harmony with others and dislike any conflicts with others does not seem to be adequate when they have to face up with foreign academics and their different communication styles.

If, as is frequently said, that languages influence cultures, then any language ability is not a mere issue about how to acquire the certain language. The more challenging issue in particular in scientific communications, where language plays such a manifold and crucial role (Weidemann/Kuhn 2005), is about how to learn other languages' cultures. In other words, if Japanese speak English with a Japanese mindset, they would inevitably have severe difficulties in lingua franca communication not coincidentally resulting in a »culture shock«, as Kalervo Oberg phrases it (Otani 2007). Language is a certainly a tool for international scholarly activities but there are obviously many cultural components incorporated in the language, which seem to have a much greater impact on scientific communications.

The Concept of Translation

Very much coinciding with the historical creation of the country based on an imported society model there seems to be a kind of faith among Japanese people (Suzuki 2006) that anything coming from abroad is positive, and this view seems also to apply to the historically import-oriented social sciences and in particular guide the design of Japanese science policies.

In Japan, the word *internationalization* is almost understood as a synonym for being able to speak, write, and read English. Translating in both directions therefore has become the major concept of Japanese international science policies.

Given the above-mentioned status of English in Japanese higher education and the challenges for Japanese communication in English, the participation of Japanese scholars in and contribution to international social science knowledge do in fact heavily depend on lingua franca abilities. This fact might have contributed to conceptualizing international collaborations as a matter of translating from and into English.

Many non-English-speaking Western countries also encourage scholars to publish in English via various incentives. In the case of Japan, the number of citations or the number of theses written in English has not even always necessarily been counted as the evaluation criterion measuring scholars' achievements, and there seemed to be little pressure for Japanese scholars to publish their research findings in English.

In Japan, it has only recently been realized that Japanese academe should urge scholars to write papers in English not only to import knowledge created elsewhere to Japan but to also export the knowledge created in Japan to the world. A recent report by the Science Council of Japan in 2002, which points out the »closed academic societies« (the Science Council of Japan 2002), particularly in social and human sciences, that take for granted that scholars only have to write papers in Japanese, exhibits a serious concern about Japanese SSH scholars' less acknowledged and less evaluated work in the other parts of the world. It is also suggested that solid English education, particularly focusing on academic writing, communicative English skills, and long-term (about three years) sabbatical programs abroad should be considered in the future to achieve the purposes suggested by the report (the Science Council of Japan 2002).

However, translation still plays a major role in either direction, and it is widely discussed that translation should serve the participation of Japanese social sciences in international knowledge (MEXT 2002).

The concept of translating knowledge always played a crucial role in the importation of the Western sciences to Japan, and it is very understandable that Japan had to rely on translated literature in the late nineteenth century to the beginning of twentieth century, considering the aforementioned historical background of the social sciences in Japan. Since then, Japan has imported most of basic scientific knowledge and theories from Europe and the United States. It still continues doing so, as Huang (2009) states that »the academic profession in Japan still maintains its basic character of being engaged in a process of catching up with advanced overseas countries, mostly identified with the English-speaking countries in Europe and especially the United States« (Huang 2009: 155), and many books are still translated from English or other foreign languages into Japanese. Now, the attention suggested by MEXT in 2002 is to translate Japanese papers into English by professional translators so that excellent studies implemented by Japanese scholars could be acknowledged by Western scholars (MEXT 2002). Such a shift toward also exporting knowledge created in Japan would certainly be useful for scholars who are less confident of writing in English.

Nevertheless, it should be noted that this would not fundamentally change the participation of the Japanese academe in international discourses. The fact that Japanese scholars are confronted with the above-mentioned communication challenges to join the discourses in the existing lingua franca would not change substantially if the communications still rely on translation as a major means for participation. The wide discussion among Japanese scholars about the technological feasibility of automatic translation options (Otani 2007) shows the ongoing view on international collaborations as a dependency on translation, which does, however, not really allow for active participation in direct scientific discourses. Also the fact that the above-mentioned suggestions made by the Science Council of Japan, of which many members are most prestigious emeritus scholars, illustrates the extent to which the very traditional concept of translating and exchanging knowledge beyond any discourses are still dominating ideas in the Japanese social science community.

English-language abilities are of course a necessary prerequisite, but cannot substitute for scientific discourses. Otani (2007) rightly criticizes the »easygoing attitudes in exchange programs set up by the Japanese government and MEXT [...]. They think of internationalization or international mutual understanding would be fulfilled if only we, Japanese, go abroad or we bring any foreigners to Japan« (Otani 2007: 169). Similarly, Aspinall (2000) who also argues about the Japanese policy approach promoting international activities, quotes Hall (1998) claiming that »the downside of this formula is that it deliberately minimises human contact—the pattern first set by the Meiji government when it sent students abroad and invited foreign teachers to Japan, putting each of the two groups back on the return steamer as quickly as possible« (Hall 1998: 174, quoted by Aspinall 2000: 18). It seem that the Japanese social science community has started to discuss the traditional approaches of sciences policies questioning the old paradigm considering the participation of Japanese social sciences in international scientific collaborations via the categories of foreign trade.

Science Policies and Higher Education

There have been various university reforms in Japan from around the 1990's to the beginning of this century (Amano 2006; Tsukahara 2008; Huang 2009). The government established a great number of committees and councils for a series of university reforms (MEXT 2009) and discussed different perspectives for improving the internationalization of Japanese higher education and the Japanese academe, including the aforementioned translation service, research programs abroad, and also exchange programs with foreign scholars. However, they all seem conventional and unelaborated, sticking to the old paradigm of importing and exporting knowledge.

Even though the government urged universities to make independent attempts toward internationalization, the extent to which they are able to change toward more internationalized academic life still depends on the capacity of individual universities. Some, for instance, well-known research universities, are very capable and flexible enough to create new programs, while others are not due to the financial situation, the human resources, and their facilities. Without major governmental support it seems hardly possible for all the universities to put the internationalization programs into practice.

As a part of the recent science policy, closer collaborations among Japan, China, and South Korea should be mentioned. Han and Shin (2008) enumerate a number of collaborative activities held after 2000 among those countries and suggest grounds for the collaboration that »as East Asian countries experienced similar economic changes and social problems due to globalization, researchers in East Asia began to develop joint research« (Han/Shin 2008). Adding to this, it is possible to consider that the collaborations among Eastern Asian countries are preferred due to cultural closeness among these countries as well as the geographic closeness. While the countries in East Asia are developing closer relationships because they are neighboring countries, they also started promoting close relations to internationalize academic activities in East Asia. Han and Shin rightly call these efforts rather an »Asianization« (Han/Shin 2008) than internationalization; however, it could be a milestone for all these countries toward a new general approach to improve internationalization of the scientific work of scholars beyond these East Asian countries.

Conclusion

Internationalization of the Japanese social sciences seems an entirely contradictory process. On one hand, the Japanese social sciences do not need to be transformed into an international science community, since even their creation was the result of an international import from other science communities in other countries.

On the other hand, unlike numerous colonies, which also imported social sciences from the Western science world, this import was neither initiated nor conceptualized as a means of the Western countries to export their knowledge into the colonies. In the case of Japan, importing social science knowledge was a means of the importing country, and this import was accordingly controlled by and for the Japanese interests of becoming as comparable, competitive, and powerful a country as those from which they most selectively imported this knowledge.

This imported knowledge and the creation of a social science community was most obviously a top-down activity clearly steered toward supporting the policy agenda of making Japan an economically, politically, and militarily powerful country. Social science knowledge was imported for this purpose and even the way it was imported via very selective translations of given knowledge and adjusted to the needs of this policy agenda was directed via this policy agenda.

Finally one can conclude that all the particular difficulties the Japanese social science community is encountering today in participating in international activities seem to be the result of precisely these very imported Japanese social sciences. Not only does the concept of imported knowledge lead to a perception of a social science community that consumes and applies knowledge that has been created elsewhere, but adjusted to the particular needs of a the Japanese society, but the manifold meaning of the notion of translating this very imported knowledge today seems to create culture hurdles for the Japanese social science community to participate in an internationalizing social science community.

Even the most recent shift in international science policies, again politically initiated, to now also export knowledge, shows the implied restrictions of a very nationally steered science community. Exporting knowledge again via translation, this time in the other direction, sticks to conceptualizing the exchange knowledge as if it were an international trade.

Thus the current status of internationalization of the Japanese academe seems still conceptually similar to the basic concept during the Meiji period at the beginning of Japanese social science history.

References

Amano, I. (2006): Daigaku kaikaku no shakaigaku (Sociology of University Reform). Tokyo: Tamagawa University Press. [In Japanese]

Aspinall, R. (2000): Policies for »Internationalization« in the Contemporary Japanese Education System. Studies in Language and Culture, 21 (2), 3-21. Nagoya: Nagoya University

Han, S.-J./Shin, K.-Y. (2008): Internationalization of Social Science in Korea. Presentation of Global SSH Project Paris Workshop held in April 2008

Hosack, I. (2005): The Effects of Anonymous Feedback on Japanese University Students' Attitudes towards Peer Review. Yamamoto Iwao kyouju taishoku kinen ronshu Kotoba to sono hirogari (Commemorative Essays for Professor Iwao Yamamoto's Retirement: Words and Their Dimensions), 3, 97-322. Kyoto: Ritsumeikan University. [In Japanese]

Huang, F. (2009): The Internationalization of the Academic Profession in Japan. Journal of Studies in International Education, 13 (2), 143-158. Sage

Ishida, T. (1984): Nihon no Shakaikagaku (Japanese Social Science). Tokyo: Tokyo University Press. [In Japanese]

Japan Society for the Promotion of Science (2009a): 2 Tokubetsu kenkyuuin (Research Fellowship for Young Scientists). http://www.jsps.go.jp/j-pd/pd_saiyo.htm. [Date of last access: 13.08.2009]. [In Japanese]

Japan Society for the Promotion of Science (2009b): Kaigai tokubetsu kenkyuuin (Postdoctoral Fellowship for Research Abroad). http:/www.jps.go.jp/j-ab/ab_gaiyo2.htm. [Date of last access: 13.08.2009]. [In Japanese]

Ministry of Education, Culture, Sports, Science and Technology (MEXT) (2002): Jinbun shakaikagaku no shinkou ni tsuite (On improvements for Human and Social Sciences). http://www.mext.go.jp/b_menu/shingi/gijyutu/gijyutu4/toushin/020601.htm. [Date of last access 13.08.2009]. [In Japanese]

Ministry of Education, Culture, Sports, Science and Technology (MEXT) (2007): Gakkou kihon chousa (Basic survey on Schools and Educational Institutes). General Views on Higher Education including the breakdown of number of students by disciplines. http://www.mext.go.jp/b_menu/toukei/001/08010901/002/001/001.htm. [Date of last access 13.08.2009]. [In Japanese]

Ministry of Education, Culture, Sports, Science and Technology (MEXT) (2008a): Kyouiku shihyou no kokusai hikaku (International Comparison of Education Index 2008). Rate of Advancement to Higher Education. http://www.mext.go.jp/b_menu/toukei/001/08030520/004.htm. [Date of last access: 13.08.2009]. [In Japanese]

Ministry of Education, Culture, Sports, Science and Technology (MEXT) (2008b): Kyouiku shihyou no kokusai hikaku. (International Comparison of Education Index 2008). Population of graduate school students per 1000 people. http://www.mext.go.jp/b_menu/toukei/001/08030520/006.htm. [Date of last access 13.08.2009]. [In Japanese]

Ministry of Education, Culture, Sports, Science and Technology (MEXT) (2009): Kokkoushiritsu daigaku wo tsuujita daigaku kyouiku kaikaku no shien. (Governmental Supports for University Reforms). http://www.mext.go.jp/a_menu/koutou/kaikaku/index.htm. [Date of last access 13.08.2009]. [In Japanese]

McVeigh, B.J. (2004): Foreign Language Instruction in Japanese Higher Education: The Humanistic Vision or Nationalist Utilitarianism? Arts and Humanities in Higher Education, 3 (2), 211-227.

Mizuki, S. (2007): Kougakureki Working Poor (Working Poor with High Academic Background). Tokyo: Koubun-sha. [In Japanese]

Nitta, S. (2006): Koutoukyouiku eno FD dounyuu to kyouiku kaikaku no genjou (The Introduction of Faculty Development into Japanese Higher Education and Academic Reform in Japan). Housei daigaku Tama ronshuu, 22, 51-69.

Otani, T. (2007): Nihonjin nitotte Eigo towa nanika (What Is English for Japanese?) Tokyo: Taishukan. [In Japanese]

Science Council of Japan (2002): Nihon gakujutu no shitsuteki koujou eno teigen (Suggestions for qualitative improvement of Japanese academics). http://www.scj.go.jp/ja/info/kohyo/pdf/kohyo-18-t979-1.pdf. [Date of last access 13.08.2009]. [In Japanese]

Suzuki, T. (2006): Nihonjin wa naze Nihon wo aisenainoka (Why do Japanese not love Japan?) Tokyo: Shincho-sha. [In Japanese]

Tsukahara, S. (2008): Koutou kyouiku shijou no kokosaika (Internationalization of the Higher Education Market). Tokyo: Tamagawa University Press. [In Japanese]

Weidemann, D./Kuhn, M. (2005): Speaking the Same Language? Lingua Franca Communication in European Social Science Research Collaboration. In: M. Kuhn/S.O. Remoe (Eds.), Building the European Research Area (pp. 85-113). New York: Peter Lang

Yamada, Y. (2005): Nihon no Eigo kyouiku (English Education in Japan). Tokyo: Iwanami-shoten. [In Japanese]

Internationalization of Social Science in South Korea: The Current Status and Challenges

KWANG-YEONG SHIN, SANG-JIN HAN

Introduction

In this paper we explore the status of internationalization of social science in Korea and discuss new challenges from the internationalization of social science there.[1]

Although East Asia first imported social science in the late nineteenth century, the discipline entered Korea mainly through Japan in the early twentieth century and America in the latter half of the twentieth century. However, full-scale social science research started in the late twentieth century in Korea. Since social science research requires institutional arrangements and trained researchers, the first interesting research results took some time to develop. In particular, the rapid increase in the number of social scientists from the late 1980s contributed to the rise of social science research in Korea. The scope of research interests also expanded from domestic to international issues with the rise of new researchers trained abroad.

Political democratization played a significant role in promoting international social science research as social scientists tried to understand

---

1 This paper was presented at the Workshop for Global Social Science and Humanities at Maison des Sciences de L'homme in Paris, France, April 23-24, 2008. We thank the participants of the workshop for their valuable comments.

economic growth and political change in East Asia from a comparative perspective. Moreover, research on labor policy and social policy in Europe has increased as state policy has become a contested issue in the public debate and policy discourse. Democratization altered the parameter of social science research as well as the political behaviors of citizens. Comparative studies and area studies became a new trend in social science in Korea.

Furthermore, when South Korea became a member of the Organization for Economic Cooperation and Development (OECD) in 1996, Korean social scientists had to broaden their scope of research to OECD countries. Topics and issues of research in Korea began to be discussed internationally as other OECD member states became reference societies. The Korean government supported this broadening of scope of social science research as it introduced the graduate school of international studies at big universities and finally supported international research through the Korea Research Foundation (KRF). The KRF also funded international academic meetings organized by academic associations in Korea and supported participation of Korean scholars and graduate students at various international congresses. In short, globalization of education and research in social science became a new trend in Korea in the late twentieth century.

Historical Background

Most modern social science in East Asia has been imported from the West since the nineteenth century. While there was some indigenous social science tradition, social science in the West overwhelmed it with new institutions such as the university and standardized curriculum. The process of importation of social science from the West to East Asia included both institution building and translation of theoretical concepts. First of all, it required an establishment of a modern university system, recruitment of professors, and choice of curriculum. In addition, social scientific theory and research practice were needed in order to talk about social science in reality. Mostly, theory and concept of social science were imported from the West.

The development of social science research has been closely related with political development and globalization in South Korea. Social science research began to be activated in some limited areas such as rural community development and population studies in the mid 1960s. The Korean government attempted to formulate policies for economic growth, and the United Nations Development Program (UNDP) tried to assist the Korean government to control population growth.[2] But other social science research has not been promoted until the 1980s, though many social science associations were already formed in the 1950s.[3] The 1980s was the period when student movements for democracy exerted significant impact on the course of political development in South Korea. Social science discourse provided theoretical resources for articulating radical discourse for students' movements.

During the long period of authoritarian rule, social science research was discouraged and sometimes suppressed. Because some social scientists were critical of the military government, it tried to block dissemination of critical ideas of professors to students. Possession and reading of books on Marxism and socialism were banned by the Anticommunist Law until the 1990s. Some social scientists were active in prodemocracy movements, challenging the military government. Sociologists who were classified as political dissidents were ousted from universities and many of them were put in jail.[4] The role of university professor as researcher was never emphasized by the government in that period. However, students in the college of social science in many universities have been also active in the student movement for democracy. Students of sociology among others played a key role in the struggle against the military regime for more than two decades.

Emphasis on social science research was connected to democratization and globalization that concomitantly took place in the late 1980s and the early 1990s. The transition from authoritarianism to democracy transformed the role of university drastically. As democratization proceeded, the role of the university began to alter from the site of the democratization movement to the site of knowledge production and education. In the late 1980s and the early 1990s, the collapse of the state socialism in Eastern Europe, which coincided with democratization in Korea and diminished the appeal of radical ideologies, contributed to transforming the role of the university from agency for democracy to agency for knowledge production.

---

2 In 1974, there was a massive fertility survey funded by the United Nations Fund for Population Activity. In its scale and precision, this survey was unprecedented in Korea.

3 Major social science associations were formed in the 1950s. For instance, the Korean Economic Association was formed in 1952 during the Korean War. The Korean Political Science Association was established in 1953. The Korean Sociological Association was launched with 17 members in 1957.

4 For example, Wan-Sang Han and Jin-Kyun Kim, professors of the Department of Sociology at Seoul National University, were expelled from the position in the 1980 and reinstated in their previous positions in 1984.

As the Korean economy became more intensely integrated into the world economy, the Korean government emphasized scientific research in the university to enhance competitiveness of manufacturing goods in the world market. Scientific research was considered as the basis of industrial competitiveness in the early 1990s. The poor research performance of Korean universities was reported in the national media. The ranking of universities in the world as an indicator of competitiveness in scientific research was regularly reported in the major newspapers. Competitiveness became a buzzword of academicians as well as laymen. Thus improvement of the research capacity of the universities was considered as one of major tasks of the government.

Consequently, R&D became one of major concerns of the universities and the government. With an emphasis on research, the R&D expenditure increased sharply at both the university and governmental level.[5] The discourse on the scientific citation index (hereafter SCI) became popularized among media and academics as a key indicator of research capacity of natural science and engineering schools. Discourse on SCI became popular even among laymen from the 1990s.[6] Nevertheless, emphasis of research by the government was on natural science and engineering in the 1990s.

In the late 1980s, as a result of a large increase in the number of social scientists, social science research began to establish itself separately from natural science. This rapid increase in the number of social scientists resulted from the popularity of social science in the 1980s. Some students entered graduate schools to continue their study of social science at home and abroad, and this enhanced social science research. New social scientists showed interests in the social and political development of Korea from a comparative perspective. An explosion of social science research on foreign countries occurred in the early 1990s as well. Social scientists began to do research on foreign countries that had already experienced democratization, a labor movement, and various social movements. That was a paradigm shift from the previous social

---

5   The R&D expenditure was 1.79 percent of the GDP in 1990. It was 2.37 percent in 1995. It consistently increased up to 2.39 percent in 2000 and 3.23 percent in 2006, respectively. Due to the financial crisis in 1997, the rate of increase of the R&D expenditure in the GDP was reduced until 2003 and resumed to the previous level from 2004. Here, R&D expenditure of social science and humanities is not included in this statistics. See the Ministry of Education and Science (2007).
6   The first discourse on SCI appeared in Korea in 1990. Dong-A Daily Newspaper reported that R&D expenditure and science education in Korea lagged behind advanced countries with regard to the number of research papers listed in the Science Citation Index (see Shin 2007: 102-103).

science in which social science research was confined to the Korean society only.

Internationalization of social science research took place in three stages in Korea. The first stage was internationalization of research interests and subjects. Traditionally, area studies have focused on understanding the society and culture of other countries. Area studies have been initiated and monopolized by the core countries that exercised hegemonic power over the world. Social science research about foreign countries has been difficult for the peripheral countries because of constraints of financial resource and number of researchers who are interested in other countries. Rapid increase of social science research on foreign countries occurred in the 1990s in South Korea.

The second stage of internationalization of social science was an increase of research collaboration among researchers across countries. Korean researchers began to show interests in doing joint research with researchers from other countries. Korean researchers began to engage in research projects coordinated with foreign researchers as well. More often, foreign researchers began to join research projects initiated by Korean researchers. International collaborative research has been rapidly increasing due to the financial support by the Korea Research Foundation (KRF) and the increase in number of Korean researchers with research networks in other countries.

However, the second stage of internationalization of social science research has limitations since most of social scientists in Korea were trained in America and research paradigms of American academics have been predominant in South Korea.[7] The dominance of American influence on Korean social science has been reinforced as publication in English-language journals and teaching in English has been highly valued by the state and university. Recently some universities in Korea began to accentuate teaching sociology in English as a part of globalization of campus. Thus some even argue that what has occurred is the Americanization rather than internationalization of social science in the late twentieth century in Korea.

The third stage of internationalization of social science research is the globalization of the education of future generation of researchers. Education for graduate students beyond the national boundary has been emphasized by universities and the government. The program of the

---

[7] For example, more than 90 percent of faculty members of some departments of social science such as economics, political science, and international relations at Seoul National University received their Ph.D. degrees in the United States. This figure is 68 percent in the Department of Sociology.

Brain Korea 21st, the program that the Korean government funded to improve education at the graduate level by supporting graduate students and postdoctoral researchers, emphasizes international experience of graduate students in research and academic activity.[8]

In the 2000s we have witnessed East Asianization of social science research in Korea in the sense that the second and the third stage of internationalization of social science research in Korea have been mostly confined to East Asian countries. It may not be properly called an internationalization of social science research yet. However, the rise of internationalization of social science research in Korea reveals that there has been a fundamental epistemic change in social science research in the 2000s. Equally important, internationalization of social science research raises fundamental questions about the nature of social science itself.

Changing Concepts of Social Science Research

Social science research as an arena of scientific endeavor was a relatively new phenomenon in South Korea. The role of professors in university had been mainly focused on education of students until the early 1990s. Prior to democratization, which began in 1987, universities did not function well neither for education nor research. Instead the university in Korea had been a site of the student movement for democracy. Universities had played the most important role as a free and rebellious sphere where radical ideas were disseminated, new discourses against dictatorship were formulated, and struggles against the military regime took place. The military regime frequently closed down universities to suppress student protest movements, and the riot police entered university campuses to squelch student demonstrations. Until 1987, when the democratization movement successfully ended the military dictatorship, the role of professors in social science was limited and the research function of the university was almost paralyzed.

As the transition to democracy began after 1987, social scientists changed their focus from domestic issues to international or comparative issues. One of the major interests of political scientists and political sociologists was the nature of democratic transition in Korea in a comparative perspective. How was democratization in South Korea different

---

8   In 1999, the Korean government launched a new program to promote research departments by providing financial support to the graduate programs of two or three departments across the nation. It comprises the single largest part of the expenditure for higher education of the Ministry of Education and Science.

from that of other newly democratizing countries? The research question itself requires comprehensive knowledge of the other newly democratizing countries in Latin America, Africa, and southern Europe. A case study of Korean transition to democracy in a comparative perspective was done by international research groups organized by scholars from many countries (Diamond 1997) and Korean scholars (Choi 1995, 1996, 2005; Han 1984, 1989, 1995). Comparative labor research also has been a major area of comparative research among social scientists in Korea. It includes East Asian countries (Jang/Paek 2005; Lee 1995, 1997; Jung/Kim 1997; Kim 2004; Paek 1996, 2001; Shin 1990b, 1999; Song 2000), America (Latin American countries) (Cho H.R. 1994, 2002; Cho D. 1994, 1996, 2003, 2005), continental European countries, and Nordic countries (Kim 2004; Lee 2006; Moon 1998; Shin 1990a, 1994, 2000; Song 1996). Social movement has been also one of the core research subjects among Korean sociologists (Han 1994, 1997; Kim 2005; Shin 2006).

The plethora of comparative research and area studies in the 1990s was based on three factors. First, the number of social scientists trained to do comparative research increased rapidly. Mostly they were trained in America and Europe in the 1980s and returned to Korea in the late 1980s and early 1990s. The dominance of comparative perspectives marked a paradigm shift from the previous research tradition, which focused on Korean society only. Comparative perspectives not only broadened the scope of social science research but also allowed social scientists to raise new questions, for further research. In sum, we might call it a paradigm shift from a Korea-centered research to comparative research in social science in Korea.

Second, the Kim Young Sam government in 1994 began to emphasize globalization as a political slogan. In 1997, the government encouraged major universities to establish Graduate Schools of International Studies (GSIS) by providing financial support for these new programs. The rationale of the support was to educate experts on international relations and area studies. The Ministry of Education selected five universities by competition and provided 25 million U.S. dollars to the newly established GSIS at each university for five years. The government initiated educational reforms to prepare for globalization understood as »marching forward to the international market«. An enhancement of international competitiveness was an explicit motivation for the globalization policy of the Kim Young Sam government. Third, the Korea Research Foundation, the government funding organization for research established in 1981, introduced three research funding programs associated with international research in the 1990s: international collaborative research in 1993, joint research with foreign scholars in 1996, and overseas area studies in 1997 (KRF 2001: 149). In addition, the Korea Research Foundation made agreements with research foundations in other countries such as the Alexander von Humboldt Foundation (AvH), Deutsche Forschungsgemeinschaft (DFG) (1987), and Deutscher Akademischer Austausch Dienst (DAAD) in Germany (1988); the Australian Research Council (ARC) in Australia (1996); the Chinese Academy of Social Sciences (CASS) in China (1999); the Japanese Society for the Promotion of Science (JSPS) in Japan (2001); and the Centre Nationale de la Recherche Scientifique (CNRS) in France. The agreements promoted exchange of researchers between Korea and foreign countries.[9] While the government played a significant role in the process of internationalization of social science, the strengthening of civil society also contributed to internationalization of social science in the late 1990s. Citizen's organizations, later known as Nongovernmental Organizations (NGOs), have exploded since the first half of the 1990s and became the most important social agency to promote democratization of the Korean politics and society. As social movements experienced globalization, the number and importance of NGOs involved in concerted activities related with democracy, human rights, global peace, environmental protection, anticorruption, and transparency increased. Thus social demands for information and knowledge about international nongovernmental organizations exploded. Suddenly demands for experts who could deal with international relations within NGOs also increased. Korean social scientists' research on new social movements in Europe in the 1990s was mainly motivated by the rise of citizen movements in Korea.

Internationalization of research has proceeded in the 2000s when the Korean economy recovered from the financial crisis. As East Asian countries experienced similar economic changes and social problems due to globalization, researchers in East Asia began to develop joint research. The Thirty-sixth World Congress of the International Institute of Sociology (ISS) in Beijing was a platform to establish academic exchanges among sociologists in East Asia. That meeting catalyzed academic exchange, joint conference, joint research, and student exchange programs among East Asian countries. As globalization proceeds in East Asia, sociologists in East Asia, relatively independent of each other, be-

---

9 Mostly the exchange of scholars has been confined to Germany. One hundred and two Korean scholars were sent to Germany and 60 German scholars came to Korea under the exchange program between Korea and Germany from 1989 to 2000 (KRF 2001: 334). The program for an exchange of scholars between Korea and China started in 2000 with only 5 Korean scholars and 1 Chinese scholar involved.

gan to recognize the necessity of expanding the scope of academic activity beyond national boundaries so as to comprehend the new dynamics of globalization in East Asia.

Internationalization of social science research occurred at three levels. At the first level, academic exchange continuously increased at the national level. Sociological associations in East Asian countries signed memorandums of understanding to promote scholar exchanges. In 2007, the Korean Sociological Association (KSA) and the Japanese Sociological Society (JSS) made an agreement for organizing joint panels at each other's annual meeting every other year.[10] Prior to the agreement of regular scholar exchange between two organizations, in September 2006, the KSA organized the international conference titled »The Global Futures of World Region: The New East Asia and Vision of East Asian Sociology« in which more than twenty sociologists from Korea, Japan, China, and various Western countries participated. Korean sociologists and Japanese sociologists participated in the annual meetings of each association from 2007. There was a special panel for a roundtable discussion for sociology in Korea and Japan in the summer meeting of the Korean Sociological Association on June 23, 2007.[11] Korean sociologists also joined the annual meeting of the Chinese Sociological Association in Changsha in 2007 and Changchun 2008.[12] The participation of sociologists from other countries in annual meetings of the sociological association of each country is a new development in East Asian sociology. Though there has been frequent participation of sociologists from different countries in the special symposiums or international conferences, it was unusual to see participation in the annual meetings of sociological association from other countries.

At the second level, there has been a rapid increase of the number of special conferences devoted to issues associated with East Asia organized by universities. With the help from the Brain Korea 21st programs, departments of sociology began to organize international conferences exclusively focusing on issues in East Asia. For example, the Department of Sociology at Pusan University organized a conference on

---

10 The agreement on regular scholar exchange between the Korean Sociological Association and the Japanese Sociological Society includes responsibility of the host organization to provide local accommodation and airfares for the delegation of the invitee organization up to two persons.

11 Four sociologists from Korea and Japan presented their papers and more than twenty sociologists from both countries joined the roundtable discussion.

12 Sang-Jin Han (Seoul National University), Seung Kook Kim (Pusan National University) and others participated in the annual meeting of the Chinese Sociological Association in Changsha in July 2008.

»Globalization and Modernity in East Asia« in which professors and graduate students from East Asian countries participated.[13] The Department of Sociology at Chung-Ang University and Yonsei University also jointly held an »International Conference: East Asia Societies in Comparative Perspectives« in which professors and graduate students in sociology from Korea, Japan, China, Hong Kong, and Singapore participated.[14] Those conferences were organized by departments of Sociology, funded by the Brain Korea 21st programs in each university. Moreover, graduate students commonly participated in the conferences. The Brain Korea 21st program that launched in 2000 to strengthen graduate programs in humanities, social science, natural science, and engineering required graduate students to present their research papers at international conferences as a part of graduate training.[15]

At the third level, collaborative research has been developed to carry out research projects by working together as a research team. This is the highest level of internationalization of social science research. This type of research collaboration has been launched among sociologists in East Asia. The Social Stratification and Mobility (SSM) project in Japan invited sociologists from Korea and Taiwan in 2005. The SSM survey has been conducted by the Japanese Sociological Society every ten years since 1955. Prior to the 2005 SSM survey, the SSM projects were confined to Japan. Perceiving the necessity of enlarging the scope of research, the SSM project team in 2005 expanded the SSM survey to Korea and Taiwan.[16] Comparative analysis of the stratification system in East Asia might generate new perspectives to understand stratification systems in East Asia. Sociologists from Korea, Japan, and Taiwan share the data set collected from three countries and explore the stratification regime in East Asia from comparative perspectives, which is not predetermined by participants.[17] Considering the relatively recent develop-

---

13 The conference was held on the campus of Pusan University, November 10–11, 2006. It was possible due to the funding from the Brain Korea 21st programs.
14 This was a joint conference organized by Chung-Ang University and Yonsei University on October 27 at Yonsei University, Seoul.
15 According to the Korea Research Foundation, an international conference refers to the conference in which scholars from more than 4 countries participate.
16 The 2005 SSM survey was organized by Sato Yoshimichi at the Center for Social Stratification and Inequality at Tohoku University. The final reports of the 2005 SSM survey were submitted to the Ministry of Education and Science of Japan in March 2008.
17 It is a relatively loose research collaboration in the sense that researchers do not necessarily share common theoretical orientation but acknowledge the exploratory nature of the collaborative research.

ment of scholarly exchange among East Asian countries, the 2005 SSM project in Japan might be regarded as a milestone in the history of internationalization of social science research in East Asia. Comparable to the Comparative Analysis of Social Mobility in Industrial Societies (CASMIN) project at the Oxford University in comparative research on social stratification and mobility in Europe, the SSM 2005 project at Tohoku University in Japan provides a rare opportunity for sociologists in East Asia to do comparative research on social stratification and inequality.

Another example is the joint project titled »The Harmonious Society and the Middle Class« in which Korean and Chinese sociologists participated in 2007.[18] It was funded by the National Research Council for Economics, Humanities, and Social Science (NRCS) in Korea. It exclusively focused on the changing dynamics of the middle class in Korea and China in order to explore the stability and social role of the middle class in the rapidly changing economies in East Asia. »The Harmonious Society and the Middle Class« project contributes to an understanding of the distinct nature of middle-class formation and the relationship between the middle class and other classes in East Asia. The middle class in Korea has been experiencing unpredicted instability of jobs and family life since the financial crisis, whereas the middle class in China has been emerging with unprecedented speed as a new social bloc.

Recently, social science research has shown new development and expansion of its scope in terms of geographic area studied and the number of researchers engaged. Massive research projects have been launched to investigate the dynamics of social change caused by globalization. The global scale of social change motivated international collaboration of social research in various ways. The new type of international research in which sociologists from different countries participate emerges with the growing interests in comparative research in social science. Internationalization of social science research in East Asia partly reflects the global trend in the East Asian region. The explosion of international social science research transforms the nature of social science research in each society, accentuating comparative perspectives in every research.

The new development will be accelerated with the increasing interdependence among countries in East Asia and homogenization of social institutions and policies due to the impact of globalization in East Asia.

---

18 In 2006, Sang-Jin Han, a professor of Seoul National University, initiated the project for the purpose of facilitating joint research of East Asian scholars on common problems that East Asian nations are experiencing.

For example, the massive capital and labor migration among East Asian countries in recent years makes it hard to analyze a society without considering East Asia as a whole. Since social change takes places at a supranational level, social scientists begin to reformulate the research agenda so that research can be extended beyond the national boundary.

Recent development of social science research in Korea indicates new ways of production and reproduction of social science knowledge. Based on self-centered research interests, sometimes depicted as particularistic interests, the traditional social science research did not pay much attention to other societies from the outset of research design. Assuming uniqueness of history and culture of the South Korean society, many social scientists exclusively confined their research interests to the South Korean society. Without reflection on the tradition social science research, the nationalist epistemology has been implicitly taken for granted by majority of social scientists in South Korea. In contrast, considering the Korean society in a comparative perspective allows researchers to keep distance from the Korean society, relatively free from nationalist sentiments in their research. The internationalization of social science research, to use Foucault's word (1972: 191), displays a sharp break with the past »episteme« of social science in South Korea.

However, internationalization of social science research rekindles questions related with the very nature of social science. Social science has been considered as half-science for being in-between humanities and natural science. Thus many contemporary social theories originate from great thinkers, including Greek philosophers or philosophers in the nineteenth century. Whether they accepted or negated previous social theories, contemporary social theories are indebted to philosophical discourse of the past generations. Social theories tended to be more universal since they dealt with more fundamental question beyond a particular person or society.

Social science research since the 1980s has tended to be more naturalistic than ever before with emphasis on hard facts rather than speculation. Even comparative researchers have developed rigorous methods to embrace comparative historical research (Brady/Collier 2004; Ragin 1987, 2000). The underlying logic of the rigorous comparative research method is to unravel causal relations from comparison of cases characterized by the small sample size (N). The small N problem of comparative research can be overcome by rigorous comparison rather than plain narratives, assuming quasi-experimental design for a comparative study. Comparative historical research also leans toward natural science model rather than philosophical discourse.

Social science in South Korea has not focused on the discovery of scientific truth. Rather the goal of social science has been implicitly assumed to improve conditions of human life by debunking and criticizing the hidden dimension of injustice embedded in social and political institutions. Furthermore, sociologists have been engaged in practical activities to promote democracy and fight against social injustice committed by the power elite. Thus many Korean sociologists have shown much different research interests from those of mainstream American sociologists, whose reference group is not the public but professional community.

When internationalization of social science research has expanded, the immediate question can be raised about the substantive nature of the international social science research. Do we need international social science research? If so, why do we need international social science research? Do we carry out international social science research to test some hypothesis? If yes, why do we need to test these hypotheses? Those questions are associated with the nature of the »social« in social science, the science of society. Regardless of political ideology, classical sociologists explicitly stated that the purpose of their research (Giddens 1986a, 1986b; Callinicos 1999: 78-178; Seidman 1983) was to discover better forms of society. Karl Marx among others advocated socialism, whereas Emile Durkheim and Max Weber vindicated democratic capitalism in different ways.[19] Critique based on research has been a long-term tradition of sociology from the outset, though it dwindled away with the ascendance of the American sociology in the interim period of the second half of the twentieth century.

It is true that although the internationalization of social science research has helped undermine the old episteme of social science, the nationalist or geocentric orientation, the new episteme has not emerged clearly yet. Three different epistemes may yet emerge. One is a return to the philosophical tradition. Postmodernism might be a good example. Rejecting the rationalist concept of science, postmodernists deconstruct the foundation of modern social science as a rational understanding of man and society. Another episteme is to emphasize the scientific nature of social science. Though the meanings of science might not be simple enough to generate consensus, critique of society based on scientific research could be considered as an alternative to the previous one. The

---

19 Wolfgang Mommsen (1989: 27) described Weber's political orientation as liberalism imbued with seemingly contradictory nationalism. Durkheim's liberalism is close to social liberalism in which the state representing the republic ideology played an important role in integrating individuals and society (see Giddens 1986: 40-45).

third alternative is the return to original thinkers of social science with scientific rigor, called international public social science.

Recently, Michael Burawoy, an American sociologist at the University of California at Berkeley, claims the necessity of public sociology by which professional sociological researchers engage in public issues.[20] To vindicate public sociology, he distinguishes four types of sociology: policy sociology, professional sociology, critical sociology, and public sociology, based on two dimensions of social research: the nature of knowledge and the target audience (see table 1). The nature of knowledge in sociology is directly related with the purpose of sociological research, »sociology for what«. The target audience is the relationship between research and audience in which sociologists implicitly assume their role with respect to the nature of the possible audience, »sociology for whom« (Burawoy 2004: 268-272). Professional training, rigorous research, and writing scholarly papers have been emphasized as a core of education of sociology at the graduate level. The research audience is limited to the professional community that shares some hidden knowledge and jargons. Policy sociology is sociology for finding solutions of problems demanded by clients that might be the state or the society. Problem solving can be effectively achieved by finding effective policy instruments. Critical sociology, as in the case of the Frankfurt School, is another type of sociology with a limited audience, interrogating the epistemological and normative foundation of society and usually appears as a form of social critique. Public sociology employs social research aimed at the public audience to examine fundamental values of social institutions and social policy. These four types of sociology are not mutually exclusive but mutually nested in a sense that professional sociology can be a part of other types of sociology. For example, professional sociology can be a baseline of public sociology. As Burawoy mentions, »Professional sociology is not the enemy of policy and public sociology but the *sine qua non* of their existence—providing both legitimacy and expertise for policy and public sociology« (2004: 267).

In spite of the difference in weight of each disciplinary category in social science, Burawoy's discussion about sociology can be extended to

---

20 The Department of Sociology at UC Berkeley states the nature of public sociology as follows: »Social science as public philosophy is public not just in the sense that its findings are publicly available or useful to some group or institution outside the scholarly world. It is public in that it seeks to engage the public in dialogue. It also seeks to engage »the community of competent«, specialists and experts in dialogue, but it does not seek to stay in the boundaries of specialist community while studying the rest of society from outside« (in http://sociology.berkeley.edu/index.php?page= forums, search March 15, 2008).

social science in general. To apply Burawoy's classification of sociology, we might consider scientific and critical international social science based on professional-cum-public social science at the international level. Thus internationalization of social science research expands not only the scope of social science from one nation to the global world, but also the nature of research from instrumental knowledge production to reflexive knowledge production. Knowledge produced by international social science research can contribute not to a specific country but to the whole world. In short, it produces and reproduces knowledge for human beings, beyond a specific country or social group.

Table 1: Division of labor of sociology (Burawoy 2004: 269)

	Academic audience	Extra-academic audience
Instrumental knowledge	Professional	Policy
Reflexive knowledge	Critical	Public

However, growing internationalization of social science research among East Asian countries has not shown any clear tendency that might be called a new episteme of social science research. There have not been any serious discussions about what internationalization of social science research really means and what goal it tries to achieve. Reflexive knowledge production at the international level might be difficult to achieve because it requires something that transcends the limits of nationalist ethos and tacit assumptions of social concepts based on thenation-state. National politics and the concept of the social based on the nation-state still dominate the episteme of most of social scientists.

Globalization makes researchers recognize increasing economic disparity and rising poverty in East Asian countries. A growing awareness of the limits of social science research based on a single nation is emerging among social scientists in East Asia. Globalization demands radically different concepts of »the social« in social theory and social research in East Asia. Now more sociologists from South Korea, Japan, and China commonly share the view that globalization can be a destructive social and economic process that generates social polarization and damages social integration beyond the nation-state, and that demographic change will radically transform the nature of the society. In the era of globalization, East Asian countries, once praised as exceptional cases of successful economic growth with a very low level of economic inequality (South Korea, Japan, and Taiwan) or an egalitarian society (China), have experienced unprecedented levels of economic inequality and social disintegration as an immediate impact of this process.[21] While the impact of globalization on work and life might be different across countries due to differences in the level of industrialization and culture, sociologists in East Asia share tacit assumption that joint research might be able to deepen our understanding of globalization in East Asia, which goes beyond a single nation.

Concluding Remarks

In the twenty-first century, we are witnessing the emergence of a new mode of production and reproduction of knowledge in social science. As globalization proceeds, social science research also becomes global in its scope of subjects and number of researchers involved. Internationalization of social science research has been accelerated by the collapse of Eastern Europe and development of electronic communication since the late 1980s. In particular, the Internet has fundamentally transformed the way in which social science research has been coordinated and carried out. Traversing national boundaries, social scientists from different countries began to formulate new research agendas and logistics to promote international social science research.

Concerns about and practice of international social science research in South Korea were really late phenomena in that international social science research was not one of the major concerns of South Korean sociologists until the late 1980s. In practice, internationalization of social science research in South Korea has been the ascendancy of social science research collaboration among researchers in the East Asian countries. Growing sociological research collaboration among social scientists in East Asian countries in recent years is remarkable. Although international social science in East Asia is in a rudimentary stage of development, research collaboration has rapidly increased in the 2000s. These impulses have been joint products of democratization and globalization, rehabilitation of the function of university as research institution, the shift of government policy for research, an increase of global-

---

21 To describe rising inequality, however, different terms have been used in each society. »Social polarization« has been used by government's officials as well as social scientists in Korea, whereas »the gap society« has been used by some social scientists in Japan. »Cleavage society« has been used by some sociologists. Those terms are used to describe newly emerging social inequality in three countries.

minded researchers, and a growing awareness of social change beyond nation-states among scholars in East Asia.

Internationalization of social science research generates a new radical episteme of social science that breaks away from the concept of »the social« restricted to the nation-state. The boundary of the society has been assumed to be equal to the geographic territory of the nation-state. However, the territory of the nation-state becomes less significant than ever for understanding the recent social change called globalization. Divergence of the national society from the nation-state paves a new way for rearticulating the concept of international social science as well.

Internationalization of social science research also generates new challenges to the paradigm of social science that has been mostly developed in a European tradition. Debates on indigenous paradigms of social science rooted in history and culture of the non-European countries not only discloses the episteme of social science research imbued with Orientalism but also promotes new theoretical endeavors to understand intricate social change at the regional and global level. However, although newly emerging paradigms of social science might be different from those developed since the Second World War in Europe, the nature of these new paradigms is not yet clear. So far, we are able to identify only a reflective and comparative perspective as a single, common distinction of the new paradigms in South Korea.

References

Brady, H./Collier, D. (Eds.) (2004): Rethinking Social Inquiry: Diverse Tools, Shared Standards. Lanham: Rowman and Littlefield.

Burawoy, M. (2004): 2004 American Sociological Association Presidential Address: For Public Sociology. British Journal of Sociology, 56, 259-294.

Callinicos, A. (1999): Sociological Theory: A Historical Introduction. New York: New York University Press.

Cho, D. (1994): Autocratic Class Rules and the Mexican Working Class in the Period of Diaz. International Regional Studies, 3, 247-281.

Cho, D. (1996): Reform and Change in Mexico: The Formation and Reproduction of the Dual Structure in the Mexican Labor Movement. Latin America Studies, 9, 51-107.

Cho, D. (2003): Neoliberal Reform and Socio-Cultural Transformation in Latin America (1). Neoliberal Economic Reform and Workers' Living Condition: Bankrupt Washington Consensus and the »Transfer of Legitimacy« Effect. Latin America Studies, 16, 93-124.

Cho, D. (2005): The Two Years of Lula Government and a Trap of Success? Economy and Society, 67, 113-136.
Cho, H.R. (1994): A Comparative Study of Labor Policy During the Transition Toward Democracy: Cases of Spain, Brazil and Korea. Society and History, 41, 125-189.
Cho, H.R. (2002): Neo-Liberal Economic Reform and the Dilemma of the Labour Movement in Brazil. Economy and Society, 53, 91-118.
Choi, J.J. (1995): A Theory of Korean Democracy. Seoul: Hangilsa.
Choi, J.J. (1996): Conditions and Prospects of Democracy in Korea. Seoul: Nanam.
Choi, J.J. (2005): Democracy after Democratization. Seoul: Humanitas.
Diamond, L. (Ed.) (1997): Consolidating the Third World Democracies: Themes and Perspectives. Baltimore: Johns Hopkins University Press.
Foucault, M. (1972): The Archeology of Knowledge. London: Tavistock.
Giddens, A. (1986a): Capitalism and Social Theory: An Analysis of the Writings of Marx, Durkheim and Max Weber. Cambridge: Cambridge University Press.
Giddens, A. (1986b): Durkheim on Politics and the State. Cambridge: Polity Press.
Han, S.-J. (Ed.) (1984): Political Regimes and the Bureaucratic Authoritarian State in the Third World. Seoul: Hanul.
Han, S.-J. (1989): A Middling Grassroots: Searching for a Model for Transformation. Thoughts, 1, 79-100.
Han, S.-J. (1994): Social Reform and a Theory of a Middling Grassroots. Thoughts, 9, 260-283.
Han, S.-J. (1995): Economic Development and Democracy: Korea as a New Model? Korea Journal, 35, 5-17.
Han, S.-J. (1997): Political Economy and Moral Institution: The Formation of the Middling Grassroots in Korea. Humboldt Journal of Social Relations, 23, 71-89.
Jang, Y.S./Paek, S.W. (2005): Reforms of State-Owned Enterprises and the Changes in Labor Relations in Shanghai. Contemporary China Studies, 7, 167-220.
Jung, E.H./Kim, J. (1997): A Study of Industrial Relations in the Taiwanese Firms. Area Studies, 6, 49-79.
Kim, I. (2004): Globalization, Flexibility and the Social Democratic Labour Market: The Case of Sweden. Korean Journal of Sociology, 38, 143-294.
Kim, M. (2005): Women's Issues in the French Second Wave Women's Movement. Review of International Politics, 45, 313-345.

Korea Research Foundation (2001): Twenty Years of the Korea Research Foundation. Seoul: KRF.

Lee, J.K. (1995): Aging and Labor Issues in Japan. Area Studies, 4, 23-51.

Lee, J.K. (1997): Transformation and Diversification of the Japanese Labor Relations. International Regional Studies, 6, 125-143.

Lee, J.H. (2006): A Case Study of Sectoral Class Compromise in Swedish Metall: Collective Egoism or Rational Choice. Korean Journal of Sociology, 40, 132-164.

Ministry of Education and Science (2007): Report on the Survey of Research and Development in Science and Technology. Seoul: Ministry of Education and Science.

Mommsen, W. (1989): The Politics and Social Theory of Max Weber. Cambridge: Polity Press.

Moon, S. (1998): Principal Orientation and Content of the Swedish Labour Market. European Studies, 8, 227-248.

Paek, S.W. (1996): The Changes of Employment Practices in China's Economic Reform Era. Studies of Industry and Labour, 2, 255-291.

Paek, S.W. (2001): Unemployment and Unemployment Policy in Contemporary China. Studies of Industry and Labour, 7, 95-115.

Ragin, C. (1987): The Comparative Methods: Moving Beyond Qualitative and Quantitative Strategies. Berkeley: University of California Press.

Ragin, C. (2000): Fuzzy-Set Social Science. Chicago: University of Chicago Press.

Seidman, S. (1983): Liberalism and the Origins of European Social Theory. Berkeley: California University Press.

Shin, K.-Y. (1990a): Swedish Social Democracy and Its Economic Policy. Social Critique, 4, 260-288.

Shin, K.-Y. (1990b): Industrialization and Labour Movement in East Asia: A Comparison of South Korea and Taiwan. Asian Culture, 6, 1-36.

Shin, K.-Y. (1994): 60 Years of the Swedish Social Democracy: Possibility and Limitations. Thoughts, 9, 40-75.

Shin, K.-Y. (1999): Industrialization and Democratization in East Asia. Seoul: Munhakkwa Jisungsa.

Shin, K.-Y. (2000): The Formation and Crisis of Class Compromise in Sweden. Korean Journal of Sociology, 34, 897-927.

Shin, K.-Y. (2006): The Citizen's Movement in Korea. Korea Journal, 46, 5-34.

Shin, K.-Y. (2007): Globalization and the National Social Science in the Discourse on SSCI in South Korea. Korean Social Science Journal, 34, 93-117.

Song, H.K. (1996): The Politics of Coordination and a Dilemma of Social Democracy: Focusing on the Swedish Unemployment Policy. Korean Journal of Sociology, 34, 951-979.

Song, H.K. (2000): A Comparative Analysis of the Labour Market Structure: Korea and Taiwan. Korean Journal of Sociology, 23, 57-87.

Internationalizing Education and the Social Sciences: Reflections on the Indian Context

PRADEEP CHAKKARATH

Introduction[1]

In an article written for the New York Daily Tribune in 1853, Karl Marx, one of the most influential thinkers in the history of the social sciences, expressed concern about the loss of Indian identity under British rule, stating:

»England has broken down the entire framework of Indian society, without any symptoms of reconstitution yet appearing. This loss of his old world, with no gain of a new one, imparts a particular kind of melancholy to the present misery of the Hindoo, and separates Hindostan, ruled by Britain, from all its ancient traditions, and from the whole of its past history.« (Marx 1853/1981: 125)

Shortly before, he had presented the following, thoroughly Eurocentric, portrait of India:

»Hindostan is an Italy of Asiatic dimensions, the Himalayas for the Alps, the Plains of Bengal for the Plains of Lombardy, the Deccan for the Apennines, and the Isle of Ceylon for the Island of Sicily. The same rich variety in the products of the soil, and the same dismemberment in the political configuration. Just as Italy has, from time to time, been compressed by the conqueror's sword into different national masses, so do we find Hindostan, when not under

---

1 The author gratefully acknowledges the assistance of Tamara Herz with the preparation of the final version of the manuscript

the pressure of the Mohammedan, or the Mogul, or the Briton, dissolved into as many independent and conflicting States as it numbered towns, or even villages. Yet, in a social point of view, Hindostan is not the Italy, but the Ireland of the East. And this strange combination of Italy and of Ireland, of a world of voluptuousness and of a world of woes, is anticipated in the ancient traditions of the religion of Hindostan. That religion is at once a religion of sensualist exuberance, and a religion of self-torturing asceticism.« (Marx 1853/1981: 125)

Thus, in one and the same article, Marx spoke about his concern for colonial India's alienation from its historical roots as a result of Westernization while at the same time perpetuating Eurocentric stereotypes of India and its culture, as exhibited by his projection of European geography and culture onto the Indian map. This tension is still palpable in the current discourse about Europe's relationship to non-Western countries, especially those that once fell victim to European imperialism and colonialism. The diagnosis of this tension and the specification of ways to resolve it is not only a political issue, but also of utmost social scientific interest. Immanuel Wallerstein, another eminent thinker and analyst in the field of social sciences, sharp-wittedly remarked that the twentieth-century social sciences, as they are still geared to nineteenth-century paradigms, are »the central intellectual barrier to useful analysis of the social world« (Wallerstein 2001b: 1). Not surprisingly, his attempt to characterize the phenomenon called »India« without succumbing to the confining borders of European thinking and geography, culminates in the skeptical question about whether India exists at all and a demonstration of how many possible answers there are, depending on the historical, political, sociocultural, *and* scientific parameters chosen in the search for an answer (Wallerstein 2001a). From a cultural psychologist's point of view, there is so much at stake when it comes to »unthinking« familiar theories and approaches that it is not surprising that most European scholars avoid such an endeavor. As Wallerstein points out, »There is resistance even to serious reflection on the underlying epistemological, metaphysical, and political issues« (Wallerstein 2001b: vii).

There is no doubt that the kind of tension I have just described also exists with respect to Europe's recent efforts to internationalize the social sciences and the humanities. This effort has different aspects and connotations depending on one's perspective, ambitions, and motives. Who is promoting internationalization, who is considered the bearer of the effort, who is considered the »recipient«, who is for and who is against it? Since it would be peculiar to pretend that questions like these can be answered without discussing the history of internationalization efforts, I will do so in the following.

From at least the nineteenth century on, it was the so-called »West« or—from an African perspective—the »North« (i.e., the industrialized, technologically advanced countries of Europe and North America) that set the standards for what is now considered »mainstream science« and has been serving as the role model for scientific communities all over the world. Nowadays, few would dispute that the contributions of the modern »West« to science, especially natural science, were groundbreaking and sometimes even breathtaking. Nevertheless, »Western« endeavors to make »Western« conceptions of science the international standard and thereby solidify claims from the West of their universality have recently met with reservations, especially on the part of non-Westerners from cultures that also once contributed groundbreaking and even breathtaking contributions to science. These reservations are not entirely new; they were already present in the late fifteenth century when one of the earlier waves of internationalization efforts started in Europe, especially in an attempt to spread Christian views throughout the newly discovered rest of the world, even by means of brute force. In those days, while the promotion of central aspects of the Western world view played an important role, Westerners' endeavors were not decisively driven by the goal of promoting knowledge or scientific standards, but rather by hegemonial and economic interests. However, history is full of examples illustrating how imperialism and colonialism—though primarily motivated by political and economic goals—have also been linked to science in one way or another. When Alexander the Great, one of Aristotle's pupils, set out to create his empire, he not only led an army, but also a group of scientists whose task it was to gather the scientific knowledge of foreign and distant regions. The Greeks' interest in Buddhist thinking and many other aspects of Indian culture and society is a product of Alexander's vision of a culturally diverse empire in which those from other cultures with other traditions were respected and invited to participate in intellectual discussion, in line with the Greek idea of scientific progress through discourse. This idealized depiction of the Age of Hellenism and its imperialistic aspirations, which found their fullest expression in the Roman Empire, is frequently found in European historiography.

The image of imperialism and colonialism has changed over centuries and when modern science was developed in the nineteenth century, the colonial powers of Europe were not expecting to learn from the scientific traditions of their colonies. In fact, most colonists did not even view non-Western traditions as scientific, as only Western traditions were seen as methodologically founded and »rational«. Therefore, science played an important role in legitimizing some European nations' exercise of power in territories so far away from their motherlands and so remote from the comfort of their technological civilization. From its early conception, Europe's modern social science with its belief in developmental stages (e.g., August Comte), evolutionary processes (e.g., Charles Darwin), social Darwinism (e.g., Herbert Spencer), and a cross culturally, biologically, and psychologically justifiable racism fostered the »Western« conviction that it was the »white man's burden« to develop the rest of the world (Jahoda 1999; Sztompka 1994). This self-proclaimed missionary role of the West along with the allegedly scientific theories to support it were essential aspects of what Edward Said (and others before him) later identified and described as the instrumentalization of science for the sake of the West's power over the rest of the world (Said 1978, 1993).

In past centuries, there was a lack of organized resistance to European efforts to internationalize the sciences, but this is changing. Former colonies, especially those like India that are aware of their rich contribution to the development of mankind (including the sciences) over centuries and millennia, are showing renewed self-confidence that is supported by their rapidly growing economic success and corresponding Western acknowledgment and concerns. Thus the »West« has a much harder time convincing the international scientific community that its efforts toward economic and scientific globalization are selfless and motivated solely by scientific reason.

In the following, using the Indian context as an example, I will highlight historical, political, and scientific aspects of the situation described above in order to shed light on problems and difficulties stemming from European globalization and internationalization efforts in the fields of education and science. Since it is impossible to discuss all the relevant aspects in one short article, I will focus on the relationship between colonialism and education and the effects of this relationship on the social sciences in India (for a more detailed discussion of colonialism and education in India, see Seth 2008). After describing important aspects of the general historical and cultural setting, I will concentrate on Western and Indian science in general and then on social-science-related specificities.

Colonialism and Indian Identity

Though there have always been various mechanisms of spreading scientific knowledge around the world, colonialism and imperialism indisputably played a crucial role in speeding the process up and securing the adoption and institutionalization of the colonizers' standards in the colonized countries. Modern discourse about colonialism and imperialism centers on the military and ideological influence exercised by European nations especially in African, Asian, and Latin American countries. Since more recent forms of similar engagement (e.g., doctrines within U.S. foreign policy and its »war against terror«) refrain from taking possession of foreign land and declaring it its own territory, the terms »colonialism« and »imperialism« are increasingly being replaced by »neoimperialism«. Structurally, however, it is not easy to differentiate between these terms, especially since the modes of justifying and legitimizing the use of force and violence to exert one's influence on another country are very similar, whether executed in the name of colonialism, imperialism, neoimperialism, or even some forms of religious proselytism. In each case, a certain set of beliefs is used to legitimize or promote colonialism, which is explicitly or implicitly based on the ethnocentric belief that the morals and values of the colonizer are superior to those of the colonized. Though belief systems of this kind are frequently linked to racist and other pseudoscientific European theories dating back to the eighteenth and nineteenth centuries, they are much older and can be traced back to examples of colonialism throughout history. Similarities with respect to structure and legitimization strategies can be found in premodern historical examples of imperialism and the creation of colonies: in ancient China, in the Greek empire under Alexander the Great, in the Roman Empire, and the early Ottoman Empire, to name a few. Interestingly enough, many of the countries that fell victim to European colonialism in modern times have a history of being colonizers themselves.

As for India, in the last 3,500 years, its territories were conquered and ruled by many foreign powers. The British were only the last of a series of invaders and colonists, among them the Aryans, Greeks, Romans, Muslim dynasties, Portuguese, French, and Dutch. Thus the discourse about the meaning of the colonial past for the Indian present has always been one about India's history and culture as a whole. In addition, this discourse has emphasized India's willingness and capacity to deal with foreign influences and threats by integrating them into the Indian culture. Thus India has beaten the invaders by »Indianizing« them. Moreover, within this discourse, (especially nineteenth-century) Indian historians have often described India as being an imperialistic power itself that for more than a millennium exported Indian standards to Southeast Asia in order to develop the politically and »culturally backward« regions and peoples. So, despite two centuries of British rule in India,

Indian scholars and politicians seem to have a relatively relaxed attitude toward colonialism. Though the experience of racism toward Indians undoubtedly left its mark on the Indian »psyche«, or what might be called the »Indian identity«, Indian discourse about (and coping with) its colonial past has always been multifaceted and to a large extent free of bitterness. This is not surprising since key elements of the British attitude and justification strategies were quite familiar to educated Indians: the conviction of belonging to one of the most influential cultures in world history; a clear sense of status, hierarchy, and authority in all domains of social life, even with the inherent idea that status comes with birth; the (often arrogantly expressed) attitude of superiority toward others (including fellow Indians); the idea of »divide et impera«, which allows one to keep societal order under otherwise chaotic circumstances. Perhaps the most intriguing example of a result of colonizer and colonized sharing cultural traits is that the British colonists did not put much effort into abolishing India's caste system. This social system has been a part of the Hindu social structure for at least 3,500 years, serves as a primary source of identity for most of its members, and is still advocated by a large portion of the Indian elite (viz., high-caste Hindus, politicians, and intellectuals).

As a consequence of this historical and pragmatic way of dealing with India's colonial past, on the one hand, the centuries of British involvement in India were considered an alienation from India's own cultural roots and heritage; on the other hand, however, the encounter with the British version of elitism confirmed many of the images that the Indian elite held of themselves. The unique and specific Indian way of finally getting rid of the colonizer even gave many Indians a feeling of cultural superiority over a Western superpower. At the same time, however, it was acknowledged that the colonial experience provided—in particular knowledge-based, value-oriented, political, and infrastructural—opportunities that India could profit from in the future. Thus Indian discourse about its colonial experience was an open discourse: open for integrating these experiences into the »Indian identity«, open for a reciprocal relationship with its former colonizers, and even open enough to trust the honest attempts of other nations to help independent India solve its enormous problems. Even the neo-Marxist movement, which was quite strong in India until recently, was not very concerned about challenging colonialism's racist attitudes or their effect on the Indian soul, but about overcoming the injustice embedded in the genuinely Indian traditions of thinking and acting. As an important outcome of this Indian way of dealing with its past, India as a nation—though continuously being concerned with the discourse about racism and colonialism—was saved from a paralyzing discourse about the traumatizing effects of colonialism. This is worth mentioning, especially against the background of African experiences with colonialism, which are still being discussed in many African countries in terms of individual and collective »traumatization«.

To summarize: There is no doubt that countries that have undergone colonization are in many respects different from other countries and face postcolonial problems with regard to societal, psychological, and economic development. However, the general discourse about the effects of colonialism should not disregard the specific meaning and consequences that colonialism has had for different countries and contexts. Actually, compared to the vast number of essays on colonialism in general, little research has been done on the impact it has had, for example, in specific and interrelated contexts within specific cultures. This is important for our topic because we should be aware that the debate about »colonialism« and its effects on cultures and countries cannot be analyzed adequately if we use concepts that are too simple and too general to describe the various phenomena and processes involved. It is especially doubtful whether the concepts developed within the scientific discourse of the former colonizing countries can grasp the whole picture which, of course, should depict the formerly colonized, too. The discourse about postcolonialism shows how fruitful discourse can be when the problems in question are not tackled by the colonizers alone, but by them and their former inferiors together. As postcolonial theorists would say, we should »let the subaltern speak« (Spivak 1988). In other words, if we want to do scientific justice to these specificities, we need to enter a collaborative scientific endeavor in which we are on par with each other and which includes a serious reflection on how to avoid any form of colonialism, including scientific hegemony. Since India's experience with its colonizers' efforts to Westernize its culture and people was also an experience of hegemony, in the next section, I will sketch some of the historical and political facets of this development.

History of Education and Science in India

The Ancient Heritage

One of the main areas in which India strove to benefit from its British colonial heritage was that of education. This area shows, in an exemplary manner, the interplay between self-confident India, on the one hand, and Britain as the colonial power, on the other hand, as well as attempts to integrate the two into the history and development of India. Since science, in one way or another, is always embedded in a country's education system, in the following, I will relate that system to the broader historical, ideological, and sociocultural framework in which the problems concerning the social sciences and the humanities in India are set.

Indian appreciation of the importance of institutionalized formal education and teaching dates back to at least the beginning of the first millennium BC (for details on ancient education in India, see Scharfe 2002). Since one of the main topics of ancient Indian philosophy and psychology was understanding and developing the self, in schools, particular emphasis was placed on self-development. The main figure in the education (*vidya*) process was the teacher (*guru*). Children were considered rather passive receivers of the guru's knowledge because, according to Hindu psychology, most human beings do not possess the cognitive abilities necessary to acquire »relevant« knowledge on their own before the age of approximately ten years. Knowledge considered »relevant« for self-development pertained to social conditions, rules and values, actions, company, food, etc. Another factor that might be of relevance to education and science in the Indian context is that the Brahmin caste has been the highest caste in India for the longest time and education has always been their main symbol. They exercised control over several crucial aspects, dictating that

- only members of the upper castes would receive training in reading and writing;
- the language of instruction was the holy language of Sanskrit, which was also the language of most of the texts studied in the schools;
- only the Brahmins would decide what to read and what discussions would be allowed;
- school subjects were scripture, philosophy, literature, warfare, statecraft, mathematics, medicine, astrology, and history;
- the main topics of ancient Indian thinking (as taught at famous centers of learning like Takshashila) were philosophical, psychological, and sociological topics, such as: understanding and developing the self, moral and societal values, social status, human relationships and interaction.

These elements of Hindu education were supplemented and promisingly modified by the Buddhist and Jainist movements starting in the fifth century BC. Buddhist education was based on the »Eightfold Path« (i.e., the goal of taking the cognitive, moral, and spiritual abilities of human

beings to the highest level) and had the following distinguishing features that, taken together, made it decisively different from the Hindu tradition:
- While retaining the Hindu topics, Buddhist education introduced new aspects to the curriculum—putting more emphasis on logic, mathematics, grammar, philosophy, astronomy, literature, Buddhism, Hinduism, the *arthashastra* (economics, politics, law), medical sciences, and the arts.
- Buddhist education combined day schools with residential education (i.e., schools with dormitories for the students).
- The door to education was open to all.
- Lessons were held in the vernacular.
- An institutionalized educational path led from elementary education to higher education at some of the earliest universities in history (e.g., Vickramasila, Odantapuri, Jagadalala, Somapura, and Nalanda universities; Nalanda, the most famous of all, at times accommodated ten thousand students and flourished for about a thousand years before Muslim invaders set fire to it in the early twelfth century).
- These universities attracted many students from other parts of Asia, made India one of the earliest centers of high-quality education, and fostered intercultural exchange between India and southeastern as well as eastern parts of Asia (e.g., China). Thus these academic institutions became an integral part of the Buddhist and Hindu mission that led to the process of cultural transmission, also known as the »Indianization« of many regions and cultures outside India.

From around 500 AD on, when Buddhist and Jainist influence in India decreased, Hinduism became the main ideological force and the country's education system became elitist again. Starting in the eleventh century, Muslim rulers introduced Quran schools and *madrasahs* (i.e., institutions for higher education in Islamic theology and law, Arabic grammar and literature, mathematics, logic, and, in some cases, natural science) that did not share the Hindu idea of elitism.

The Quality of Indian Education and Science Before the Arrival of the British

Before turning to the development of India's education system under British rule, let us interrupt the historical chronology for a moment and take a brief look at some of the achievements of Indian science within the framework of the ancient education system.

It is a truism that India, along with Mesopotamia, Egypt, and China, is one of the cradles of human and urban civilization. Indian philosophy, science, and technology contributed substantially to development in Eastern and Western parts of the world and at the same time established traditions of thought and practice that remain among the few alternatives to mainstream Western approaches. The previously mentioned emphasis of Indian schooling on self-development is a prime example of what is often called the »spiritual orientation« of Indian thinking, but it could also be called »Indian psychology«. Although modern-day academic mainstream psychology considers itself a product of Western origin (and in many aspects rightly so), it is still true that theories about the self, its anthropological function, and its social meaning abound in Indian literature and, for millennia, have had an enormous effect on Asian intellectuality and social scientific thinking. Since I have touched on the quality of Indian theories about the self and their social scientific relevance elsewhere (Chakkarath 2005, 2007, in press), I will only quote Max Mueller, one of the most knowledgeable Western scholars of Indian academic literature, here:

»If I were asked under what sky the human mind has most fully developed some of its choicest gifts, has most deeply pondered on the greatest problems of life, and has found solutions of some of them which well deserve the attention even of those who have studied Plato and Kant—I should point to India. And if I were to ask myself from what literature we, here in Europe, we who have been nurtured almost exclusively on the thoughts of Greeks and Romans [...] may draw that corrective which is most wanted in order to make our inner life more perfect, more comprehensive, more universal, [...] again I should point to India.« (Mueller 1883: 24)

There is a certain degree of romanticism in almost all of Mueller's assessments of the Indian culture's achievements and he often seems to repeat the well-established stereotype of India being the most spiritually oriented country of all (see Chakkarath 2007). However, unlike many other Western scholars, he knew that in Indian thinking, reflection on different aspects of the self was not an isolated philosophical and metaphysical endeavor, but closely related to psychological, sociological, and political analyses. Thus, from early on, many of these theories, which are documented in a vast number of academic texts, had what today would be called a »social scientific« concern (Ames/Dissanayake/Kasulis 1994; Bharati 1985; Chakkarath in press; Paranjpe 1998). For this reason, we even find many aspects of these self-theories in Indian politics textbooks that treated the art of governing and diplomacy, like the ancient *arthashastra* or the medieval *shukra niti*. While Western scholars often see a sort of »Machiavellian realism« in these works, this kind of political realism should really be termed »Chanakyan realism«, after the author (also known as Kautilya) of the *arthashastra*, which was written a millennium earlier than his Venetian successor's *Il principe*.

Though it may be true that social scientific topics like these were often discussed within the framework of Hindu, Buddhist, and Jain philosophy (i.e., in the field of the humanities), the widespread assumption that Indian thinking was mainly religious and spiritual is wrong. Apart from the fact that the concept of »religion« is Western in origin, it does not aptly characterize the worldviews and conducts of life favored by Hinduism and Buddhism. Moreover, there have always been various schools of thought in India, arguing for or against any of the positions held in European classical philosophy. Accordingly, the universities provided an open academic culture of reasoning and argumentation that resulted in fruitful scientific pluralism and interdisciplinary perspectives (Sen 2005) as documented by eminent thinkers like Asanga, Dignaga, Siddhasen Diwakar, Gautama Buddha, Nagarjuna, Kapila, Vasubandhu, Ramanuja, Sankara, and many others. This was the basis for the ability of Indian thinkers to contribute largely to almost any domain of scientific interest. While, for example, the achievements of some of the Indian mathematicians are well-known around the world (especially the groundbreaking development of the decimal number system and the concept of »0« as a number), the names of Baudhayana, Aryabhata, Brahmagupta, and Bhaskara II, and many others who advanced the disciplines of mathematics and astronomy in important ways (e.g., by introducing the concept of a mathematics-based astronomy to Arabia) are unfamiliar even to most educated scholars in the West. Panini (fourth century BC) and Bhartrhari (sixth century AD) were two other exceptional scholars who are still considered main figures in the history of logics and linguistics, with Panini considered one of the founders of the disciplines. Outside the sphere of Indian scholarly knowledge, it is also little known that Indian medicine, therapy, and pharmacy did not rest exclusively on the pillars of holistic systems like *ayurveda* and *yoga*. The *susrutasamhita*, a compendium of textbooks by Susruta, a physician in the sixth century BC, documents the early and impressive knowledge about various branches of the discipline (e.g., anatomy, surgery, toxicology, and anesthesia). Among the surgical procedures described by Susruta, there are detailed accounts of plastic (e.g., rhinoplasty) and cataract surgery. From very early on, Indian scholars also developed the branch of veterinary medicine and even established veterinary hospitals.

Of course, this short enumeration of scientific contributions could go on for several chapters and include achievements in the application of sciences, in technology, architecture, the arts, etc. Here, however, it is not and cannot be my aim to give a detailed account of how effectively the early Indian education system was capable of producing scientific knowledge and contributing to the development of the sciences. Here, too, I can only remind the reader of the lasting historical meaning Indian scholarship had for large regions in the rest of the world. Especially Southeast Asia was influenced by Indian philosophy, psychology, political theory, architecture, and the arts for almost a thousand years before the European colonizers took over. Those who are interested—like Max Mueller was—in overcoming the cultural knowledge barriers that, for the vast majority of European scholarship, are marked by the ancient Greeks, the Romans, and various national thinkers can find related information elsewhere (e.g., Arnold 2000; Baber 1996; Kulke/Rothermund 2004; Sen 2005). For the purposes of this article, I simply want to draw attention to the fact that one can easily and convincingly show the effectiveness of the ancient and medieval Indian culture when it comes to pathbreaking, sometimes breathtaking scientific insight and standards. We should keep this in mind before turning to quite a different assessment of the Indian education system and scientific capabilities given by the British colonizers and their sympathizers.

Colonial Education in British India

Starting in the seventeenth century, the British reintroduced the (in India well-known and formerly Buddhist) idea of a school system that was open to all. During the British rule between 1858 and 1947, elementary education was made mandatory in India. However, the British system had its own elitist traits and therefore did not interfere with the Hindu social system of caste and status in that the British system also favored those who could afford a better education, namely, the Indian elite. They were considered the »pool« from which the British wanted to recruit staff for administrative matters, in other words, the skilled and loyal paladins that were to be trained for government service in controllable numbers. No one found clearer words to convey the colonizers' intentions and justification strategies than Thomas B. Macaulay, the historian, essayist, and politician who, serving on the Supreme Council of India, remarked, »It is, I believe, no exaggeration to say, that all the historical information which has been collected from all the books written in the Sanskrit language is less valuable than what may be found in the most paltry abridgments used at preparatory schools in England« (Macaulay 1835/1972: 241). Therefore, he demands, British knowledge and the British language as its primary medium had to be introduced and taught to a selected few who would be capable of completing the British mission in India:

»It is impossible for us, with our limited means, to attempt to educate the body of the people. We must at present do our best to form a class who may be interpreters between us and the millions whom we govern; a class of persons, Indian in blood and colour, but English in taste, in opinions, in morals, and in intellect. To that class we may leave it to refine the vernacular dialects of the country, to enrich those dialects with terms of science borrowed from the Western nomenclature, and to render them by degrees fit vehicles for conveying knowledge to the great mass of the population.« (Macaulay 1835/1972: 241)

This clearly demonstrates how colonialism, issues of cultural and individual identity, education policies, and the problem of internationalizing the social sciences interact in a complex relationship that I will discuss further below. It is worth mentioning, though, that the British attitudes expressed toward Indians and their history were not exclusively British but were shared by intellectuals all over Europe. One of the most famous among them was surely Hegel, a leading figure in the history of Western philosophy, who found few compliments when assessing the intellectual and scientific achievements of India before British rule. Ignoring the fact that Indian astronomy was in many regards more accurate than Greek Ptolemaic astronomy, which was cutting-edge for European astronomers until the sixteenth century, he bluntly used that discipline for his general verdict on all Indian thinking, »If we had formerly the satisfaction of believing in the antiquity of the Indian wisdom and holding it in respect, we now have ascertained through being acquainted with the great astronomical works of the Indians, the inaccuracy of all figures quoted« (Hegel 1837/1995: 125). Based on Indian scholars' alleged incapability of doing proper science and as a justification for Western scholars to rewrite the history of India and Indian science, he continued, »Nothing can be more confused, nothing more imperfect than the chronology of the Indians; no people which attained to culture in astronomy, mathematics, etc., is as incapable for history« (Hegel 1837/1995: 126). Summing up his analyses, Hegel concluded that a culture that had never come up with the idea of a solid and stable individual self and/or the idea of a person also could not be qualified for self-government. Thus he considered it natural that the British became the rulers of India, it being the »fatal destiny« of Asia to subject itself to the Europeans.

In India, the British took on the burdening task of bringing up a new breed of properly educated Englishmen of Indian origin. Thanks to Macaulay's successful intervention, one important first step of internationalizing what he and Hegel considered the only valuable knowledge system was the establishment of the English language as the language of the educated. This was effectively pushed through in 1935 when the British government decided that only schools and colleges teaching English and European knowledge and science would be entitled to receive government aid. This had an adverse effect on the traditional educational system and laid the foundation for the development of post independence education in India.

Education in Independent India

After decades of failure by most Indian governments when it came to educating the masses, today, the perception of India, especially in the West, is rapidly changing. While there were only 20 universities and approximately 500 colleges on the Indian subcontinent (including Bangladesh and Pakistan) in 1947, the year India gained independence, today there are about 375 universities and about 18,000 colleges in India alone, many with an infrastructure of the highest standard. With a population of more than a billion, India has the second largest education system in the world (after China) and produces nearly 3 million college graduates every year. Since more and more of these college graduates are going on to successful careers in the West and increasingly in their home country, too, the Indian education system is considered the engine of the country's impressive technological and economic growth as can be witnessed in the industries of cinema and music, fashion, steel production, agriculture, genetics, pharmacy, software engineering, nuclear science, as well as information and space technology. Many link this success to India's excellence in the field of higher education. In fact, according to European and U.S. sources (e.g., Times Higher Education Ranking) the Indian Institute of Technology, Indian Institute of Science, Indian Institutes of Management, All India Institute of Medical Sciences, Indian Statistical Institute, National Institutes of Technology, Birla Institute of Technology, Tata Institute of Fundamental Research, and the Indian School of Business rank among the best academic institutions in the world in their respective fields of specialization.

But counter to this image—which is as stereotypical as was that of spiritual India—on average, males in India complete only three years of schooling, females even less than two years, and only 10 percent of the relevant age group is enrolled in higher education. An estimated 30 percent of children between the ages of 6 and 14 do not attend school at all. This partially explains why India's illiteracy rate is still around 55 percent, and even higher among the elderly and women (see Rhines Cheney/Ruzzi/Muralidharan 2006). State reports as well as internationally conducted surveys show that the quality of instruction varies widely from school to school (also depending on region and state) and is especially unsatisfactory at state-run primary and secondary schools. Studies concerning the quality of textbooks for Indian schoolchildren have shown that most still transport perspectives that can be found in canonized ancient texts (Kumar 1989): For example, poverty and failures are said to be the consequences of (inborn) personality traits; artistic freedom, creativity, and spontaneity are not held to be as important as the need to follow orders from superiors. The Indian texts are linked to symbols of the traditional, feudal norms of behavior that still prevail in society. Consequently, training for primary-level teachers specifies that the teacher has the power to show approval or disapproval (even corporal punishment), but is compelled to minutely follow the curriculum guidelines set by the education authorities. Teachers may pose clarification questions, but not challenge the authorized version of knowledge. When India was governed by the Hindu national BJP party (from the late 1980s to the 1990s), many schoolbooks were even revised: Chapters on »Hindu sciences« like astrology, fortune telling, and other magical practices were added. In some states, schoolbooks even introduced justifications and praise for the caste system. The continuing influence of the traditional Indian elite and their standards can also be seen in the fact that the vast majority of students that make it through middle school to high school are still from higher castes and upper-class urban families (Rhines Cheney/Ruzzi/Muralidharan 2006).

Since a couple of years, the National Alliance for the Fundamental Right to Education (NAFRE), an alliance of some 2,400 education-related grassroots-level voluntary agencies, has been issuing reports on the status of education in India. In short, these are some of the main problems addressed in the report: since India gained independence, its government has failed to fulfill its constitutionally anchored task to abolish illiteracy and make primary and middle school education accessible to and free for all children between the ages of six and fourteen. Moreover, it has failed to stimulate the cognitive skills of Indian schoolchildren and college students in order to lay open the vast potential of creativity in India's youth, which must be considered the groundwork for the country's future.

Against this background, the world-renowned Indian institutes of technology, business, mathematics, and medicine (see above) seem to

profit from the fact that they have created an education culture that is quite different from general Indian standards:
- Although the recruiting procedures of these institutions encourage students who take the admission examinations to memorize huge amounts of material (thus following an old Indian tradition), they especially differ with respect to their international orientation, which can be seen in exchange programs that include the exchange of students as well as instructors.
- In the same vein, the institutes were able to recruit Indian scholars from the Indian diaspora who until lately did not show much interest in returning to their country of origin.
- The interaction with foreign, especially Western, students and instructors has also introduced a variety of teaching styles and modes of communication.
- The institutes are places of intercultural communication and thus prepare the Indian students for their future working environments which are largely in a global context.
- Students at the prestigious institutes do research on and solve problems in supervised work groups which they mainly organize themselves, starting with the formulation of research questions, assessing adequate methods for the research procedure, finding and discussing alternative approaches, and developing solutions.
- On the administrative level, the institutes enjoy almost complete autonomy from the government and the boards have recently tried to improve the quality of education by increasing the percentage of practical and project courses.

The Indian government (under Prime Minister Manmohan Singh) has been pushing for liberation from traditional obstacles since the 2004 election. A newly created committee on the state of education in India found that Indian social scientists had failed to do thorough research on the complex relationships between factors contributing to this unsatisfying situation, which negatively affects the country's poor majority in particular. Meanwhile, the first state-sponsored research programs have been established to investigate the intrinsic and interrelated problems on the individual, societal, and cultural level.

To summarize: India has been the venue for Western internationalization efforts for many centuries. The results have been manifold. India has attempted to hang on to its own cultural heritage in order to preserve indigenous achievements that are the foundation of Indian self-confidence. These attempts have proved to be partially successful as can be seen in their impact on the education system in India from 1947 until today. Unfortunately, it is especially the idea of elitism that survived the ages while the quality of indigenous traditions of scientific work that were established before British rule does not seem to matter anymore. As for a stable and protective national identity, one may call these attempts positive, keeping in mind that the British endeavor of creating a new breed of anglicized Indians was successful, in part because many educated Indians saw a resemblance between the British colonizer and themselves In other respects, however, Indian traditions as well as British colonial interests also contributed to failures in many fields due to an uncritical assessment and promotion of traditional and British standards that especially served the traditional Indian elite. Thus innovations were hindered, time and generations wasted, and overcoming the deeply rooted obstacles became even more difficult than before. All this is true at least for the field of education and would also be true in all societies that hang on to the wrong, in other words, inadequate standards.

We have now reached the crucial point of the discussion: Does the internationalization of education and scientific culture provide the means necessary to enable all the world's nations to find culturally and psychologically adequate ways to gain the knowledge they need to provide their citizens with a better future? What should we accept as an »adequate« way and what should be the indicators of adequacy? In an attempt to contribute to the resolution of these questions, let us draw some lessons from the Indian experience.

The Dilemma of the Social Sciences and the Humanities in India

Disregard of India's Role in the Internationalization of the Sciences

From its early beginnings, Indian science had an international or, more precisely, intercultural dimension. Scholars from various foreign cultures—among them the Greek, Roman, Arab, and Chinese—came to India for trade, proselytizing, imperialistic goals, or for scientific exchange. They left their mark in India, but also brought Indian knowledge back to their own cultures where it influenced the development of international science. For many centuries, some of the scientific and technological knowledge produced in India was trendsetting and cutting-edge. While ancient and medieval Western scholars relatively rarely acknowledged the quality of Indian intellectuality, its scientific outcome, and historical role, their Chinese and Arab counterparts, for example, did so readily (Sen 2005). When the Western nations needed legitimization for their imperialistic and colonialist policies, it was mainly the Western romanticists who admired Indian knowledge. However, most of them called »Indian knowledge« »Indian wisdom« as if Indian intellectuality was related more to philosophy and spirituality than to science. Thus even many of those (e.g., Max Mueller) who admired at least some of the achievements of Indian culture indirectly surrendered to those (e.g., Hegel and Macaulay) who denied that these achievements had *scientific* value. By claiming that only Western scholars possessed the philological skills necessary for a sound understanding and interpretation of the classical Indian authors and their textbooks, they even attempted to exclude Indian scholars from »proper« scientific discourse (as can be seen very clearly in the political aspects of philological disciplines like Indology, where mostly only Westerners achieve any sort of eminence).

It is fair to say that Western scholars' assessment of non-Western scientific achievements (e.g., in historiographies of science) are to a large extent ethnocentric. They are influenced by culture-specific cognitive schemas that create the impression that the development of Western science was a continuous process leading from the Pre-Socratic ideas of early Greek thinkers to the scientific foundations of the Industrial Revolution. In addition, they tend to ignore the fact that non-European achievements (e.g., by Indian scientists) had an impact on what is often considered an almost exclusively European show with almost exclusively European actors on the stage. Many aspects of this notion are fatal for an accurate assessment of the history of science. The development of European science was not a self-made, homegrown success story, but depended on the import of foreign ideas and foreign knowledge. In this regard, it was intercultural and »interculturalized« from the start.

Thus Western theories about scientific and cultural development need to be read with caution. For example, according to Max Weber (1904/2002), the difference between Western and Indian traditions of thought is that the former resulted in European »rationalization« while the latter resulted in Indian »world renunciation«. Relying on what Western »experts« on Indian intellectuality had chosen to translate and building on their interpretations and judgments, Weber was almost completely unaware of the variety of Indian philosophical systems and thinkers mentioned above. He was thus likewise unaware of the important role of scientific realism, empirical and logical analysis, and rational argumentation in India's scientific traditions, which makes it difficult to endorse many of the key points of his own argumentation. Nonetheless, Weber's theory, which has been criticized by many Indian scholars for the reasons given, is still one of the most influential theories in Western sociology and historiography of the sciences. In more recent assessments of the question concerning when modern science began and why it began in Europe, we can find even more ignorance regarding non-European contributions. Reijer Hooykaas (1987) is just one of many Western historians of science who did not deem it necessary to include the influence of non-European traditions when trying to answer this question. These various forms of ethnocentrism and ignorance were the grounds for Wallerstein's critical remarks on current social science's ongoing dependence on theories and concepts borrowed from nineteenth-century theories (see my related remarks in the introduction).

Failures such as these of Western science to provide an accurate assessment of India's contributions to the development of science, not to mention the colonized's experience of colonization, through which ethnocentric attitudes surely found their most arrogant expression, help us understand why the latest Western efforts to internationalize the sciences has been met with reservations by many, including many Indian scientists.

Internationalized and Dissatisfied: Social Scientists in India

One upshot of an internationalized science would be internationalized scientists. According to mainstream science's understanding of internationalization, an »internationalized« scientist would deem it true that:
- there are universal laws governing nature (including the nature of the social, cultural, and individual realm) and they can be discovered by means of science;
- there should be universal standards in methods and methodology (e.g., agreement about comparability);
- there is a need to assimilate academic institutions (schools, universities, technological facilities) in order to provide some degree of standardized scientific education;
- all members of the international scientific community should have open access to knowledge and training (including open data access, data sharing, exchange programs)
- the papers of all members of the international scientific community should have a fair chance of publication (i.e., especially in the leading scientific journals);
- there should be a lingua franca that allows all members of the international scientific community to communicate efficiently across lan-

guage barriers and that serves the goal of standardizing scientific practice.

If we take these basic characteristics of the internationalization of science and scientists as a measurement for the degree of internationalization already acquired by Indian scientists, the proponents of the effort can be satisfied. Many Indian scientists are so to speak grandchildren of »Macaulay's children«, the term used to characterize Indians who embodied Macaulay's vision of a new Indian elite, indigenous in blood and color, but English in tastes, opinions, morals, and intellect. They are trained in Western scientific methods and methodology, advocate the paradigms of their respective mainstream scientific communities, are well-read in the fundamentals of Western scientific literature, and even manage to publish in Western handbooks and journals every now and then. A growing number of Indian scientists receive their postgraduate training in Western countries and an increasing number decide to make their careers in the West. This phenomenon, known to many developing countries as the problem of »brain drain«, is currently being discussed in India as that of the »rising non-resident Indian«. The term draws attention to the question about why Indian scholars, including scientists, are so much more creative and successful in the West than they are in their country of origin.

There are probably several answers to this question, but one simple and nonetheless plausible answer is that there is a crucial difference between internationalizing India's scientists and internationalizing the sciences in India. As we have seen, large parts of the education and science environments in India are still to some extent influenced by traditional perspectives and sociocultural constraints. Even for many of the scientists who received their training in one of India's Centers of Excellence, it is difficult to find a work environment that allows them to live up to the scientific potential that their training was meant to inspire. Leaving India and exercising the scientific skills in the place from which they were exported to India seems to be a plausible decision. Moghaddam and Taylor (1985: 1145) described this situation as the result of a historically grown »dualism«, resulting from two main factors: first, the shaping of colonized nations primarily in response to the needs of Western European and North American powers; second, the unequal distribution of resources within most developing nations.

Those who have become skeptical as a result of the historical experiences with Western education efforts consider the latest internationalization efforts by Europe and the United States a new wave of colonialism and plausibly argue that the currently steadily growing numbers of emigrating Indian scientists is the latest step of continued exploitation by the West. They also point out that the leading scientific journals are Western journals published by Western companies and supervised by scientific boards and editors that are mostly Westerners who favor certain research topics, contents and methodologies, thus consolidating Western scientific hegemony.

However, there are other forms of skepticism, objections, and dissatisfaction that do not simply blame the unfortunate situation on colonialism and postcolonial outcomes, but rather focus on the insufficiency of Western scientific concepts and approaches. Not surprisingly, this critical assessment does not come from the natural sciences, but mainly from the social sciences. While most scientists would agree that the *laws* of physics, chemistry, and biology do not lose their validity when crossing national borders, the *approaches* taken in order to obtain data or solve scientific puzzles vary sometimes. This is especially true for scientists who study phenomena that are inherently imbued with historical, sociocultural, linguistic, contextual, and situational specificities. Thus, in the social sciences and the humanities, the effort to internationalize faces especially great domain-specific challenges and resistance. In India, the two-pronged attitude of many social scientists (i.e., a willingness to adopt Western approaches, on the one hand, with a critical stance toward their usefulness, on the other hand) can be traced back to the beginning of the twentieth century when the first university departments for social sciences (initially, mainly sociology, anthropology, history, economy, and psychology) were established. From early on, research was closely linked to the British goal of governing and controlling the colony with the aid of quantitative and qualitative data (gathered through censuses, analyses of cast inventories, ethnographic field work, etc.). However, at the same time, Western-style sociology had an outspoken interest in pushing social reform, an aspect that was willingly taken up by many Indian social scientists of the first generation. Thus their goal was not to serve the British, but rather the Indian independence movement. The dilemma of wanting to learn from the Westerners while, at the same time, avoiding assimilation of their overly narrow assumptions and principles is best described and documented in the work of the late M. N. Srinivas (2002), the eminent Indian sociologist and anthropologist of the twentieth century. As the current discussion among social scientists shows, this dilemma still needs to be resolved (Patel/Bagchi/Raj 2002).

At a seminar on the »Future of Social Sciences: Search for New Perspectives« in April 2001 (organized by the Indian Ministry of Education), leading Indian social scientists resolved to set the following goals,

viewing them as crucial for evaluating and improving the quality of social sciences in India (see Natraj/Majumdar/Kumar Giri/Naidu 2001):
- achieve autonomy (for the researcher) juxtaposed with accountability
- overcome the persistent elitist elements in Indian academia
- replace the authoritarian and hierarchical structures in higher education
- take academic and political measures to solve gender- and minority-related problems
- learn academic lessons from postcolonial studies
- develop new methodologies
- revitalize and integrate indigenous approaches
- achieve transdisciplinarity and disciplinary grounding
- strengthen the teaching-research-nexus
- solve institutional issues (e.g., funding, infrastructural improvement)

This list of measures tellingly shows how Indian social scientists are trying to solve two main problems at the same time: they are striving to overcome the culture-related, especially social, obstacles to a successful modernization and internationalization of their profession and their institutions (for a more detailed assessment of inequalities and injustice within the social sciences in India, see Guru 2002). In addition, they feel the need to find political and scientific ways to do more justice to the specificities of the Indian environment as expressed in their wish to include insights from postcolonial studies in their own research, to seek new methodologies, and to integrate indigenous traditions of science into their research programs. Since a growing number of Indian social scientists (in and outside of India) feel the need to find a middle road between internationalizing and indigenizing their research work, I will now go into this growingly important topic within the social science community in India.

As I aimed to show above, compared to other nations that also suffered from the negative impacts of colonialism, India has largely managed to preserve a cultural identity built on a strong awareness of its manifold contributions to the development of mankind over millennia. Moreover, there are Indian intellectuals and scientists who feel that Western scholars have not taken due notice of these contributions, especially in view of their meaning for the history of the sciences. The Indian capabilities that Westerners consider more impressive are related to the latest successes of the country's economy and the role Indians are beginning to play on the political and economic world stage. For many Indian intellectuals, the two phenomena have something in common: the growing international interest in India's economic and political rise is strengthening cultural and national self-confidence, including a national awareness of India's achievements in its precolonial past. For some, the latest developments even seem to ring in a return to former Indian glory, a time in history when Europe was lagging behind the Asian civilizations in many ways.

While these are important aspects of the sociocultural and cultural psychological framework within which the need for an integrative science (i.e., a science that allows for modernization and indigenization at the same time) is felt more strongly than before, it is also the century-long experience with Western social scientific approaches that has led many Indian scientists to question standardized Western science's adequacy for examining the heterogeneous cultural contexts that make up India. Since psychology is the field in which many of the topics discussed in this chapter are especially salient, it is no coincidence that the movement for an indigenization of the social sciences is particularly strong in Indian psychology.

At the »National Conference on Yoga and Indian Approaches to Psychology« held in Pondicherry in 2002, 160 of the participants—among them, some of the most prominent Indian psychologists—issued the »Pondicherry Manifesto of Indian Psychology« (2002) with the following assessment of their discipline:

»We believe that the state of psychology in India is none too flattering. In fact, we find psychology in India unable to play its necessary role in our national development. It is widely believed that this unfortunate state of affairs is largely due to the fact that psychology in India is essentially a western transplant, unable to connect with the Indian ethos and concurrent community conditions. Therefore, it has been said repeatedly that psychological studies in India are by and large imitative and replicative of western studies, lacking in originality and unable to cover or break any new ground.«

The authors then express their surprise concerning this situation, claiming that psychological topics like the self, identity, emotions, cognition, social and moral development, coping as well as methodology and various forms of psychological therapy have played a significant role in classical Indian thought:

»Our culture has given rise to a variety of practices that have relevance all the way from stress-reduction to self-realization. Rich in content, sophisticated in its methods and valuable in its applied aspects, Indian psychology is pregnant with possibilities for the birth of new models in psychology that would have relevance not only to India but also to psychology in general.«

Interestingly, in the view of this group of Indian psychologists, the indigenization of the discipline in order to make it more adequately fit Indian subjects and their culture-specific environments does not run contrary to internationalization efforts, but rather supports them: »Judicious introduction of Indian psychology at various levels in our universities and colleges would help (a) to promote indigenous psychology in India and (b) to develop new psychological models, which may have pan-human relevance.«

In a similar vein, the recently launched peer-reviewed *Psychological Studies,* the official journal of the National Academy of Psychology (NAOP) India, aims to develop an »inclusive psychology« that considers the diversities, complexities, specificities, and continuity of human experiences. In the editorial to the first issue, the Indian cultural psychologist Girishwar Misra notes that culturally informed research of the last few decades has shown »that the Euro-American perspective is a special case of culturally rooted indigenous psychology« (Misra 2009: 1). The aim of the journal is therefore to encourage contributions that help to overcome one indigenous psychology's hegemony over others, and enhance our conceptual repertoire, symbolic resources, and methodological knowledge by drawing on diverse cultures and employing interdisciplinary perspectives.

To summarize: In this section, I have proposed answers to questions concerning the usefulness of some aspects of Western internationalization efforts and the adequacy of Western scientific concepts when conducting research in non-Western contexts. Do these efforts and concepts provide us with the means to gain the scientific knowledge we need to improve living conditions all over the world? What indicators of their adequacy should be selected and who should be considered qualified enough to answer these questions? Presumably, it is the social scientists with the profoundest knowledge about the people and the environments and conditions they are living in. In addition, those qualified enough to answer these questions should be the best-skilled scientists who are able to use their knowledge in order to do the finest research possible. Although we may not have found the final answers to our central questions, at least we can say the following: according to a growing number of Indian social scientists, internationalization can be useful in some respects (e.g., if it helps overcome social inequality in national and international scientific communities and provides a culturally and scientifically unbiased evaluation of scientific work in non-Western countries). However, internationalization can also be detrimental to scientific progress (e.g., if the training scientists receive is standardized such that it builds on theories, concepts, and methods mainly developed in the West, mainly for Western purposes, mainly tested on Westerners, and mainly tested by those without the expertise to assess their adequacy in non-Western contexts). If internationalization is taken as a serious effort to improve scientific practice and knowledge and not as a disguised continuation of colonial interests, its proponents need to overcome the theoretical and practical limitations of current mainstream science. Accordingly, proponents of internationalization must aim to investigate alternative (e.g., indigenous) scientific traditions and the expertise of those indigenous scientists who possess the knowledge and skills to do their work effectively in environments in which the range of mainstream Western approaches is limited. Of course, approaches and results varying from the mainstream must also continue to stand up to (international) scientific debate. Such a debate, however, requires that alternative scientific traditions and approaches are recognized and that the voices of their proponents are heard.

Concluding Remarks

In this chapter, I have described the broader historical, sociocultural, and political framework in which the Indian discussion about the internationalization of the social sciences is taking place. I have tried to show that, in order to avoid overly simple and schematic judgments about the relationship between colonization, education, and internationalization, we need to consider the broader context, the historical developments, and the cultural specificities that have led to the current state of the social sciences in India and elsewhere. In other words, even an accurate assessment of the question concerning what needs to be taken into consideration before successfully executing the project called »internationalization« requires a thorough and therefore *collaborative* scientific effort in order to understand why internationalization is perceived so differently across nations and cultures. Moreover, only an open, collaborative, respectful, and well-informed effort can reveal the reasons why the prospect of internationalization has been welcomed enthusiastically in some countries and by some scientists, but met with reserve, and even resistance in other countries and by other scientists.

In the case of India, we have seen how the interplay between very specific experiences with Western interference, on the one hand, and culturally embedded coping strategies, on the other hand, has resulted in Indian social scientists' attempt to find a middle road to internationalization. Using a more recent concept from the social sciences, we can describe this strategy as a »glocalization« strategy, in other words, a strategy that reacts to globalization efforts by propagating global localization (i.e., indigenization). The reasons for this reaction are manifold and include subjectively negative experiences with Westernization in the past as well as experience-grounded scientific arguments against the blind adoption of mainstream Western social science. In addition, in order to acquire a more complete picture of specific Indian attitudes toward internationalization, we also need to identify the heritage of the indigenous scientific traditions that shows up in the current suggestions made by Indian scientists for resolving the dilemma they are in. Demanding interdisciplinary approaches, for example, reflects the conviction that the narrow demarcation lines between the modern scientific disciplines signify a Western trademark, rooted in ancient Greek efforts to systematize the sciences accordingly. The result, highly specialized experts working in narrow scientific niches, proved to be successful for some time. In the social sciences, however, it is becoming increasingly evident that the complex, diverse, and often specific patchwork that we call the »social sphere« needs collaborative interdisciplinary efforts if we want to understand it more adequately. For the Indian advocates of the revival of indigenous scientific achievements and their fruitful integration into current science, this signifies a return to the roots, since classical Indian science was interdisciplinary from the beginning (and, as I tried to show, that is one of the reasons why so many modern Western historiographers have such a hard time finding the »sciences« in classical Indian culture). Finally, the indigenous approach as a whole can be understood as an Indian tradition in that, historically speaking, »Indianizing« foreign invaders' imports has always been a successful cultural coping strategy in India.

It is evident that an internationalization of the social sciences and the humanities must be preceded by an evaluation of aspects like those discussed above. Of course, they need to be supplemented by additional aspects depending on the questions we ask, the answers we need, and the national and cultural specificities we want to consider. Whatever kinds of internationalization goals and strategies come out of this, this much is clear: in the long term, any fair internationalization effort requires a culture of power-sharing that ends scientific hegemony and prevents new scientific monopolies. That is a prerequisite for more voices being heard, for a less culturally and scientifically biased selection of papers by leading journals, and as a result, for tougher competition between opinions and theories, which has always been a key postulate by Western science for constructive scientific discourse.

References

Ames, R.T./Dissanayake, W./Kasulis, T.P. (1994): Self as Person in Asian Theory and Practice, Albany: State University of New York Press.

Arnold, D. (2000): The New Cambridge History of India, Vol. 3, Part 5. Science, Technology and Medicine in Colonial India. Cambridge: Cambridge University Press.

Baber, Z. (1996): The Science of Empire: Scientific Knowledge, Civilization, and Colonial Rule in India. Albany (New York): State University of New York Press.

Bharati, A. (1985): The self in Hindu thought and action. In A. J. Marsella/G. Devos/F.L.K. Hsu (Eds.), Culture and Self: Asian and Western Perspectives. London: Tavistock, 185-230.

Chakkarath, P. (2005): What Can Western Psychology Learn from Indigenous Psychologies? Lessons from Hindu Psychology. In W. Friedlmeier/P. Chakkarath/B. Schwarz (Eds.), Culture and Human Development: The Importance of Cross-Cultural Research to the Social Sciences (pp. 31-51). New York: Psychology Press.

Chakkarath, P. (2007): The Stereotyping of India: Spirituality, Bollywood, and the Kamasutra. Sietar Journal, 1, 4-7.

Chakkarath, P. (in press): Indian Thoughts on Psychological Human Development. In: G. Misra (Ed.), Psychology and Psychoanalysis in India. New Delhi: Sage.

Guru, G. (2002): How Egalitarian Are the Social Sciences in India? Economic and Political Weekly, 5003-5009.

Hegel, G.W.F. (1837/1995): Lectures on the History of Philosophy (Vol. 1, transl. by E. S. Haldane). Lincoln: University of Nebraska Press.

Hooykaas, R. (1987): The Rise of Modern Science: When and Why? The British Journal for the History of Science, 20, 453-473.

Jahoda, G. (1999): Images of Savages. London: Routledge.

Kulke, H./Rothermund, D. (2004): A History of India (4th ed.). Oxford: Routledge.

Kumar, K. (1989): The Social Character of Learning. New Delhi: Sage.

Macaulay, T. B. (1835/1972): Minute on Indian Education. In: T. Macaulay, Selected Writings (ed. by. Clive and T. Pinney). Chicago: University of Chicago Press.

Marx, K. (1853/1981): The British Rule in India. In: K. Marx/F. Engels, 1853-1854, 12, Collected Works (pp. 125-133). New York: International Publishers.

Misra, G. (2009): Towards an Inclusive Psychology (Editorial). Psychological Studies, 54, 1-2.

Moghaddam, F.M./Taylor, D.T. (1985): Psychology in the Developing World: An Evaluation through the Concepts of ›Dual Perception‹ and ›Parallel Growth‹, American Psychologist, 40, 1144-1146.

Mueller, F.M. (1883): India—What Can It Teach Us? New York: Funk & Wagnalls.

Natraj, V'K/Majumdar, M./Kumar Giri, A./ Naidu V.C. (2001): Social Science: Dialogue for Revival. Economic and Political Weekly, 3128-3133.

Paranjpe, A.C. (1998): Self and Identity in Modern Psychology and Indian Thought. New York: Plenum.

Patel, S./Bagchi, J./Raj, K. (Eds.) (2002): Thinking Social Science in India: Essays in Honour of Alice Thorner. New Delhi: Sage.

Pondicherry Manifesto of Indian Psychology. (2002). Psychological Studies, 47(1-3), 168-169.

Rhines Cheney, G./Ruzzi, B.B./Muralidharan, K. (2006): A Profile of the Indian Education System. Paper prepared for the New Commission on the Skills of the American Workforce, Washington, DC: National Center on Education and the Economy.

Said, E. W. (1978): Orientalism. New York: Pantheon Books.

Said, E.W. (1993): Culture and Imperialism, New York: Vintage Books.

Scharfe, H. (2002): Education in Ancient India (Handbook of Oriental Studies, Sect. 2, Vol. 16). Leiden: Brill.

Sen, A. (2005): The Argumentative Indian. In: A. Sen, The Argumentative Indian: Writings on Indian History, Culture and Identity, London: Penguin Books, 3-33.

Seth, S. (2008): Subject Lessons: The Western Education of Colonial India, New Delhi: Oxford University Press.

Spivak, G. C. (1988): Can the Subaltern Speak? In: C. Nelson/L. Grossberg (Eds.), Marxism and the Interpretation of Culture (pp. 271-313). Urbana, IL: University of Illinois Press.

Srinivas, M. N. (2002). Collected Essays (with a foreword by A. M. Shah). New Delhi: Oxford University Press.

Sztompka, P. (1994): The Sociology of Social Change. Oxford: Blackwell Publishers.

Wallerstein, I. (2001a): Does India exist? In: I. Wallerstein, Unthinking Social Science. The Limits of Nineteenth-Century Paradigms (2nd ed.) (pp. 130-134). Philadelphia: Temple University Press.

Wallerstein, I. (2001b): Unthinking Social Science. The Limits of Nineteenth-Century Paradigms (2nd ed.). Philadelphia: Temple University Press.

Weber, M. (1904/2002): The Protestant Ethic and the Spirit of Capitalism. London: Penguin Books.

Indonesian Experiences: Research Policies and the Internationalization of the Social Sciences

I KETUT ARDHANA, YEKTI MAUNATI

Introduction

The development of the social sciences in Indonesia has been strongly influenced by the development of the social sciences in the West, particularly by developments in England, France, Germany, and the United States. The social sciences that mainly developed in these four countries later spread to other regions, including Indonesia (Balan 1982: 212). The discussion of the development of social sciences in present-day Indonesia can therefore not be separated from considering the developments in the West. Yet studies on the state of the social sciences in Indonesia or even reflections on the extent to which the social sciences contribute to solving some of the crucial issues in Indonesia are still rare. Such a study would not only concentrate on the scientific community but also investigate policymaking and its contribution to the social, political, and economic relevance of social research. It would also have to take the limited resources of social sciences in Indonesia (as in other parts of the Third World) into account.

Our following reflections emphasize the existing link between history and the social sciences. This emphasis rests on the premise that social sciences are targeted at the explanation of social phenomena that are embedded in historical context (Bien 1978: 84). Despite being an academic discipline, the social sciences are therefore interlinked with social, political, and ethical considerations and demands (Nachmias/Nachmias 1981: 317; Ladd 1975: 177).

Our analysis will focus on several issues: first, on the role of historians and social scientists from the early development until the decline of the New Order Regime; second, on the place of history and social sciences in the Indonesian universities; and, third, on the challenges of internationalization of the social sciences and humanities in present-day Indonesia.

Indonesia certainly faces the need to solve the negative impacts of modernization and globalization. We will therefore focus part of our attention on the contributions of social scientists to anticipating the current situation and consider accusations that social experts in Indonesia have not been capable of foreseeing developments quickly enough.

The Emergence of the Social Sciences in Indonesia

In his articles, »The Social Sciences in Indonesia: Performance and Perspectives« and »Trends and Perspectives in Historical Studies«, Taufik Abdullah (1979) analyses the background of the emergence of social sciences in Indonesia after the Second World War. Whereas prior to the Japanese occupation in 1942 there were only two colleges of social sciences and humanities (SSH), there was an increase of faculties of social sciences and humanities after the war. In 1975 there were no less than 340,000 students in all fields of study in state universities plus 120,000 students in private universities. Out of this total there were 53,700 students enrolled in SSH courses at the state universities. Abdullah rightly calls this »an impressive gain«. Based on this large number of social scientists he expected substantial benefits for the future of social sciences in Indonesia (Abdullah 1979: 11–13). However, if we compare the situation of the 1970s with the current development in Indonesia, we observe marginalization trends and decreasing numbers of student enrollment in the classical disciplines of the social sciences and humanities.

The Early Years

Social scientists played an important role during the period of Dutch colonization as well as in the era of postindependent Indonesia. In the early days, they often contributed valuable expertise on local social realities that was then used by the colonial administration. During the interlude of British rule, for example, Raffle in his book, *The History of Java*, explores the structure of land ownership in Java. His research was later very useful for the colonial government to effectively rule the indigenous people (Hong 1995: 48). Snouck Hurgronje's work on Aceh and Islam likewise greatly contributed to the formulation of colonial policies.

Based on their knowledge of Indonesian society, social scientists even prepared ground for a military strategy to solve the internal conflicts and tribal wars among the various ethnic groups, such as in Timor, Flores, Sumba, and other regions, and to integrate all islands of the Indonesian archipelago. In the Timor Residence and the surrounding islands (now called, Nusa Tenggara) most of these social scientists were ethnologists and anthropologist such as H. G. Schulte-Nordholt (Ardhana 2000: 8). Their ideas were followed and applied by Dutch rule. Finally, they succeeded in establishing the colonial government. The social scientists had, indeed, contributed to the understanding of how to unite all small regions into one larger unit, which came to be called the Netherlands Indies. On the other hand, in the following years, social scholars began criticizing the government policy on the indigenous peoples. During that time, social scientists often acted as critics or even antagonists of colonialism.

In postindependence Indonesia, particularly in the period of Guided Democracy, academic quality eroded, especially after the expulsion of the Dutch professors from the older universities in 1958. Abdullah (1979) notes that out of the forty-one state universities and teachers' training institutes thirty-two were newly established during the Guided Democracy period (1958–65) and only two were originally part of the older universities. This situation was not really conducive for the social science faculties (Abdullah 1979: 15). Abdullah describes:

»Not only the disciplines themselves were rather suspected, training program for young lecturers abroad, were, to say the least, also discouraged. In the early 1960`s the Minister of Education even instructed to the extent that no social science and humanities students would be sent abroad for further studies. President Soekarno himself in several occasions attacked some branches of social science, notably economics and political science, as being outdated and not in tune with ›the spirit of revolution‹.« (Abdullah 1979: 16)

If during the colonial period social sciences played a significant role serving the interests of the Dutch colonial policy, during the reign of Sukarno (1945–67), the government distrusted the social sciences. However, as professors and lecturers came to be treated as technocrats, many social scientists (especially economists) got directly or indirectly involved in the activities of the provincial as well as central planning agencies (Abdullah 1979: 16). This situation basically also partly existed for the more than thirty years of Suharto's rule (1967–98).

Nevertheless, under the New Order regime there was a new atmosphere. Reacting to the weakness of the Indonesian social sciences, Clifford Geertz proposed the establishment of a social science training center to be funded by the Ford Foundation in cooperation with the local administration. The first center was founded in the Syah Kuala University, Banda Aceh in 1974, and the second was established in Ujung Pandang (Makassar) on the campus of Hasanuddin University in 1975. The low qualifications of the teaching staff indicates a still unsatisfactory situation: for instance, a local university with a student population of more than two thousand only had four teachers with MA or MSC degrees. Furthermore, at the national level there was no integrated plan for developing the social sciences (Abdullah 1979). This partly reflects the judgment of policymakers that the social sciences have not been very supportive of the development program (ibid.: 18–28).

Social Sciences under the New Order Regime

During Suharto's New Order regime (1967–98), the social sciences faced both government involvement and a general trend toward marginalization. The discipline of history shall serve as an example to highlight the interplay of governmental interest and scientific research and as a point of reference for discussing some of the reasons for social sciences' increasingly marginal position.

Government Involvement: The Case of History

In the 1950s the dynamics of Indonesian historical science entered a period that Abdullah calls »the search of meaning«. This search for meaning became apparent in the first national seminar of history that was held in 1957. One of its results was the idea of establishing a commission for the writing of the national Indonesian history. But the resolution came to nothing. The second national seminar of history was held in 1970 in Yogyakarta. A major decision of the seminar was to rewrite the Indonesian history from national perspective. Elected as the chairman of team, Sartono Kartodirdjo was aided by Marwati Djoened Poesponegoro and Nugroho Notosusanto. Taufik was also involved in the 1970 discussion section.

Nugroho Notosusanto expected the book *The National History of Indonesia* (*Sejarah Nasional Indonesia*) to be based on the ideology of

Pancasila.[1] He expected that this book would not only be a standard book for the university level but also a reference for history books at lower school levels. However, Sartono Kartodirdjo, one of the outstanding historians in Indonesia, did not agree to have the book, *Sejarah Nasional Indonesia* (Indonesian National History) serve as general standard, because it lacked relevant material and needed further research (Klooster 1985: 122–123). However, it seems that until the present no substantial changes have ever been made, although the books were republished during the regime of the New Order (see Poesponegoro/Notosusanto 1984).

In several countries in Southeast Asia (such as Indonesia, Brunei Darussalam, and Vietnam), there is strong government involvement in the development of history and other social sciences (Ahmad/Tan 2003). Historical writing cannot be separated from social and political dynamics of the state. All of the research that has been carried out by the government in the field of history and other social sciences strongly reflected political events in the regions (Ahmand/Tan 2003: xiv), and the governments funded such research.

Yet, in general, during the New Order period, the government did not pay much attention to the social sciences. The New Order regime prioritized economic growth over social issues. Some people argue that this is the reason why history and other social sciences have played a marginal role. In general, these disciplines were not thought to contribute significantly to social and economic developments. However, other reasons may also have played a role. Since entering a transition period from authoritarian into a democratic society in 1998, it seems that social scientists have failed to anticipate several crucial issues. In different parts of the Indonesian archipelago divergent types of social conflict broke out. For instance, in Aceh and Papua, Indonesia is facing disintegration threats, in Ambon, Sambas, Pontianak, and Poso ethnic conflicts, and in Bali and other regions of Java incidents of international terrorism. Before and after this chaotic situation, so-called *pemerhati budaya* (cultural commentators) or *pengamat sosial* (social observers), have emerged, who have spoken up about the reasons behind such conflicts. Those cultural commentators and social observers are often members of NGOs who do not have high academic backgrounds and whose analysis

---

1 »Pancasila« is the philosophical basis of the Indonesia state. The five principles: 1. Belief in the One and only God; 2. Just and civilized humanity; 3. The unity of Indonesia; 4. Democracy guided by the inner wisdom in the unanimity arising out of deliberations amongst representatives; 5. Social justice for the whole of the people of Indonesia.

is shallow. Yet social scientists did not play a major role in anticipating and solving these events.

Marginalization of the Social Sciences

We may agree with Abdullah that in many ways Indonesia can be proud of its achievements in the development of social sciences in the last decades. Many efforts such as seminars, research, and conferences have been successful and further programs have been started. Additionally, many of the social scientists do have international reputations. Yet, we may also ask in which respects social sciences have been successful beyond academia. To what extent have social scientists contributed to better solutions of social, cultural, economic, and political issues?

Despite being confronted with important social and political issues in the past—which would have offered historians and social scientists ample opportunity to demonstrate their expertise—many scientists have stayed clear of these issues. Of course, there are presumably several reasons behind this avoidance.

In the field of history, for instance, there is a need for theory in order to be able to offer clear explanation and to sharpen analysis. However, historical sciences in Indonesia have been weak in the adoption and development of theory. Winichakul (2003) observes a general reluctance of historians to pay significant attention to social theories. This is in stark contrast to other branches of the social sciences. Some historians think that if they applied social theories their work would be more comprehensive and scientific. However, this ignores the fact that theories and social concepts in historical writings have become more elaborate and challenging (Winichakul 2003: 3). The low degree of theory adoption may be considered one reason why historical research has failed to give better academic explanations of social and historical development. As a consequence, research results cannot be disseminated to the wider society.

Another reason may be seen in the fact that during the New Order Regime, social scientists ran the risk of being considered as opponents of the ruling government if they criticized the political, cultural, or economic situation. As a result, academics (as the people in general) were afraid to express their arguments and also afraid to be labeled as destroyers of the »national stability«, which was a important slogan for the New Order Regime. Also, often, there was no a channel to express academic views. However, there were also some scholars who supported the government's policy deeply.

The New Order regime of that time contributed to the marginalization of the social sciences in Indonesia. The research community could not expect much assistance from the government, but looked for private institutions, such as research foundations for funding and support. In the 1980s, there was considerable willingness to compete with other countries for funds. After initial successes, since the turn of the century there has been stagnation.

Chris Manning (1984: 178) offers additional reasons why the social sciences have been marginal in Indonesia. First, he states that most of the social scientists in Indonesia are government employees who are obliged to support government policies. In addition, the rest of them who were academic scientists have not been involved in the process of the decision-making. Second, social scientists did not have freedom to express their aspirations. There were, indeed, sociologists or anthropologists who did some research, but there were considerable obstacles to applying their research findings to decision-making, and therefore their results had no impact on problem solving.

At the end of the twentieth century Indonesia has entered a transition period from authoritarian to a democratic society. The year 1998 was the time when the New Order era was superseded by the reformation period. At this transition period many people hoped for profound reform in terms of achieving a better life and sound nation building in the future.

Social Sciences in Contemporary Indonesian Universities

The role of the universities in developing social sciences has increased with the creation of the main academic principles of certain universities, called »*pola ilmiah pokok*«, or core competencies. This principle has involved the creation of new study programs, such as cultural studies, tourism studies, maritime studies, and others. These study programs aim to confront and take advantage of the impacts of globalization. They also influence the development of history and social sciences as academic disciplines. There are now an increasing number of students who enroll in the departments of history and social sciences. Gadjah Mada University, for instance, sees increasing numbers of students in the department of history, especially in the newly established field of tourism studies.

Table 1: The number of students enrolled in social and cultural studies in the Faculty of Cultural Studies at the Gadjah Mada University

Program	1998	1999	2000	2001
Indonesian language	31	33	29	29
Literature	14	12	15	14
West Asia	19	9	15	21
English literature	33	31	33	40
Romans literature	22	24	20	25
Japanese literature	27	20	23	32
History	30	23	17	28
Archaeology	26	30	25	37
Anthropology	30	30	25	28
Tourism studies	24	95	110	94
Archival Studies	1	89	122	94
English Language		33	38	45
Arabic Language		23	34	37
French language		27	39	44
Japanese language		36	41	42

(Source: Statistik Mahasiswa Semester II Tahun 2001/2002: S1 dan D3, Fakultas Ilmu Budaya Universitas Gadjah Mada)

Enrollment in history, however, remains stable at a low level, due to bleak employment perspectives. Tourism has proved a more promising alternative. The same development can also be observed in other universities (e.g., the University of Udayana in Bali). In addition, there are increasing numbers of students who enroll in master programs in cultural studies, since most of the students have already got the jobs secured not only in private business but also in the government.

Table 2: The number of students in the Faculty of Social Political Science at the University of Indonesia

Program/number of students enrolled	97	98	99	00	01	02
Extension Program (Society Relations)	6	22	43	88	64	73
Indonesian Politics	1	6	9	6	7	4
Comparative Politics	2	2	11	16	6	7

(Source: UPT Komputer Universitas Indonesia)

Table 2 above shows that except for the Extension Program (Social Relations) the programs of Indonesian Politics and Comparative Politics have enrolled insignificant numbers of students. This situation is not only limited to Indonesia but also occurs in other countries. Talib Ahmad and Tan Liok Ee (2003: xiii) write about the weakness of the historical material subjects in Southeast Asia:

»School textbooks, perhaps, are most heavily permeated by the political and didactic functions of history. There was a long, but interesting, discussion into how history textbooks, at the university as well as school level, tend to be »ideological, repetitive and mantra-like«, containing little beyond the »officially sanctioned versions of the national narrative«. Textbooks are an important genre of history writing that affects entire generations of young Southeast Asianist. There was much concern that »mantra like« textbooks accompanied by uninspired teaching are killing any seeds of interest in the study of the past that might still be left in a generation already uninterested in the histories of their families and societies. It seems that history not only has to be rescued from the nation, but also from bad teachers and boring textbooks!«

Nevertheless, the development of history and the other social sciences in Indonesia cannot be separated from the role of wider society, especially those who experienced important historical matters in Indonesia. Abdullah (1979) highlights that some issues such as the quality of teaching and the spread of social scientists are two factors that must be addressed. There are ways in which the objective can be achieved. First, the quality of social science teaching has to be improved and the capability of the social scientists has to be further nurtured. It is certainly the top challenge that has to be met if we really want to accomplish advances in our social science. Second, progress in our social sciences depends very much on a more widespread distribution of qualified social scientists. Third, there should be a smoother communication among social scientists. In other words, a better academic community should be encouraged. Fourth, the social scientists in Indonesia have to be deeply aware of their two commitments—the academic and the social commitments. At least both of them share a fundamental requirement, namely, uncompromising integrity. Without this attitude, the social sciences and their custodians will be doomed. Fifth, the place of the social sciences in Indonesia is both promising and precarious at the same time. It is promising because the usefulness of the social sciences has been widely recognized. This recognition also makes its place precarious. Sixth, the precariousness is due to lack of sensitivity of the social scientists toward the manipulative potential of social science. Seventh, intellectual integrity as a central question cannot be separated from the social commitment of the social scientists.

Nowadays, the challenges for social scientists are getting bigger, and because of the rapid process of globalization, which has increased the speed of information exchange internationally, the need for building international networks has become a must.

Patterns of Internationalization

There are many patterns of internationalization of social sciences and humanities that could be discussed with respect to the Indonesian experience. The first pattern is a collaboration of Indonesian universities and other universities outside Indonesia in the fields of social sciences and humanities in producing degrees. As a second pattern, universities could offer an international graduate school, opening for students regardless of their nationalities. A third pattern is research collaboration, while the fourth pattern is international collaboration in workshops, seminars, and conferences. Finally, there is another type of international collaboration in the form of capacity building, including trainings and exchange of researchers and lecturers.

Toward International Degrees and Classes

In Indonesia, getting a higher degree at the international level has a long tradition in the field of business (via a Masters in Business Administration [MBA]). For example, the University of Gadjah Mada (UGM), one of the oldest universities in Indonesia, has collaborated with many universities in Australia and the United States to provide an MBA degree. Other state and private universities have also sent students overseas to study for few months. This is to attract more students by offering overseas' experiences. In the Faculty of Cultural Sciences, an international class was also established in the field of Indonesian languages. It has a long collaboration with Monash University in Melbourne, Australia, and universities from the United States and Japan. UGM has also promoted its international postgraduate school (UGM 2009b).

Other international study programs at UGM in the social sciences and humanities include arts performance and art studies, American studies, public administration, Middle East studies, comparative religious studies, tourism studies, policy studies, culture and media studies, and peace and conflict resolution studies.

In the fields of social sciences and humanities, UGM is one of the pioneers in offering international postgraduate programs. UGM has been recognized internationally, especially because it has been ranked fifty-sixth amongst universities in the world in the field of arts and humanity by the Times Higher Education published in England (Sairin 2006: 212)

International Cooperation in Area and Cultural Studies

The establishments of area studies at universities and research institutions has promoted many international collaborations. In universities, this type of collaboration is formed in the curriculum, research, and capacity building. Meanwhile, for research institutions international collaborations usually occur in the forms of research collaborations, seminar/workshop/conferences, and capacity building (training and exchange researchers).

Area studies and research centers have mostly been developed in prominent universities like (UGM, University of Indonesia, and Universitas Hasanuddin (UNHAS), to mention a few. In UGM of Yogyakarta, for example, undergraduate programs for Japanese and Korean have been established in the Faculty of Cultural Sciences. It has also established Mandarin studies for a graduate diploma. Recently it has also established American studies both for master and Ph.D. degrees. Meanwhile research centers that are independent from faculty have also been established in UGM, including the Research Center for Japanese Studies, Research Center for Korean Studies, Research Center for Asia Pacific, Research Center for Social and Southeast Asia, Research Center for Cultural Studies, and Research Center for Population and Policy. For example, the Center for Social and Southeast Asia, UGM, collaborates with a university in Thailand to exchange lecturers in teaching languages and area studies on Southeast Asian Studies. The Research Center for Social and Southeast Asia collaborates with universities in Netherlands and Norway. According to its director the establishment of international collaborations has been the priority of this center (UGM 2009a).

The Research Center for Asia Pacific collaborates with the Ford Foundation, while the Research Center for Population and Policy has collaborated with many international institutions in different countries, including Australia.[2] Indeed, the establishments of area studies at universities under the auspices of social sciences and humanities have opened opportunities to establish wider international networking.

---

2 Based on interviews with several lecturers of UGM.

Apart from area studies, a study by a IPSK-LIPI team notes new developments of cultural studies within Indonesian universities, especially within prominent universities, which promote international collaborations.

»At the Gadjah Mada University, for instance, since the 1990s, there has been opened up Cultural Studies with two programs: Religious Studies and Cross-cultural Studies. At the Sanata Dharma University, there are Religious Studies and Cultural Studies as well. UIN has established a Centre for the Study of Religious and Social Cultural Diversity focusing on religious studies and diversity of social and cultural norms. Now, the Centre for the Study of Religious Studies and Social Cultural Diversity has on-going collaborative research with the Oslo coalition New Direction in Islamic Thought and Practice: Exploring Issues of Equality and Plurality.« (IPSK-LIPI Team 2005: 181)

Similar to UGM, the University of Indonesia has also established area studies, including Korean, Japanese, American, and European studies. International collaborations have been made to develop such studies, including with the Japan Foundation for Japanese studies.

In South Sulawesi, Hasanuddin University has also established research centers which have international collaboration:

»Being the biggest state university in South Sulawesi, Hasanudin University has many research centres coordinated by the Research Institute (Lembaga Penelitian—Lemlit), including the Research Centre for Gender Studies, the Research Centre for Population Studies, and the Research Centre for the Environment. Indeed, the research centres at Hasanudin University can be categorized as good because of their continuity in terms of carrying out studies and their collaborations with other institutions both national (including RISTEK, DIKTI) and international (Ford Foundation, Sumitomo Foundation and Toyota Foundation). For instance, the Research Centre for Gender Studies has had a long collaborative relationship with an institution in Canada.« (IPSK-LIPI Team 2005: 167–68)

International networking has been growing in the field of social and humanities. This can partly be explained by the rapid process of globalization as well as the need to develop internationally recognized universities. Another important issue is that many lecturers have been studying abroad and when they return to Indonesia they continue their existing networking and embrace the way in which universities in the developed world have directed their universities.

Research Center for Regional Resources (PSDR-LIPI)

Apart from research centers at universities, the establishment of area study research centers, such as the Research Center for Regional Resources at the Indonesian Institute of Sciences (PSDR-LIPI), contributes to the process of internationalization of social sciences and humanities. The center which has three divisions—Southeast Asian Studies, Asia Pacific Studies, and European Studies—was established in June 2001 and has continued to make an effort to wider international collaborations in many forms.

After the economic crisis in Southeast Asia beginning of 1997, it was realized that the understanding of our neighboring countries or other countries in general was very limited. As a result scholars and government officials saw a crucial need for research centers on area studies so that Indonesian researchers could study other nations. The embryo of this center was the Program of Southeast Asian Studies, founded by Taufik Abdullah, former chairman of LIPI, in 1983. Currently, the center has twenty-five researchers and undertakes comprehensive and comparative studies on the dynamics of society, culture, and economy in Southeast Asia, the Asia Pacific, and European countries. In relation to the development of each division, PSDR-LIPI not only established a network with its counterparts in Indonesia but also with other foreign countries, and national and international foundations.

International Projects and Seminars

PSDR-LIPI has carried out several collaborations and it has relations with international organizations, institutions, and foundations, Southeast Asian Studies Regional Exchange Program (SEASREP), International Federation of Social Science Organizations (IFSSO), Asian Public Intellectuals (API) Nippon Foundation, Japan Foundation, Toyota Foundation, Leiden University, Royal Netherlands Institute of Southeast Asian and Caribbean Studies (KITLV).

Since its establishment, there are many international research collaborations and international seminars/workshops/conferences held by this center in collaboration with such international institutions. The following list introduces some of the topics addressed in these collaborations:
- The (Re)Construction of the »Pan Dayak« identity in Kalimantan and Sarawak, Jakarta: PSDR -LIPI, PMB-LIPI, SDI, and SEASREP, the Toyota Foundation (research collaboration).

- Martinus Nanang and G. Simon Devung (eds.) (2004), *Local People in Forest Management and the Politics of Participation*. Indonesian Country Report. Kanagawa, Japan: IGES Institute for Global Environmental Strategies (research collaboration).
- *Border of Ethnicity and Kinship: Cross Border Relations between the Kelalan Valley Sarawak and the Bawan Valley, East Kalimantan*, Jakarta: PSDR-LIPI, Sarawak Development Institute, SEAS-REP, the Toyota Foundation (research collaboration).
- Forestry products and income of indigenous people in East Kalimantan, ADBI, Tokyo, Japan (research collaboration).
- The Chinese and the Japanese Economic Activities in Eastern Sabah during the Prewar 1930s–1940s, The Sumitomo Foundation (research collaboration).
- Politics and business, Bangka-Belitung in Post Suharto Era, KITLV the Netherlands (research collaboration).
- The XVI General Conference of International Federation of Social Science Organizations (IFFSO) on Environmental Protection and Regional Development (EPRD), Jakarta, Research Center for Regional Resources, the Indonesian Institute of Sciences (PSDR-LIPI), October 3–5, 2003.
- International Seminar on »Identity, Multiculturalism and the Formation of Nation States in Southeast Asia«. Jakarta: PSDR-LIPI and the Japan Foundation. This seminar aimed to better understand Southeast Asia from the Southeast Asian perspectives. The question of cultural and ethnic identity, multiculturalism in Southeast Asia has become prominent in public debate, particularly in response to the increasing attention to ethnicity and religion as sources of often violent conflict among different ethnic, religious, and linguistic backgrounds. The seminar aimed to see the development of the Southeast Asian Studies in terms of the practices, methodologies, and theories, particularly the understanding of identity, multiculturalism, and the formation of nation states in Southeast Asia. There were thus three immediate aims of the seminar: to have an exchange of opinions and experiences among Southeast Asianists; to publish a book on *Identity, Multiculturalism and the Formation of Nation States in Southeast Asia*; and to promote the understanding of Southeast Asia. This work has been published by PSDR in collaboration with the Japan Foundation in 2004.
- The International Workshop »Towards Sustainable Development in Southeast Asia: From Forest Management to Eco-tourism«, PSDR-LIPI in Collaboration with the Japan Foundation, Jakarta, August 9–10, 2005. The seminar was expected to stimulate new ideas in order

to understand the recent social, cultural, and political development in Southeast Asia. The objectives were to examine the impact of globalization on the transition from traditional to modern life; analyze the policies on forestry and tourist sectors in Southeast Asia and its implementation; understand challenges and chances in developing the forestry and tourist sectors, particularly in Southeast Asia; sharing among participants, the experiences of forest management in Southeast Asia; look at the changing policies in forestry and tourist sectors, not only in micro level, but also in macro level; explore the understanding and status of ecotourism in Southeast Asia; share among participants the experiences of ecotourism in Southeast Asia; suggest strategies for the promotion and enhancement of sustainable ecotourism in Southeast Asia; search for local wisdom on sustainable development in Southeast Asia; and offer any collaborative programs with scholars from Southeast Asian countries in particular and Asia in general in managing the sustainable development.
- Second workshop on »Local Scholarship and the Study of Southeast Asia: Bridging the Past and the Present Location«, PSDR-LIPI, NUS Singapore, SEASREP, the Toyota Foundation.
- Seminar on »Japan's International Environment around the Time of Meiji Restoration of 1868«, PSDR-LIPI, PMB-LIPI, the Japan Foundation.
- The Second Korea Forum on »Recent Political and Economic Developments in Korea«, February 14, 2006.
- International Seminar on »Stranger-Kings in Southeast Asia and elsewhere«, PSRD-LIPI, PMB-LIPI and KITLV, 2006.
- International Seminar on »Tracking Development«, PSDR-LIPI and Leiden University, January 2007.
- International Workshop on »Questions of Nationalism and Cultural Identities in the Present Day Asia«, PSDR-LIPI in Collaboration with the Japan Foundation, Jakarta, November 5–6, 2007. The four objectives of the international workshop were to have an understanding of complex array of »authorities« involved in the formation, definition, and representation of »ethnic groups« in Asia, including religious and local leaders, government officials, politicians, anthropologists, and so forth; to understand the notions of nationalism and cultural identities in Asia; to contribute to understanding of the process of formation of identities in relation with other issues, including nationalism, ethno-nationalism, ethno-development, decentralization, and social conflict; and to promote the proceeding of the international workshop to the stakeholders and public by publishing it.

Participants included researchers from Australia, Vietnam, Malaysia, Korea and Indonesia.
- International Conference on »Comparative Multidiscipline Approach in Socio-Cultural Studies Marginalized People«, PSDR-LIPI and IFSSO.
- Annual API Country workshop, PSDR-LIPI and the Nippon Foundation, Tokyo, Japan.

International Funding Organizations

Important funding organizations of PSDR-LIPI international cooperations include SEASREP, API the Nippon Foundation, and the Japan Foundation. For illustration, we shall briefly explain the aims of these organizations.

SEASREP, established around fourteen years ago, is funded regularly by the Toyota and Japan foundations. SEASREP has three main programs: language training, Luisa Mallari fellowship, and comparative and collaborative research program. Apart from this, it has also had other activities, such as a training workshop in collaboration with SEPHIS (South-South Exchange Program for Research on the History of Development) and a seminar in cooperation with KISEAS (Korean Institute of Southeast Asian Studies). SEASREP will revitalize its scheme and will shift from the core grants program to training and capacity-building and global collaboration, while the focus remains on promoting Southeast Asian studies. Even as support for Southeast Asians will continue, the revitalized program entails a major change: the expansion of SEASREP's target audience to include Southeast Asian specialists and scholars outside the region. Participation in training and capacity-building program and the global collaboration program will thus no longer be exclusive to students and researchers based in the Southeast Asian region.

Another important source is API Fellowships, the Nippon Foundation. The first API Fellowships, batch 1, were given in 2001–2. Currently funding is in the batch 7, 2008–9, cycle. The API Fellowships Program, which operates in five countries (Japan, Malaysia, Thailand, the Philippines, and Indonesia), is purposed to meet the political, economic, and social challenges that transcend national boundaries. It is believed that »the region needs a new pool of intellectuals who are willing to be active in the public sphere and can articulate common concerns and propose creative solutions« (see *Nippon Foundation Fellowships for Asian Public Intellectuals (API Fellowships) 2008–9*, The Nippon Foundation and PSDR-LIPI). Themes of API Fellowship are changing identities and their social, historical, and cultural contexts; reflections on the human condition and the search for social justice; and globalization: structures, processes, and alternatives. The notion of the fellowships is for a recipient to study countries other than his/her own. An Indonesian, for instance, cannot study in Indonesia, but other member countries (Thailand, Japan, Malaysia, or the Philippines), while Japanese cannot carry out research in Japan, but in the other four member countries. This program has promoted the building of regional community and its roles in Asian society.

The Japan Foundation has also promoted an international exchange program in which PSDR-LIPI has applied for and received grants several times. The Japan Foundation is one of the outstanding institutions that is concerned with how Southeast Asian Studies can be developed in Southeast Asia, particularly by conducting collaborative work either through research or seminars, workshops, and conferences with the main institutions regarding the development of the studies in the region. The Japan Foundation has been very important for opening and widening networks amongst Southeast Asianists and Asianists from Asia and other countries like Australia.

Conclusion

From the above description it can be concluded that a better understanding of the development of historical studies as well as the other social sciences in Indonesia can be reached if we can understand their development in Europe as well. Difficult challenges have been experienced not only by social scientists in Indonesia but also in the Western countries. However, research on the social sciences continues to be conducted for the sake of maintaining academic truth as well as developing social sciences.

On top of these difficulties, each time a social scientist conducts research on a sensitive issue, he/she will be viewed as a threat to the ruling-class or the government. He/she will be considered to be against the status quo. For this reason, we can see the social scientists' double roles in Indonesia. There are those social scientists who act both as scientists and bureaucrats, and they usually pay more attention to power than to academic interests. This phenomenon tends to prevent the development of social sciences. In this case, it can be understood why the decision-makers are not interested in using the results of social science studies. This is one reason why the social scientists in Indonesia are very often considered as incapable of solving the crucial social-cultural, economic, political issues in the present-day Indonesia.

Maintaining the existence of the social sciences in Indonesia will require hard work, particularly for historians and social scientists. In other words, higher-quality research based on the academic and social commitments, with a high degree of dedication and intellectual integrity, is needed. With these strategies, we can expect that the development of historical studies and the social sciences in Indonesia will enjoy a much better prospect in the future. The development of international networking is also important if Indonesia is to catch up with the development of social sciences and humanities worldwide.

Joint degrees and international classes have been opening up in Indonesia in social sciences, although these have been limited to a few state and private universities. UGM, a prominent university, has promoted both joint degrees and international classes. The Times Higher Education, published in England, has ranked field of arts and humanity of UGM to be number 56 worldwide. This has become a kind of promotion for international classes in social sciences and humanities.

Another feature is that in terms of internationalization of social sciences and humanities, regional studies seem to be dominant, and Japanese institutions and foundations have played important roles in promoting Asian or Southeast Asian studies or collaborations in research and seminars, workshops, and conferences. This can be observed from PSDR-LIPI's collaboration with Japanese institutions in strengthening the regional cooperation in the region, and could also be seen in a few universities. It is expected that in the near future there will be more institutions worldwide participating in the process of widening the internationalization network in the fields of social sciences and humanities in Indonesia.

References

Abdullah, T. (1979): The Social Science in Indonesia, Performance and Perspective. In A.K. Kenkyūjo (Ed.), A Study on Trends and Perspectives of the Social Sciences in Indonesia. Tokyo: Institute of Development Economics.

Abdullah, T./Surjomihardjo, A. (1979): Trends and Perspectives in Historical Studies. In Ajia Keizai Kenkyūjo (Ed.), A Study on Trends and Perspectives of the Social Sciences in Indonesia. Tokyo: Institute of Development Economics, 1979.

Ahmad, A.T./Tan, L.E. (2003): Introduction. In A.T. Ahmad/T.L. Ee (Eds.), New Terrains in Southeast Asian History (pp. ix-xxiv). Singapore: Singapore University Press.

Ardhana, I.K. (2000): Nusa Tenggara Nach Einrichtung der Kolonialherrschaft 1915 bis 1950. Passau: Richard Rothe.

Balan, J. (1982): Social Sciences in the Periphery: Perspective on the Latin America Case. In L.D. Stifel/R.K. Davidson/J.S. Coleman (Eds.), Social Sciences and Public Policy in the Developing World. Canada: Lexington Books.

Bien, J. (1978): Phenomenology and the Social Sciences: A Dialogue. The Hague: Martinus Nijhoff.

Bourchier, D./Hadiz, V.R. (2003): Indonesian Politics and Society: A Reader. London: Routledge Curzon.

Campbell, D. (1975): The Social Scientist as Methodological Servant of the Experimenting Society. In S.S. Nagel (Ed.), Policy Studies and the Social Sciences (pp. 27-31). London: Lexington Books.

Coleman, J.S. (1982): Policy, Research, and Political Theory. In William H. Kruskal (Ed.), The Social Sciences: The Nature and Uses. Chicago: The University of Chicago Press.

Gadjah Mada University (UGM) (2009a): Lebih banyak membangun kerjasama. http://www.ugm.ac.id/index.php?page=headline&artikel =346edisi=Edisi%20No.%2074/TAHUN%20IV/31%20Maret%202 008. [Date of last access: 24.08.09]

Gadjah Mada University (UGM) (2009b): The Graduate School. http://www.ugm.ac.id/eng/content.php?page=1&display=2. [Date of last access: 24.08.09]

Hong, L. (1995): History. In Mohammed Halib/Tim Huxley (Eds.), An Introduction to Southeast Asian Studies. London and New York: Tauris Academic Studies.

Horowitz, I.L. (Ed.) (1974). The Use and Abuse of Social Science. New Brunswick, New Jersey: Transaction Books.

IPSK-LIPI Team (2005): The Development of Social Sciences and Humanities in Six Cities in Indonesia: Case Studies of Denpasar, Yogyakarta, Medan, Samarinda, Surabaya and Makassar. In Reflections on Social Sciences and Humanities Research in Southeast Asia, Jakarta: MOST-UNESCO (Management of Social Transformation) and Social Sciences and Humanities Division—the Indonesian Institute of Sciences (IPSK-LIPI).

Klooster, H.A.J. (1985): Indonesiers Schrijven Hun Geschiedenis: De Ontwikkeling van de Indonesische Geschiedbeoefening in Theorien en Praktijk, 1900-1980. Dordrecht: Foris Publications.

Ladd, J. (1975): Policy Studies and Ethics. In S.S. Nagel (Ed.), Policy Studies and the Social Sciences (pp. 177-184). London: Lexington Books.

Marwati D.P./Notosusanto, N. (1984): Sejarah Nasional Indonesia, Vol. I–VI. Jakarta: PN Balai Pustaka.

Manning, C. (1984): Peran Ilmu-ilmu Sosial dan Teori Ekonomi. In Krisis Ilmu-ilmu Sosial dalam Pembangunan di Dunia Ketiga. Yogyakarta: PLP2M.

Nachmias, C./Nachmias, D. (1981). Research Methods in the Social Sciences. London: Edward Arnold.

Rüland, J. (1998). Politische Systeme in Südostasien. Lanberg: Olzog.

Sairin, S. (2006): Penutup. In Refleksi Penelitian Ilmu Sosial dan Kemanusiaan, Jakarta: Most-UNESCO and IPSK-LIPI.

Steele, I. (1982). The »New« History in Schools: From Content to Understanding. In B. Dufour (Ed.), New Movements in the Social Sciences and Humanities. London: Maurice Temple Smith.

The Nippon Foundation and PSDR-LIPI (2008): The Nippon Foundation Fellowships for Asian Public Intellectuals (API Fellowships) 2008–2009.

Winichakul, T. (2003): Writing at the Interstices: Southeast Asian Historians and Postnational Histories in Southeast Asia. In A.T. Ahmad/L.E. Tan (Eds.), New Terrains in Southeast Asian History (pp. 3-29). Singapore: Singapore University Press.

The Current Internationalization of the Social Sciences in Latin America: Old Wine in New Barrels?

HEBE VESSURI

Introduction

In this first decade of the twenty-first century, the social sciences in Latin America are split over debates about what problems they should study and what theories and methods should be used to study them. Added to this are the perennial *loci classici* of state, sovereignty, and nationalism, and the enduring interest in the researcher's autonomy juxtaposed with accountability, relevance, and the search for new methodologies, transdisciplinarity, and disciplinary grounding; the teaching-research nexus; and institutional aspects such as funding. But dynamic cross-currents in society are redrawing the contours of the social sciences and new approaches seek to unveil the problematic relationship between state and market; poverty, violence, and social conflict; ethnicity, multiculturalism, and cultural diversity; governance and the search for a new sense of community and/or sociality; universality versus particularity; environmental and social sustainability.

The recognition of a major divide between those who argue that globalization is the landmark of the new age and their critics is very much at the forefront of debate. In many ways, this dichotomy seems no more than another instance of polar metanarratives that obscure as much as they illuminate.

In the not-so-distant past, the social sciences discourse about science was predominantly Eurocentric, and in many ways it continues to be so.

However, in Latin America as in other peripheral regions of the world, this intellectual hegemony of the West was challenged by individuals and thought traditions way before the recent Northern »post-colonial« theory of science became fashionable. Science, and the social sciences in particular, »opened up« very early in the south to subjects, themes, actions, and formerly neglected groups, and it has provided a scope for critically revaluing non-Western knowledge forms as part of a rediscovery of hybrid traditions, identities, and cultures (Nandy 1995; Pecujlic/ Blue/Abdel-Malek 1984).

The current internationalization, or globalization, of the social sciences as seen from the vantage point of the so-called peripheral regions appears as a heterogeneous and multicentric process, endorsing the idea of multidirectional flows of knowledge across civilizations and aiming at creating a global social science based on genuinely epistemological egalitarianism. This is not a novel idea, however. A discipline like anthropology, for instance, has long embraced the notion of cultural diversity as the way to approach the universality of the human species and human society, thus prefiguring a form of global society.

The social sciences have been involved in international activities since their inception. Indeed, they have been part of the process of complexification and internationalization of society. In a region like Latin America, disunited by economics and politics and united by culture and linguistic affinity, globalization implies, among others, understanding the role of knowledge and technology in the region, the impact of migrations and diasporas, and analytically assessing some possible paths toward the future, and the further transformation of the social sciences in the new conditions.

Latin American Higher Education and the Internationalization of the Social Sciences

A way of looking at the current transition toward more intense globalization is to describe changes of cultural and intellectual habits, organization, and accountability in Latin American higher education, asking how the regional structures responded to and supported internationalization at different times, and what the institutional means or constraints are for international research collaboration in the present time. Throughout three quarters of the twentieth century, the university in Latin America was seen as one of the principal if not the main tool of modern nation-building. The central rationality of governments was grounded in the notion of »investment in human capital«, whereby the population was understood as a national resource to be harboured and developed. Among the areas to be supported, were the humanities and the arts, as befitted civilized nations; psychology, economics, sociology, and other social sciences were also deemed necessary to administration and public order. The nation-building university enjoyed strong popular resonance. An ever expanding capacity in higher education multiplied opportunities for children from professional and white-collar families, and some working-class families as well, in a synergistic spiral of growth.

As an integral component of this process, since the 1930s, at a juncture of important political and social change in Mexico, Brazil, and Argentina, the earliest institutions linked to the social sciences were created; and in the 1950–60 period, driven by public concerns with modernization, the process of effective institutionalization of sociology as a discipline through teaching and research began.[1] However, this second institutionalization period ended in the disruption and crisis of existing projects, mainly due to the irruption of authoritarian regimes (Brazil, 1964; Argentina, 1966 and 1976; Uruguay and Chile, 1973).

Curiously, the expansion of higher education and of the social sciences occurred especially since the 1970s in the midst of political turbulence and repression, after the closing down of the public social science schools. Social criticism gave place to instrumental social technology. As the number of students in the social sciences swelled, many sought an ill-defined professionalization that the university was unable to supply in satisfactory measure; for a much more structured supervision was required than what the university was prepared to offer. The majority of students who enrolled in courses with high drop-out rates and conferring dubious professional status came from social backgrounds manifestly less privileged than those seeking to enter more competitive social professions. The range of courses in the social field has not been homogeneous either between countries in the region or within one and the same country. Within a motley and variegated picture of the social field, the main systems producing the greatest numbers of graduates with first degrees have been those of Brazil, Mexico, Colombia, Argentina, Venezuela, Perú, and Costa Rica.

As a result, the long-standing governmental project of the nation-building university fell into a deep crisis in most countries of the region

---

1 Among others, the Centers for Historical and Social Studies (1943) in the Colegio de Mexico; the Institute of Sociology of the University of Buenos Aires (1947) and after Perón's demise, the program of Sociology (1957) and the Institute of Economic Development (1958) in Argentina; in Brazil, the Paulista school of sociology at the University of Sao Paulo in the early 1950s, and the Joaquin Nabuco and Bahia institutes.

since the 1970s, with mutually reinforcing but distinct elements. There is a crisis of academic identity brought about by the »corporatization« of university structures and cultures, and by the lack of alternative solid models to the Euro-American one capable of better responding to the problems of the national contexts; there is also a crisis of government commitment to the public role of the universities, expressed mainly in a weakening of the resource base; a crisis of positioning and strategy vis à vis the current international restructuring of higher education, accepting the imposed division of labor by which cheap engineers and technicians are produced (and not only workers as in the previous model) for the worldwide spread of research and development (R&D) and marketable engineering products (and thus a legal framework that supports strong intellectual property rights (IPR). This has been visible even in the processes of transition toward democratic regimes that since the mid-1980s have coincided with the reemergence of alternative approaches.

The historical continuity of the university as a social institution and the coexistence of new and ancient forms in its bosom, however, can make us lose sight that the university is undergoing a great transformation. Under the drive of an emerging global market for higher education, most universities have entered, willingly or by force, into conflictive processes of conformity, subordination, homogenization, and unequal competition toward the hegemonic global model of the North American university, losing their distinctive character and impact in the local and national contexts to which they responded to a greater degree in the past.

The global era creates a more extensive and intensive engagement. In many disciplines there have long been world circuits in which particular knowledges circulate, are augmented, and reformed. These circuits are now larger, »thicker« in the traffic the carry, and more immediate and determining in their local effects. Thus disciplinary research groups affiliated with universities are a strong pressure force from below, with divided loyalties to their international professional communities and the local knowledge institution to which they belong. Changes in the institutional governance structure, the opening of international relations offices, and the multiplication of cooperation agreements are intervening factors. Online education across national borders hastens the cultural interpenetration of nations and higher education institutions. However, as we have already recalled, this is not a free exchange of equally weighted cultures. It is often noted that globalization is associated with two contrary trends: one to convergence and homogeneity, and another one to diversity via more extensive and complex encounters with cultural »others«. One could imagine working toward a situation in the social sciences in which every initiative about research on online communication and e-science begins with the recognition of diversity and specificity. The infrastructures produced would therefore be »specific« or »situated« rather than »generic«. Common elements, if present, would arise from practices—emergent, but not presupposed (Beaulieu/Scharnhorst/Wouters 2008).

In recent years the number of Latin American social science graduates rose spectacularly, jumping from almost 400,000 in 1997 to almost 900,000 in 2005 (i.e., from 54 percent to 61 percent of total graduates). The university sector in the region is strongly oriented toward graduate education of professionals in the social sciences. The master's degree level has many professional qualifying programs in administration, law, taxation and finance, and psychology, but already shows the presence of programs aimed at training for research, for example in economics and development, social sciences, and communication sciences.

With all the changes, the university is the only existing institution in contemporary society able to establish a bridge between specialized knowledge and society as a whole in the new global context. It is most valuable in the re-creation and construction of shared contemporary values and social understanding, as well as an essential space for the formation of diverse groups for a broad range of interactions in society and with the environment. And it is a fundamental site for knowledge production, attending to a very broad range of social concerns, demands, and problems in different domains. In the current context of increasing homogenization, the challenge for Latin American universities is preserving and re-creating the diversity of traditions and responsibilities starting from a fundamental axis: a commitment to society in its broadest sense.

The Institutionalization of Research in the Region

Latin American Science Bodies

An early setup for regional and international collaboration in the social sciences was the establishment of the Latin American Faculty of Social Sciences (FLACSO), in 1957 and the Latin American Social Sciences Council (CLACSO), in 1967 (Vessuri 1999). The governments of Chile and Brazil called a meeting in Rio de Janeiro in 1957, which decided to establish FLACSO with its headquarters in Santiago, Chile. FLACSO emerged as a regional and autonomous international organization from a cooperative initiative between UNESCO and the governments of the region aimed at promoting education, research, and technical cooperation in social sciences throughout the subcontinent. Its basic functions have been providing training in the social sciences through postgraduate and specialization courses; performing research in the social science field on Latin American problems; disseminating, with the support of governments and appropriate institutions, advances in the social sciences, particularly its own research results; and collaborating with university institutions as well as with government and private international, regional, and national bodies to encourage development in the social sciences. Right from the start, thirteen Latin American countries were parties to the agreement, and since 1993 others showed an interest in becoming members. The regional and autonomous nature of FLACSO is ensured both by the participation of all member countries and eminent intellectuals in its governing bodies and also by the Latin American origins of the student and administrative bodies, which carry out activities in its ten academic units and in the General Secretariat. Its Latin American nature is likewise strengthened by the content and scope of its teaching and research programs, which are geared to the region's scientific and social needs. Assistance also comes from the financial contribution by governments of member countries and from an extensive network of cooperation agreements with various institutions in the public and private sectors of this and other continents.

Since its creation in 1967, CLACSO has become the most extensive coordinating body for social science research centers in Latin America, currently including 159 member centers. Its Executive Secretariat has always operated in Buenos Aires. CLACSO emerged partly as a counterpart to the Social Science Research Council of the United States. A group of Latin American social scientists believed in the importance of having a council that might coordinate the pioneer social science centers and institutes, among which were the Colegio de Mexico and the Universidad Nacional Autónoma de Mexico's (UNAM) Instituto de Sociologia in Mexico City; the Centro de Estudios del Desarrollo (CENDES 2007) in Caracas; the Instituto Torcuato Di Tella (ITDT) and the University of Buenos Aires's Instituto de Sociologia; the Center for the Study of National Reality (CEREN) at the Catholic University of Chile in Santiago, and the Instituto de Economia of the Universidad de Chile and other centers in the capital city and other places in the hinterland; the Brazilian Center of Analysis and Planning (CEBRAP) in Sao Paulo; and the program of Sociology of the Universidad de la Republica in Uruguay. Also the connections with the United Nations' Economic Commission for Latin America (ECLA) created in 1949 and FLACSO were very close.

From its early years CLACSO built up relationships with international agencies and research institutes in the advanced countries and the Third World, particularly those carrying out research on Latin America, as in the case of the Latin American Studies Association (LASA) in the United States (1966), the Canadian Association of Latin American Studies (CALAS-ACELA) and the European Council for Social Research about Latin America (CEISAL) created in Westphalia in 1971. After several previous attempts, it also engaged in a policy of rapprochement with social scientists from other Third World regions with the creation of the Council for the Development of Economic and Social Research in Africa (CODESRIA) (1973). Since 1972, when it joined UNESCO as an international nongovernmental organization that offered information and advice, its links with that agency as well as with the International Social Science Council have been very close.

CLACSO has developed a work program to foster greater integration of Latin American social sciences and defend the working conditions of social scientists at member centers and other institutions in Latin American countries ruled by authoritarian regimes. Its postgraduate program was drawn up to deal with two major areas: the Southern Cone Research Program, which, with financial support from the council, provided aid in the countries of the Uruguay, Paraguay, Argentina, and Chile subregion to researchers experiencing work difficulties because of their political and/or theoretical views; and in cooperation with the United Nations Development Program and UNESCO, the Young Researchers Training Program, since it had become apparent that the main problem in the region were lack of funds for research and the difficulties experienced by young graduates of universities in obtaining funds from international agencies. Its working groups and commissions have a membership of some five thousand researchers in a program of academic exchange, debates, and publications.

Both FLACSO and CLACSO have had great significance in the institutionalization of social science research and the development of a regional research and intellectual agenda that spread to the institutions of higher education in Latin America. This development has occurred even though military dictatorships knocked down the Southern Cone social sciences and civil society as a whole in the 1970s, and in spite of the authoritarianism that persisted in other countries of the region until recent times. The extension of the process of »deinstitutionalization« of social research in the Southern Cone has divided in two the process of building up an independent and autonomous trajectory for the social sciences in Latin America; its repercussions are still noticed today in the functioning of academic contexts (Bayle 2007). Nevertheless, in ensuing decades one of the main aims of FLACSO and CLACSO remained the endogenization of the social sciences, trying to explore the specificities and peculiarities of the region and its countries, its people, and its social and political dynamics. The interplay between the development of a local/regional *thématique* and international/universal concepts and notions as well as theoretical frameworks has marked the growth of the several disciplines.

This international support provided by CLACSO and FLACSO helped the new crop of social science research centers in the region during the last forty years. The bulk of current centers were established since 1970, especially between 1970 and the early 1980s. Part of the reason for this can be found in the impetus given to the whole of higher education throughout those years. It is also important to note, however, that a high proportion of the existing centers have been set up in the last twenty-five years, with an emphasis on education, economics, anthropology, and sociology centers. Other centers were also set up during this period for history and administration. The proliferation of centers seems to be inspired by needs of various kinds. Many were established as an aid to teaching others because of the need to explain local or regional realities or on account of some particular social or cultural problem or new thematic interests. Probably the smallest numbers were created in order to promote theoretical, methodological, and instrumental progress and innovation. The pace and level of geographical concentration of research units in the several countries, particularly the largest ones, have tended to fall over the last twenty-five years, resulting in a more balanced distribution, although a strong concentration persists in metropolitan areas. In Mexico, for example, all states have research centers in some social or humanist discipline, which is not the case in other knowledge fields; although more than half of all research capacity continues to be concentrated in the metropolitan area of Mexico City.

Publishing in Latin American Social Sciences

Although books continue to have great importance in the social sciences, their diffusion and circulation are basically restricted to the local community and in some countries with new incentive systems for research, contributions in the form of Spanish-language books are not as highly regarded as in the past. On the other hand, publication of English-language papers in international journals, as is usual among the physical sciences, is promoted by important segments of the social sciences. Papers become increasingly visible as the Information Technologies make them available through more efficient and comprehensive databases.

The Latin American research journal is increasingly perceived as satisfying four main functions:
1. It affords a means of communication with interested colleagues,
2. It helps to guarantee quality through the process of peer review,
3. It allows authors to show the originality and value of their thought, and
4. It facilitates the credit distribution in the academic community.

The internationalization index for general scientific literature rose significantly in the last three decades, in a trend shown also to be valid for Latin American social sciences (cf. Narváez-Berthelemot 1995), and we may infer that in recent years it has intensified as a consequence of incentives and recommendations by science councils and funding agencies. However, the profile of Latin American journals has continued to be judged as poor (Sancho/Morillo/De Filippo/Gómez/Fernández 2006) when considering the dynamic growth experienced by Latin American science in the same period. And although the science citation index (SCI) and social science citation index (SSCI) have significantly increased their coverage of Latin American journals, the changes are unlikely to have altered to any considerable extent the heavy underrepresentation of Latin American journals in the mainstream databases (Collazo-Reyes/Luna-Morales/Russell/Pérez 2008). In connection with this, there has been a conscious effort by government agencies and local scientific communities that has led to great improvement in the formal features of Latin American social science journals and in databases making use of new information and communication technologies. The emergence of services such as Latindex, Scielo, and Redalyc have helped overcoming the restriction to local users of most existing institutional databases that was still noticeable in the 1990s.

The international flow of communication continues to be highly asymmetrical. In this sense, translations are an interesting indicator of asymmetrical fluxes. Even though we do not have specific figures for the social sciences, but rather general ones for the literary field, we have reasons to believe that the situation is rather similar. In any particular year translation from English into, say, Spanish, Portuguese, Bulgarian, Dutch, or Arabic vastly outnumbers Spanish, Portuguese, Bulgarian, Dutch, or Arabian works translated into English. In fact, translation in the literary fields has been said to currently comprise about 2–4 percent of the annual output for British and American publishers. This contrasts with 25 percent in Italy, and the astonishing figure for Brazil, where 60 percent of new titles consist of translations, 75 percent English (Venuti 1998). But the issue is not only one of translations. Changes have been found in the preferred languages for publication due to the continuing loss to English of the two local languages in the Latin American region in some social sciences. For countries with little scientific tradition, participation in the international scientific community has been considered particularly relevant, also for the social sciences. Collaborations of Latin American scientists in mainstream papers with intraregional colleagues of two or more Latin American countries rose 2000 percent from 1975 to 2004 (Russell/Ainsworth/Narváez-Berthelemot 2006), while there was a 36 percent increase in ISI papers published between 1999 and 2002 by Latin American scientists in international collaboration generally (Sancho et al. 2006).

A study from the beginning of this decade, observing the UNESCO DARE Database from 1991 crossed with the economic index of the World Bank, showed that in Latin American social sciences journals tended to be multidisciplinary, with several of the specialized journals belonging to the humanities. This coincided with the observation that high-income countries tended to publish on all subjects and also in specialized fields like psychology, social security, applied science, and information sciences, while middle-income countries, a category which includes the majority of Latin American countries, tend to publish more titles in the humanities. Spanish was the language of the majority of journals (68 percent), with 17 percent for journals in Portuguese, 9 percent bilingual, 5 percent trilingual, and journals in English being only 0.6 percent and French 0.3 percent (Narváez-Berthelemot/Russell 2001).

A study about social sciences publishing behavior among Venezuelan social scientists in the National System for the Promotion of Researchers explored the categories adopted by the program: book or book chapter, peer reviewed article of international circulation, and indexed journal (Vessuri/Martinez/Estévez 2001). Contrary to the common expectation that social scientists only publish books addressed to a local audience, which would lead to a slower growth of knowledge, results showed that the most frequent publication outlet for Venezuelan social scientists (among whom were humanists, for they are all in the same commission) is the article in a research journal. Journal articles and papers in compilations were 6.6 times more frequent than books in the sample. Nonetheless, a certain preference of historians for publishing books was confirmed, while psychologists were seen to prefer articles as outlet.

Results such as this suggest that the Latin American social sciences are not exclusively aimed at a local audience and, therefore, can be analyzed in some of their components, with bibliometric tools valid for international comparison. In practically all fields and specialties there are authors and groups that publish not only in their own country but also abroad and even in languages different from Spanish or Portuguese, although some fields are clearly more internationally oriented in their habits of publishing research results. Nonetheless, a fuller and fairer analysis would require bibliometric exploration of Latin American databases, which register the large majority of journals published in the region.

Regional Collaboration

In the last few years a renewed climate of knowledge cooperation has set in. As illustration we may mention the Sixteenth Iberoamerican Summit of heads of state and government, which met in Montevideo in 2006, approving the creation of the Iberoamerican Knowledge Space (in Spanish, EIC). The EIC was defined as a domain for promoting regional cooperation in the generation, diffusion, and transfer of knowledge. Its aim was to generate mechanisms to improve higher education, scientific research, and innovation oriented to the region's sustainable development (Aintablian/Macadar 2009).

The following year, the seventeenth summit of heads of states and government met in Santiago and approved the creation of the Pablo Neruda Iberoamerican Initiative of Academic Mobility, aiming at strengthening masters and doctoral programs in the region, particularly through regional and subregional exchanges within a framework of multilateral cooperation. In 2008, in the UNESCO Regional Conference for Higher Education (CRES 2008) held in Cartagena, the Final Declaration included a special chapter on regional integration and internationalization of higher education. It stated the fundamental need to build a Latin American and Caribbean Meeting Space of Higher Education (ENLACES in the Spanish acronym), forming part of the agenda of governments and multilateral agencies of a regional nature (CRES 2008). This is basic for reaching higher levels aiming at fundamental aspects of regional cooperation: deepening its cultural dimension; the development of academic strengths that may consolidate regional perspectives in the confrontation with the most excruciating world problems; creating synergies at regional scale through capacity building; overcoming the gaps in knowledge availability; considering knowledge from the point of view of collective well-being; and the creation of competences for the organic connection between academic knowledge, the world of production, work, and social life.

Table 1: Aims of ENLACES

> The Latin American and Caribbean Meeting Space of Higher Education (ENLACES) aims:
> a. To renew the educational systems in the region, with the aim of achieving a better and greater compatibility between programs, institutions, modalities, and systems, integrating and articulating the cultural and institutional diversity.
> b. To articulate the national information systems about higher education in Latin American and Caribbean (LAC)AC to foster [...] the mutual knowledge among the systems as the ground for academic mobility and as an input for adequate public and institutional policies.
> c. To strengthen the process of convergence of the national and subregional accreditation and evaluation systems, aiming at regional standards and procedures of quality assurance of higher education and research to project their social and public function.
> d. The mutual recognition of studies, degrees, and diplomas, on the basis of quality guarantees, as well as the formulation of common academic credit systems accepted in the entire region.
> e. To foster the intraregional mobility of students, researchers, teachers, and administrative staff.
> f. To undertake joint projects of research and the creation of multi-university and pluridisciplinary research networks and teaching.
> g. To establish instruments of communication to favor the circulation of information and learning.
> h. To give impulse to shared distance education programs, as well as supporting the creation of institutions of a regional character combining online and on-site education.
> i. To strengthen the learning of the languages from the region in order to favor a regional integration that incorporates as wealth the cultural diversity and plurilinguism.

An example of a research network is that of the Iberoamerican network of researchers about Globalization and Territory (RII) which has some eight hundred members from a hundred institutions and twenty-three countries who have joined this network for the critical analysis of the spatial-temporal dimension of the globalization process. The diversity of approaches about territorial phenomena is one of the main advantages; territorial problems are analyzed from these multiple perspectives, privileging aspects such as economic, social, and political-institutional, among others. This convergence of diverse approaches has made RII a critical space particularly apt for interdisciplinary analysis and debate. Linked to the meetings of RII are those of the workshops of Red Iberoamericana de Editores de Revistas (RIER), a group of academic journals specializing in globalization and territory. There are over twenty such journals in the region (CENDES 2007).

Thematic Concerns

Given the variety of institutional, intellectual, and professional models, contemporary social sciences in the region are numerous, varied, contradictory, incorporating elements of both the humanistic and technoscientific traditions, linked at one extreme to secondary education and at the other to the foundation of emerging professions, and subdivided into ever-greater specializations.

National Identity and Modernization

A dominant theme in the social and intellectual history of the region has been the issue of national identity and, linked to it in contradictory fashion, the issue of modernization. The reception of technological modernity was intimately intertwined with the experience of cultural penetration and subordination, not necessarily unwillingly. The researcher joins a (scientific) subculture that is doubly foreign to her/him: as esoteric and special development of modernity and as a historical product of a particular cultural tradition—the European one—not easily transferable from one place to another. In this double heteronomous adscription, the Latin American researcher can be an important agent of cultural change in his country, for science continues to be an effective symbolic bridge between competitive ideological and political universes. But this is not a unitary category. Academics, within universities, often do not care much about what goes on beyond the institutional walls. Another very large segment works for the state and/or government and has a bureaucratic functionality. There is also a critical social sciences segment. The actions of the academic groups have been moving from a conservative political orientation before the Second World War to a progressive one in the 1960s and early 1970s, and back into conservative molds after 1990. This goes for the majority, although there are always exceptions. Usually, in either position of the political spectrum, their actions and views were linked to the protection of their own interests as a social group.

Development Challenges

The analysis of Latin American underdevelopment has focused on concentration of income in the region (Furtado 2000). This was a structural feature that tended to reproduce and perpetuate itself in the different development models that were adopted by the Latin American economy. Reports from UNDP, UNCTAD, ECLA, World Bank, and other multilateral agencies have repeatedly shown the deep inequality existing in the region. Latin America is the most unequal region in the world, in which poverty and extreme poverty do not decrease and only some spaces of the economies become positively articulated with the international economy. There has been economic growth without the creation of sufficient formal employment, and the segment with the highest income has maintained its power.

Throughout the modern period two main economic models prevailed: national development and international competitiveness. Key components of the early scheme of economic development were the role of national government as the main promoter of economic growth and the protection of local industry against foreign competition through high tariffs and regulations. This "import substitution model" reached its climax in the 1970s. Economic growth led to improvements in income for all social sectors, but income inequality also increased. By contrast, the economic model of international competitiveness was based on the assumption that the economy should be allowed to grow unhindered, and with increased productivity and higher income, people would be able to take care of their own needs of health, education, and retirement, with as little help from governments a possible. This assumption, however, has been questioned. Political scientists have debated whether, within the framework of a capitalist society, underdevelopment and democracy, inequality and governance, and development and distributive injustice may coexist. It is possible to conceive of a scenario in which the state becomes very efficient and the economy very competitive, while maintaining, simultaneously, high levels of income inequality and large pockets of poverty.

The social sciences became less critical of government—even the liberal government. It was common in the 1990s to talk about the »exhaustion of the development model« as if the previous fifty years of state planning had been a mess when in reality all the industrialization of these countries took place during that period! It was natural to behold the criticism against the government—even if they were neoliberal and promoted the Washington consensus—because they favoured academics' salaries: an example of this is the instauration of the Sistema Nacional de Investigadores (SNI) in Mexico.

The idea of the area of studies on development and the reproduction of underdevelopment in the dominant academic circles of universities in the advanced countries and even in Latin American universities and research centers have been largely abandoned. Dominant ideas are now about emergent markets, and solutions in economics are understood as a function of markets. With globalization, the power of large firms grows, driving accumulation, incorporating techniques, and further concentrating wealth. For Latin American countries, as well as for other peoples in the developing world, there is a double crisis: that of industrial civilization itself, derived from the progressive advancement of instrumental rationality, and the specific crisis of peripheral economies, whose situation of dependence tends to become deeper.

Universalism versus Nationalism

For a long time, the tension between the search for universal understanding and the desire to solve pressing national needs led the Latin American scientific community through a difficult path that was often perceived as a dilemma between cosmopolitanism and populism/provincialism. The transnational, or cosmopolitan, character of the researcher in a peripheral context has aroused as much critical as apologetic comment (Varsavsky 1975). Some have praised the endogenization of university education according to the standards of the United States or Western Europe as a demonstration of modernity. Others have criticized what they perceived as »brainwashing«, foreignness, and intellectual dependence. The ideologically positive tradition of cosmopolitanism in Latin America has been important, reflecting the wish of educated groups for an ecumenical, open culture, one with a broad approach contrasting with the often closed, provincial, sectarian, and inquisitorial world in which they lived. But a negative meaning of the term has been especially present in connection with the struggle against colonial domination (cf. Martí 2007; Fanon 1961), considering it as an intellectual stand destructive of the national essence by the passive assimilation of exotic and foreign influences. According to this line of thought, cosmopolitanism and »individualism« appear as the two black beasts that threaten the cultural identity of postcolonial nations. Thus, side by side with the idea of cultural opening to the entire world, the term also carries the connotation of unpatriotic uprooting.

The national issue and the social issue got entangled in a complex equation in which history challenged intellectuals and politicians in the most varied contexts, from the literary pages, through the political platform to, in less conspicuous fashion, the social researcher's office. The force driving the international scientific community is centripetal and exerts a strong attraction toward the centers, where most resources are concentrated, independent of the national origins of individual researchers. The central dynamic moving the system was explained quite some time ago by Merton (1957) in terms of a basic motivation of researchers: the search for the widest possible professional recognition, the appreciation of colleagues, from the quotation in a footnote to the Nobel Prize. Social science research in the region has repeatedly illustrated of this phenomenon (Kreimer/Meyer 2008).

From »Brain Drain« to Transnational Policies

Until recently, Latin America was largely understood in terms of communities, even in the midst of intense urbanization processes. The notion of community, including that of scientific community, entails strong and long-lasting ties, proximity, and a common history or narrative of the collective. It involves stability of links, coherence, embeddings, and belonging. The growth of scientific communities in Latin American societies involved a mixture of settlement with institutional affiliation and hence copresence within the confines of a specific territorially based society (Schwartzman 1991). Migration has been a thematic concern of the social sciences in the region for several decades now, probably related to the fact that the emigration of scientists, engineers, and highly qualified professionals has been a phenomenon long registered in the literature linked to development, politics, science and technology, and higher education (Oteiza 1971). Particularly since the sixties, migration has been seen as damaging to community building and therefore one of the obstacles to development strategies.

Today, however, there are other interpretations. Growing numbers of knowledge workers are becoming »nomadic« in their personal and work biographies. More people from Latin America are being recruited internationally (Charum/Meyer 1998; Pellegrino 2001; Lema 2007). This reedits, with an inverted sign, a situation long past when in the nineteenth century and early twentieth century most university faculty in Latin America were foreigners. As the twentieth century progressed, the institutionalization of science in the region meant to a large extent that staff positions became increasingly filled with nationals. Today this is changing again: although there is still a small but certainly growing number of foreign post docs, researchers, and teachers in Latin American institutions, the mobility of Latin American researchers to other countries in the region and internationally has become a more common feature. This presence abroad does not mean exclusively emigration, but also visits of variable duration, through sabbatical, post doc programs, short stays, conferences, and the like. No doubt, there continue to be significant attempts at strengthening the community feelings of identity, but this tends to occur through novel mechanisms, since the changed structural conditionings facing developing nations and the emergence of new international norms are leading toward convergence among states' transnational policies.

From »Closed« to »Open« Sociality

It is increasingly common for Latin American researchers—social scientists as well as natural scientists—to be involved in relationships that some of them consider »quasi/social,« in the sense that they distinguish them from what they traditionally identified as »true« social relationships, associated with long-lasting ties and a common history. But they more and more engage in brief intense encounters with other scientists, decision-makers, and publishers, and particular sociality becomes more frequent, occupying a greater portion of the time of the individual researcher. To this can be added a set of new factors, which include even components of a more psychological nature, such as the escapism that underlies a certain amount of disgust and anxiety with the immediate social context because of the syndrome of violence and insecurity that pervades daily life. The current process may be interpreted as a shift away from regimes of sociality in closed social systems and toward regimes of sociality in open social systems. Both communities and organizations are social systems with clear boundaries, with a highly defined inside and outside. Networks, however, are open social systems.

Technological Sociality

In recent years a rapidly growing body of literature that seeks to explore the social implications of information and communication technologies has emerged. E-mailing, online chatting, web surfing, and other interactive practices have given rise to *individualization,* and a high degree of mobility, by translocal communications, by a high amount of social contacts, and by a subjective management of the network. People are, so to speak, transplanted from their contexts and reinserted in largely disembedded social relations, which they must at the same time continually construct. In Giddens's (1990) words, this kind of sociality is both distantiated and immediate. Although I have only anecdotal experience, I am certain that it would be rewarding to analyze the mailing address books of Latin American social scientists, for we would surely find the most unusual long-distance linkages in the kinds of contacts they engage in.

Network sociality is a technological sociality insofar as it is deeply embedded not only in communication technology but also in transport technology and technologies for managing relationships. It is a sociality based on the use of cars, trains, buses, and metro lines, of airplanes, taxis, and hotels, and it is based on phones, faxes, answering machines, voicemail, video-conferencing, mobiles, e-mail, chat rooms, discussion forums, mailing lists, websites, and databases. Transportation and communication technologies provide an infrastructure for people and societies on the move. Network sociality is delocalized, which is most convenient at a time when the collapse of urban infrastructure, violence, and insecurity make life almost unbearable in many urban areas in the region. It is a sociality on the move, a sociality over distance. We are increasingly experiencing an integration of long-distance communication to our realms of face-to-face interaction.

For scientists in the peripheries like those of Latin America, travel becomes essential to shorten distance and defeat the feeling of isolation. Most science councils in the region have a special window to support attendance to congresses, and at least one trip per year has been the common practice for several decades now. In a country like Mexico, UNAM researchers have operational funds that allow them several trips per year as well as bringing guest researchers to deliver conferences and seminars. Establishing and maintaining ties in networks are not cost free, partly because of the importance of travel. Meetings involve a massive amount of physical travel. In order to continue to be participants in academic networks there are obligations to travel and meet up that cannot be evaded, at least within particular periods of time.

Meetings and travel are central to the character, effect, and consequences of scientific networks as they self-organize in the world. I posit that today, although face-to-face interaction organized by norms or comingling continues to be the most common social activity, a substantial change has occurred in academic sociality and a significant number of Latin American social scientists know an increasing number of foreign people through a growing number of meetings of various sorts. I think that empirical research would prove that many of them spend more time in maintaining certain distant contacts because there is less chance that casual, quick meetings occur, as would happen when there was a greater overlapping among social networks. I suggest that researchers have to spend considerably more time planning and keeping meetings with a rather small number of those who are »known,« communicating, and also traveling from afar to »keep in contact«.

Discussion

Do all these changes related to current networking sociality and specific international research agendas mean that the old questions characteristically present in the old social science research agenda in the region, concerning sovereignty, legitimacy, and power, have been quietly forgotten? It does not seem so. By the middle of the first decade of the new century, with the ascent to power of several center-left and left-wing governments in the region, the political landscape changed, again giving way to a strong resurgence of a social concern with the very unequal distribution of powers in today's world, and movements of regional integration in which social, economic, and political thought have a fundamental role in trying to fill the gap on the political theory flank of Latin American social sciences.

Thus in the decade that is about to end, we have attended to a change in many of the schemes that ruled social sciences in the 1990s. We witness a sui generis return to some of the ideas guiding the regional social sciences in the 1960s and 1970s. Old theoretical perspectives, such as the subjectivities of indigenous and other marginalized social groups, contestations by feminism, cultural studies, and science studies, have been vindicated. Among the themes being retaken and/or reformulated are social movements, social participation, multiculturalism, endogenous development, Latin American identities, education, and urban violence. At the same time, new topics have emerged such as those related to the role and power of the media, information and communications technologies, the deepening of democratization, sustainable development, and climate change.

The ethnic or racial issue has acquired some visibility through mobilization and activism, particularly in view of the fact that a large proportion of some of the larger Latin American countries, such as Brazil or Mexico, is either black or of native stock, and racial prejudice seems widespread. As with race, the gender issue has developed in the last decade, in a context of significant gender-related differences in income, occupation, and work opportunities. The environment question, still largely restricted to groups of intellectuals and middle-class activists and to nongovernmental organizations concerned with issues such as the destruction of the Amazon forest, the extinction of animal species, and the loss of biodiversity, is gaining attention in recent years.

Current processes of social change occur in the context of political instability and conflicts in some countries. It seems clear that the definition of new (or alternative) research themes by the endogenous-oriented governments of Chavez, Morales, Lula, Correa, and Lugo are an original political phenomenon and represent a real democratic advance. They talk in the name of the people from where they come, which is something new for Latin America. It is new for the social movements, and it may eventually be new for the social sciences of the region, if they approach the study of these social processes in an unbiased way. This is not easily done, however, as reflected in the observed resistance to analyze what is really going on, the mistrust on all sides at the others' views, and/or the rejection of established hierarchies and academic privileges, in what appears as a blunt denial questioning of the entire former institutional setup, which risks preventing a richer exploration from multiple vantage points.

The conventional understanding of the social sciences in Latin America has to do with the way they have grown in Europe and North America. In their expansion to other regions, they have imposed concepts, theories, methods, outlooks, and validation mechanisms, and not only institutional blueprints. But of course they cover only a minor portion of reality, both at the theoretical and practical levels, and there is a vast domain of knowledge and thinking on society and social issues that remains outside the reach of the social sciences narrowly understood. Formal training in the social sciences made a systematic effort to exclude from scientific consideration those bodies of knowledge that did not respond to the canons of social science scholarship, very often without recognizing in specific areas the primacy of the knowledge produced by other means or the presence of well conceived social or political institutions resulting from ways of life not derived from the Western tradition.

The challenge implied in integrating and assimilating unrepresented knowledge domain remains overwhelming. Throughout history intellectual contributions from Euro America were affected by a reality that was not theirs and could only be understood once they were realized to have ceased to be such in order to become something different from their origins. Some people in the region seek to uncover the local realities in different contexts, trying to find in them clues to the understanding of the broader reality. This involves pursuing fieldwork methods, participant observation, ethnographic work, microsociological approaches, asking new questions, mixed methods, multiple case studies, data gathering from micro to macro levels, and a variety of analytical techniques linked to particular questions, discrete levels, and units of analysis in unprecedented rates.

The ideas of popular sovereignty, regionalism, and local powers, direct democracy and municipalities, transfer to society of functions today assumed by a hungry state, together with all the others that go in the same direction of an advanced social democracy, should constitute a trial bench for the existing intellectual traditions, which ought to become fused in this crucible of matrixes. Problematizing the future study of the social sciences in the region by interrogating the work we currently do raises exciting new questions for research. The theories used in conjunction with new actors, new approaches, and novel practices push us to reframe our understanding of social research. This implies revising what knowledge is taken to be social science, what comes out of the relationship of theory and practice, how social science translates into practices of a very varied nature, or which parts of it do. Drawing on new questions, new theories, new approaches, we would be able to better follow the fault lines that demarcate new knowledge/power regimes that encompass what has become a globally situated social science. The future of the social science is rich with promise and difficulty.

References

Aintablian, G./Macadar, O. (2009): La cooperación internacional en Ciencia y Tecnología. Educación Superior y Sociedad 14 (1), 17-26.

Bayle, P. (2007): Emergencia académica en el Cono Sur: el programa de reubicación de cientistas sociales (1973-1975). Iconos. Revista de Ciencias Sociales, 30, 51-63. Quito: FLACSO-Ecuador.

Beaulieu, A./Scharnhorst, A./Wouters, P. (2008): Not Another Case Study: A Middle-Range Interrogation of Ethnographic Case Studies in the Exploration of E-Science. Science, Technology and Human Values, 32, 672-92.

CENDES (2007): Eventos. Red Iberoamericana de Investigadores sobre Globalización y Territorio (RII). Red Iberoamericana de Editores de Revistas (RIER). Cuadernos del CENDES, 24 (66), 135-37.

Charum, J./Meyer, J.-B. (Eds.) (1998): Hacer ciencia en un mundo globalizado. La diáspora científica colombiana en perspectiva. Bogotá: Colciencias/Universidad Nacional/Tercer Mundo Publ.

Collazo-Reyes, F./Luna-Morales, M.E./Russell, J.M./Pérez Angón, M.A. (2008): Publication and Citation Patterns of Latin American and Caribbean Journals in the SCI and SSCI from 1995 to 2004. Scientometrics, 75, 145-61.

CRES (2008): Plan de Acción. Conferencia Regional de Educación Superior UNESCO-IESALC, Cartagena de Indias.
Fanon, F. (1961): The Wretched of the Earth. Harmondsworth, Penguin.
Furtado, C. (2000): Brasil: Opciones futuras. Revista de la CEPAL, 70.
Giddens, A. (1990): The Consequences of Modernity, Cambridge: Polity.
Kreimer, P./Meyer, J.-B. (2008): Equality in the networks? Some are more equal than others. In: H. Vessuri/ U. Teichler (Eds.), Universities as Centres of Research and Knowledge Creation: An Endangered Species? Rotterdam/Taipei: Sense Publishers.
Lema, F. (2007): Migraciones calificadas y desarrollo sustentable en América Latina. Educación Superior y Sociedad, 12 (1), 107-124.
Martí, J. (2007): América para la Humanidad, Havanna.
Merton, R.K. (1957): Social Theory and Social Structure. New York: The Free Press (revised edition).
Nandy, A. (1995): Alternative Sciences: Creativity and Authenticity in Two Indian Scientists. New Delhi: Oxford UP.
Narváez-Berthelemot, N. (1995): The Distribution of Latin American Scientific Periodicals. In: M. Koenig/A. Bookstein (Eds.), Fifth International Conference of the International Society for Scientometrics and Informetrics Proceedings (pp. 383–92). Medford: Learned Information.
Narváez-Berthelemot, N./Russell, J. (2001): World Distribution of Social Science Journals. A View from the Periphery. Scientometrics, 51, 223-239.
Oteiza, E. (1971): Emigración de profesionales, técnicos y obreros calificados argentinos a los Estados Unidos: análisis de las fluctuaciones de la emigración bruta, julio 1950 a junio 1970. Desarrollo Económico, 10, 429-454.
Pecujlic, M./Blue, G./Abdel-Malek, A. (Eds.) (1984): Science and Technology in the Transformation of the World. Tokyo: St. Martin's/United Nations University Press.
Pellegrino, A. (2001): Trends in Latin America Skilled Migration: Brain Drain or Brain Exchange? International Migration, 39, 111-132.
Russell, J.M./Ainsworth, S./Narváez-Berthelemot, N. (2006): Colaboración científica de la Universidad Nacional Autónoma de México (UNAM) y su política institucional. Revista Española de Documentación Científica, 30, 180-198.
Sancho, R./Morillo, F./De Filippo, D./Gómez Caridad, I./Fernández, M.T. (2006): Indicadores de colaboración científica inter-centros en los países de América Latina. Interciencia, 31, 284-292.

Schwartzman, S. (1991): A Space for Science. The Development of the Scientific Community in Brazil. University Park, Pennsylvania: Pennsylvania State University Press.

Varsavsky, O. (1975): Ciencia, política y cientificismo. Buenos Aires: CEAL.

Venuti, L. (1998): The Scandals of Translation: Towards an Ethics of Difference. London: Routledge.

Vessuri, H. (1999): National Social Science Systems in Latin America. UNESCO, World Social Science Report. Paris: UNESCO/Elsevier.

Vessuri, H./Martinez Larrechea, E./Estévez, B. (2001): Los científicos sociales en Venezuela. Perfil bibliografico e implicaciones de politica. Cuadernos del CENDES, 18, 89-121.

The Americanization of Argentine and Latin American Social Sciences

TOMÁS VÁRNAGY

Contemporary Internationalization Practices in Latin America

Globalization establishes a dominant paradigm that, combined with the reforms in higher education in Latin America and the intervention of funding agencies, helps to create a social science community that serves the needs of the richer countries in the North instead of its own in the poorer South. Whatever globalization means, it is primarily a homogenizing process, and according to Ron Kassimir, »the active agent of this process is an omniscient center imposing itself on a passive periphery« (Kassimir 1997: 155). Cultural globalization is not only the democratic dimension via the massive access to Internet, but at the same time, it is also a process by which powerful communication media promote American values and tastes. Globalization is *McDonaldization,* the establishment of a single dominant paradigm and franchising of this paradigm; it is the suppression of alternative cultural approaches and has mostly profit motivations.

On the other hand, internationalization, although it has many meanings and facets, could allow for participatory intervention among equal partners. In such an idealized version »internationalization« could be interaction, cooperation between countries and educational institutions, and could act to preserve diversity. There is, in fact, an interchange of students and staff, the sharing of other cultures, a transfer of skills and of methodologies.[1] But even such collaborations on an equal basis between advanced countries and developing countries do not always ensure equal treatment among the partners. Internationalization of social sciences in Latin America and Argentina thus mainly means for most academics the »Americanization«, »North-Americanization«, or more precisely, the »USAfication« of social sciences.

In fact, the participation of scholars in the development of global research networks is most uneven and unequal, as social scientists in the developing countries know quite too well (Kassimir 1997). In these countries, academics have seen their research and training capacities impeded because of budget cuts related to the economic crisis, and sometimes because of political repression. The political instability related to the economic crisis has also caused a brain drain as scholars leave the country.

If there is already a scientific gap between the United States and Europe, one can imagine the size of the gap between the developed world and the »developing world«. Some simple figures speak a clear language: the production of knowledge is concentrated in OECD countries. With only 19 percent of the world population they account for 85 percent of the world expenditure in research and development (R&D), while Africa—for instance—with 13 percent of the world population accounts for less than 1 percent. Will the current internationalization process of research and development change this situation?

Indeed, the number of foreign scholars working in the United States has been growing by 4.6 percent per year during the last fifteen years. Academics coming from foreign countries represent around 30 to 40 percent of university researchers in the United States. Foreign-born workers represented around 10 percent of the highly skilled employment in the United States, 20 percent in Canada and 25 percent in Australia. About one third of the researchers born in developing countries are working in OECD (Organization for Economic Cooperation and Development, i.e., the most developed) countries (Chassériaux 2005).

Although international collaborations could be an exchange of people, knowledge, and money, gathering a critical mass that should be a public good that contributes to the production of other public goods such as health, environment, and education, the privatized production of social knowledge responds to market pressures rather than to social demands; for example, only 0.2 percent of the world expenditure of health

---

1 See IAU Durban Conference, August 20–25, 2000. 11th General Conference: Universities as Gateway to the Future. Working Group VI: Internationalization.

is devoted to diseases linked to water which accounts for 15 percent of diseases worldwide (Chassériaux 2005).

Universities and Higher Education in Latin America

Compared to their outstanding role in the past, which varied according to the particular conditions prevailing in different Latin American countries, Latin American universities today are facing a severe crisis. In the last two or three decades, major changes shocked them to their foundations and, after more than a quarter of a century of democratic recovery in the political realm, the social, economic, and educational debts of the Latin American governments are as impressive as they are unforgivable.

In a region of the world where there are more than eight hundred universities, any generalizations are problematic, even though one could give a general picture, though descriptive, that is at least partially valid. With the decline of the former welfare-state paradigms, the new neoliberal governments in Latin America shifted to hegemonic policy paradigms, also forcing the local universities to reconsider their political missions, their social visions, their academic priorities, and their organizational structures (for a comparative perspective see Torres/Schugurensky 2001).

The universities, together with all the other institutions in Latin America, have been challenged by the wave of the so-called »market-friendly reforms« that reshaped in socially regressive terms the very structure of our societies. These »reforms« carried out under the unchallenged inspiration of neoliberal policies to privatize our universities and introduce policies of »structural adjustment«, resulted in weakening the scientific capacities of our universities to respond to the challenges occurring in this global region, which are substantially different from those in other parts of the world (Borón 2008).

The transformation of higher education was subsumed under the economic priorities of globalization, setting in motion the neoconservative policies toward the consolidation of multinational corporations and the redefinition of the role of the state in Latin American countries toward these goals. The impact of these changes on the universities brought, as mentioned earlier, severe budget cuts, increasing dependence on private financing, parallel growth of private universities, deregulation of the working conditions of all level of academics, allowing an increased influence of the global market players and of the Latin American political elites in university affairs.

Changing Mission of Universities

In the debate of the 1960s in Latin American universities, many defended the idea that the universities should become the critical consciousness of society, the engines of new knowledge, and the guardians of the permanent interests of the communities. In order to perform these roles, the university had to guarantee a unique combination of scientific excellence and outstanding humanistic education, that is, advanced analytical skills joined to a vision of the good society. These ideas, rooted in the theories of great Latin American educators such as Rodolfo Mondolfo and Risieri Frondizi in Argentina, Darcy Ribeiro and Paulo Freire in Brazil, Barros Sierra in Mexico, and many others, provided the theoretical background and doctrinal foundations for the progressive reforms experienced by Latin American universities in the 1960s.

The beginning of the dictatorships in the 1970s stopped suddenly this discussion and made the idea of the university as a critical consciousness of its epoch dangerous subversive. This paradigm was banned, and those who supported it were persecuted, jailed, or killed or »disappeared«. Unfortunately, the democratic recovery that started at the beginnings of the 1980s failed to reintroduce the theme, and the question of the role of the university was forgotten and remained largely in the shadows.

With the transformations that took place in the 1990s the importance of the question reappeared once again: what should be now, in the era of globalization of capital, the purpose, role, and mission of the university? According to the neoliberal (neoconservative) theorists and practitioners the answer was quite clear: the university must train the students in the kind of professional skills required by the market; it is the market that provides the incentives (material) for the development of the different professional occupations.

Latin American governments did the same as their counterparts in Europe, Africa, and Asia: they reduced subsidies drastically, compelling universities to depend on private financing and promoting radical changes in all aspects of university life, like priorities in the research areas. The combination of the Latin American crisis prior to the 1980s and the radical regressiveness of the neoliberal reforms carried out in the last decades has created a series of major problems for Latin American universities, which are discussed below.[2]

---

2   On these issues see the outstanding chapter 2: »Las ciencias sociales en la era neoliberal: entre la academia y el pensamiento critic« by Borón (2008: 77-110).

Institutional Effects of Neoliberal University Reforms

Following the major recommendations of the Washington Consensus, beginning in more or less the early 1980s, the educational system was redefined as an »educational market« in which private providers were welcomed. Previously, education was considered as a right of every citizen and a social investment, an idea that was displaced in favour of market mechanisms. Deregulation and privatization of university education followed from this philosophy. Today's enrollment in private universities in Latin America accounts for some 40 percent of the total student body. A closer look at country cases show that in Brazil, Chile, Colombia, the Dominican Republic, and El Salvador the majority of the students are enrolled in private universities, while the opposite is true in Argentina, Cuba, Guatemala, Mexico, and Venezuela (Holm-Nielsen/Thorn/Brunner/Balán no date).

While there are in our region a very small number of good private universities, the overwhelming majority have been just commercial enterprises profiting from the continuing expansion of educational demands and taking advantage of the decreased state capacities to establish and enforce strong standards of academic quality. On the opposite side to what happened with the handful of very good private institutions, the overwhelming majority of the rest specialized in the creation of »chalk and blackboard« careers, or in short courses which, supposedly, ensure an easy insertion in the labor market (finance, marketing, public relations, social communication, tourism, etc.). Their contribution to educational and scientific development is zero.

Former Harvard president Derek Bok expressed, in recent years, his deep concern with this trend and the dangers posed by what he calls the »commercialization of university education« to the core academic values of the university (Bok 2003). The ideological ascendancy of neoliberalism in our region is reflected in many changes introduced in university life. This orientation is the triumph of the idea that education is a market, a service that—as any other in the market—must be purchased at a given price by those who can afford to do it.

One of the consequences of the neoconservative predominance in the so-called »educational reforms« of the eighties and nineties has been the generalized acceptance gained by the idea that universities should be regarded as profitable, as money-making institutions able to live on their own incomes without the intervention of the state. The main concern of these reforms was to assure that universities would be able to function with financial resources generated by themselves, reproducing in the educational sphere the general trend toward privatization.

As a consequence, the »university reform« in our countries must be mainly considered as a collection of savage budgetary cuts: massive faculty layoffs, introduction of fees and abolition of gratuity, closing down departments and research institutes. This financial crisis is related to the gradual and steady abandonment by governments of their essential responsibility in education, notably of their financial responsibilities.

This region of the world has one of the smallest amounts of money invested for each student in the tertiary level per year: some U.S. $650, against a figure almost four times as big in the Asian countries while in the United States and Canada the annual investment per university student is around $9,500, or fourteen times bigger than in Latin America (Rhoads/Torres 2006: 151). It is not by chance that this is one of the regions of the world with the most unequal distribution of income and wealth.

Latin American countries experienced a fast expansion of university enrollment in the second half of the twentieth century. The number of university students rose from some 270,000 in 1950 to almost 9 million at the beginning of the twenty-first century. These figures are impressive, but far from being a major achievement of our countries, they only reflect a universal trend that is even stronger in the developed world. Massiveness at universities is good; the bad thing is that there are not enough financial resources allocated. The growing massiveness of the universities thus added to the institutional problems already mentioned, caused by the neoliberal reform activities.

Adding to this, the rapid speed of these changes must be mentioned. University reforms that earlier took place over centuries now happen in the life span of a single generation. This situation is not only valid for the so-called »hard sciences« but also in the much less flexible social sciences and the humanities as well, where as shown in the Gulbenkian Report (a team work coordinated by Immanuel Wallerstein) the traditional separation between anthropology, economics, history, political science, and sociology has become completely untenable (Wallerstein/Juma/Keller/Kocka/Lecourt/Mudimbe/Mushakoji/Prigogine/Taylor 1996).

Deteriorating Quality of Teaching

Another factor of the quality crisis in our countries is the low level of training of a great part of the faculty, explained by the lack of adequate policies to upgrade the qualifications of all university professors and the absence of material and intellectual stimuli to undertake the academic improvements.

Before the impact of the neoliberal reform on the quality of teaching, military dictatorships had already devastated universities in Argentina because they perceived them as »nests of subversives and guerrillas« that needed to be annihilated. There was a famous »Night of the Long Sticks« in Argentina in 1966, when the police beat students and professors and fired a number of professors, many of whom eventually left the country. The bloody dictatorship of 1976–83 produced not only thirty thousand *desaparecidos* (missing people) but also a great migration of academics and researchers to other countries.

Adding to this situation in Latin America are the facts that despite of all other efforts being made, still only 7 percent of the faculty hold doctoral degrees, and another 20 percent hold barely some years of graduate studies. In addition, full-time teaching is far from widespread, except in some elite programs at the graduate level. At the Universidad de Buenos Aires, the largest in Argentina with more than 300,000 students, to cite just an example of the trend prevailing in the area, full-time faculty do not even reach the 10 percent mark, while 90 percent of professors have second and third jobs (Várnagy 1992).

Combined with the effects of neoliberal institutional reforms, the low level of skills of the teaching stuff reinforce the crisis of our universities.

Impacts on the Contents of Academic Thought

However, the transformation did not only affect teachers and staff in Latin America. The ideas of »neoliberalism«, which emphasize an economic reductionism that glorifies the prioritization of finance, replaces collective actors with a radical individualism, and in terms of its scientific paradigms, praises mathematical formulations as the only valid criteria of any social science knowledge, also had and have a very strong influence on academic thinking in the social sciences in the region. This epistemological trend in the social sciences could be called an outbreak of »neopositivism«. Many Latin American scholars see this growth of neopositivism as an assault on critical thinking because it postulates that there is only one, unique, valid way of scientific thinking, undermining utopias and even claiming the end of history.

The neoliberal reformers insist that the markets are the ones who decide what to teach and what to investigate, leaving aside wasteful courses, disciplines, and research agendas. After all, who cares for a philosopher or a theologian? What is the market value of wisdom and understanding? Do we really need astronomers or social workers? Should we waste our limited resources teaching our students about ancient Greek history, or Medieval political thinking?

I will come back to this point after having a brief look at the reforms in higher education in Argentina.

University and Higher Education Reforms: The Case of Argentina

Argentina is a country that gave birth to the progressive university reform of 1918 that shook the archaic university system of all Latin America to its foundations and established one of the fundamental principles of Latin American universities for decades, that is, autonomy and autarky. The case of Argentina is also interesting because it is the country in which university enrollment has reached the highest mark in the region.

The institutionalization of social sciences in Argentina started in the second half of the 1950s, after Peron's fall in 1955, which coincided with deep social, political, and cultural transformations after a decade of Peronism. Sociology, economy, psychology, and anthropology—among other disciplines—were created as careers at the Universidad de Buenos Aires in 1957. At the same time, laws allowing the founding of private universities were enacted. Yet, in spite of this, unlike other Latin American countries such as Chile and Brazil, Argentina never had more than 20 percent of its students in private universities, although this tendency is changing slowly after the crisis of 2001 (Neiburg/Plotkin 2004).

With the rise of dictatorships, in a process openly or covertly supported by the United States, academic freedom was replaced by political and ideological control, which ranged from—as mentioned above—the wholesale purge of disaffected professors to their being mercilessly beat up, as happened with the professors of the Faculty of Exact Sciences of the University of Buenos Aires shortly after the coup d'etat of June 1966, or the kidnapping, torturing, and »disappearance« of hundreds of students and academics with the dictatorship that started in March 1976.

Since the second half of the 1970s, but mainly during the 1990s, there was in Argentina a real cultural revolution strongly related to what the historian Tulio Halperin Donghi defined as »the long agony of Peronist Argentina«.[3] As we mentioned above, deep transformations affecting public and private spheres took place not only in Argentina, but

---

3 It is the title of his book: La larga agonía de la Argentina peronista. Barcelona: Ariel, 1994.

in the whole world in a process known as »globalization« guided by a hegemonic neoliberal ideology.

In a country in which »privatization« was a four-letter word, thanks mainly to the »Third Position« Peronist ideology, and in which »national sovereignty« was confused with the possession of enterprises and services by the state, privatization has swept Argentina to such an extent that it is one of the few countries in the world that does not have a national post service (Beccaria et al. 2002). The financing of higher education diminished drastically and some »enterprising« aspects of the universities started to be promoted, like the selling of professional services. In this way public education lost its legitimacy as a value in itself (as it was some decades before).

The new private universities created in the 1990s still try to adapt and copy the U.S. education model as much as they can. Some people think that the implementation of graduate studies (specially master's degrees) in our country is related to the »Americanization« (»USAfication«) of higher education and also the privatization of it, because all master degree courses, even in public universities (in which undergraduate studies are free), require students to pay tuition. The situation probably is more complex but it is true that there is a certain mimicry in the hegemonic model of education worldwide.

In the past few years there has been a strong urge to get graduate degrees in Argentina, which did not have master's degrees in the 1970s. Instead, many universities offered five- or six-year career degrees, which was considered terminal and copied the French educational model. Following the completion of that initial degree, and after many years of teaching, you could write a doctoral thesis. At that time not many professors had a Pd.D., while today students of twenty-five years of age are in a hurry to get their Ph.D. before they turn thirty years old.

In fact, today in Argentina, you must have a graduate degree in the academic field, and this is a great change in the local university culture. This new »culture of graduate studies« produced a shortening of careers; usually they took five or six years if you dedicated full time to it, and now the tendency is four years, like a bachelor's degree in the United States.

For decades, the salaries of professors and researchers in Argentina has been kept at low levels, or straightforwardly frozen for years, and this helped the role of the foreign influences to become more and more crucial in determining the research agenda of our universities.

Our salaries are much better now and a full time professor, as mentioned earlier, less than 10 percent of the total faculty, gets a bonus (*incentivo*) for research four times a year, which is not part of the salary and therefore does not count for our retirement payments. It is, however, not only the salaries that have deteriorated—Chilean, Brazilian, or Mexican professors get paid much better than Argentine professors do—the infrastructure, specifically the learning facilities, are in bad condition too. Argentine professors rarely have a decent room, they have no computers, poor libraries, and the classrooms are always too small for all the students, creating a poor learning environment.

There is chronic labor insecurity at the university in Argentina: near 80 percent of the faculty enjoy a precarious status because they are not tenured (*concursado*), their designation is provisional (*interino*), or as in many cases, because their work is done *ad honorem*. At the University of Buenos Aires about one third of the professor work for free. This means that work stability is uncommon and that every six months or so scholars have to be reappointed. Even with a very modest salary, in some cases professors work just to have medical insurance.

In Argentina, at least in theory, research scholars are an institutionalized element of our universities. However, the financial restrictions and political situation inside the universities present severe obstacles to the effective implementation of research scholarship. Teaching and research openings are supposed to be filled by a public contest, the *concurso de antecedentes y oposición,* which is set to create a public and scholarly space in which several candidates compete for one position under the scrutiny of their colleagues, graduate professionals, and students. However, this procedure does not always guarantee the fairness of the competition.

Professors in other Latin American universities certainly have better working conditions than the Argentines. In any case, even for full longtime faculty with assured institutional positions, salaries are still below the official poverty line. This means that the professors look for and find alternative ways of survival, in many cases in activities not related to their academic job, thus preventing them from reinforcing their scholarly education and upgrading their skills in the classroom or in the labs.

As far as publications are concerned, Professor Atilio Borón (2008) explains that the university publication activities are extremely weak throughout the whole region. And if they are not weak, as in the case of Universidad Autónoma de México (UNAM) or the Universidad de São Paulo (USP, Brazil), they are very slow in getting books out and, if this happens, they do not ensure an adequate distribution of the books. Due to this situation, academic writers are forced to seek other publication means in small publishing units of the diverse departments or institutes, however, with very little chance of having the work really introduced into the public debate.

On the other hand, with the widespread criteria in use for the evaluation of »productivity« which could result in bonuses, faculty are forced to publish quicker and more often. Since the local evaluators tend to underrate journals or publishing houses based in Latin America, there is a strong tendency to publish—preferably in English—in the North (the United States or Europe), because these papers are more highly regarded than those published in Spanish in the Southern hemisphere.

As a matter of fact, evaluating agencies give a higher premium for any paper or work published in the North. The result is that the research agenda of our universities is increasingly dependent on the practical and theoretical priorities set by the North, and we know very well that their priorities are not always the most convenient for our societies.

Social Science Research under the Control of Foreign Funding Agencies

There seems to be a dominant paradigm in social sciences and an apparently irresistible tendency to imitate the intellectual fashions of the developed North,[4] which has led to a situation in which the exceptional contributions made by Latin American social scientists in the second half of the past century is today almost completely ignored by the younger generations who, on the other side, are quite familiar with the latest papers produced in the field, for instance, of »rational choice«. We now have a new professional disease: the uncritical reproduction and imitation of anything produced in the rich North, especially if it is written in the English language.

Some years ago, research in our countries was a project that was usually carried out by a senior professor with students in a continuing plan with great success and remarkable results especially in the 1950s and 1960s, which were supported by the public universities; in this period, the Universidad de Buenos Aires produced the only three Nobel prices in science in Latin America. This kind of research has been brushed off by neoliberal policies and »state reforms«, implying the reduction of the states financial engagement to a minimum level. The funds for research have thus been constantly reduced and the university gave up part of its institutional autonomy over the sphere of research. If faculty now want to carry out research, they must apply for funds at newly emerged consulting institutions.

---

4   Even the muchused term cientista social in Spanish is a bad use of the English term »scientist«, whose correct translation is *científico*.

These consultancy organizations combine the financing of research with the conceptualization of public policy strategies, sometimes also carrying out research. They are all under the supervision of the World Bank. This new private complex making public policies was called by Joel Samoff »the intellectual-financial complex of foreign aid« (Samoff 1992). This complex imposes and defines the local policy objectives, even the restructuring of educational policies, not only in Latin America, but all around the world, all showing strikingly similar agendas, methods of working, and activities in all countries and regions in which they are active, in spite of the very different social, political, and historical characteristics of these different global regions.

Despite these direct influences of the World Bank on research-lead consultancy agencies, there are other more indirect paths of dependencies from foreign resources. As the university budgets have little capacity to finance scientific research, it has become common practice for Latin American social scientists to compete for funds from governmental agencies, which in many cases are provided as grants to the various governments by international financial institutions supporting their neoliberal reform agendas. Thus Latin American policy-oriented research is held hostage to the financing schemes and strategies of organizations like the World Bank (WB), the Inter-American Development Bank (IDB), and the International Monetary Fund (IMF).

Via these funds these international institutions are the ones who decide about the policy agendas of most Latin American countries—how to define what their problems are, and what they are not. For instance, income distribution or economic inequalities are issues that are not to be researched. Not only do these foreign-based international institutions decide about the topics which deserve to be researched, but the WB and the IDB also define the theories to be used, the methodologies to be applied, the research areas in which they have to be carried out, and even what is the »politically correct« way of phrasing research of issues and what is not.

These agencies define how, what, who, where, and when to research and which are the satisfactory expected outcomes (Borón 2008). It would be, however, naïve to believe that there are direct mechanisms of intervention to define the scientific research agendas in Latin American countries. The interventions are very subtle and circumlocutory, working via the acceptance or rejection of papers for publications, the appointment of scholarships, and the distribution of funds for research.

Via these mechanisms research in the social sciences is controlled by the financing agencies and by the editorial committees of U.S.–based journals who decide if an article written by a Latin American social scientist is to be published or not. This is not to say that there is any direct censorship made by the U.S. academic community, but via their belief that their scientific paradigms represent the highest standards of social sciences, both in terms of the topics as in the terms of methodologies and approaches, they »naturally« expect that articles written by Latin Americans should have the very same standards, constructed via their dominant paradigm of what is considered as scientific knowledge. And since it is well known that an article published in the United States is more valuable than one published in a book in any Latin American country, the circle of these publication mechanisms confirms the validity of their standards.

In this way our social scientists are forced to give up any pretension of developing a research agenda of their own, or of conducting research as it had been done in the 1950s and 1960s. It is inconcievable that any researcher could work with a particular theoretical and methodological framework that is not agreeable by the donors. Researchers are expected to produce consulting reports serving the needs of the donor and not real, independent social science research. To guarantee this service, even the methodological and theoretical framework is carefully specified beforehand in a contract that cannot be modified by the researcher. Thus the findings of research are largely built into the basic premises of the theoretical work of funded research.

The result of such research can only be considered as deficient social science, no matter the millions of dollars spent for this peculiar kind of »social research«. In fact, the parallels with late-era Sovietology are striking: a system that loudly proclaimed its scientific paradigm, that, assisted by pseudoscience, vetted and legitimatized a narrow set of politically selected projects and policies, nonetheless was unable to predict the fall of the Berlin Wall or the implosion of the Soviet Union. As in the Soviet system, now in Latin America under the current knowledge regime no valuable knowledge is produced to alleviate some of the more critical problems faced by ordinary people: poverty, insecurity, health, education, and so on.

It is well known that multinational enterprises sponsor professorial chairs at the faculties of medicine, biology, pharmacy, and odontology among others. Any student of these careers could tell that, for instance, the »anatomy« or »pathology« chair is supported by such and such laboratory, sometimes with financial help or goods (medicines, instruments, and others). It is not difficult to imagine the lines of research of these chairs . . .

This kind of distortion is typical in Latin American universities: for example, the limited funding and support given to the study of the Chagas disease, which affects millions of poor people in the region in our Latin American schools of medicine, compared to the unreasonable attention lavished on studies of physical illnesses typical of affluent societies; or the training of our medical students using highly sophisticated electronic instruments disregarding the basic tools of clinics needed to serve the overwhelming majority of the population in Latin America (Matsuura no date).

This is not different in the social sciences. The research studies on poverty may serve as an example. The World Bank, for instance, is convinced that poverty is a problem that should only be addressed focally, an approach that might be appropriate for countries like Sweden, Iceland, or Norway, but certainly is of little use for countries in which 50 percent and even 80 percent of the people are poor, in which the problem of poverty is structural and thus cannot be tackled via the concepts of »focal« solutions. In countries with such a percentage of poverty, the issue of poverty can only be approached via macroeconomic variables grasping the redistribution of wealth on the level of a national economy and not via any micro situations, in which the real problems poverty do not appear and which can thus not solve the real problems of Latin American countries.

References

Beccaria, L./Feldman, S./González Bombal, I. (2002): Sociedad y Sociabilidad en la Argentina de los 90. Buenos Aires: Universidad Nacional de General Sarmiento-Biblos.

Bok, D. (2003): Universities in the Market Place: the Commercialization of Higher Education. Princeton: Princeton University Press.

Borón, A.A. (2008): Consolidando la explotación. La academia y el Banco Mundial contra el pensamiento crítico. Córdoba, Argentina: Espartaco.

Donghi, T.H. (1994): La larga agonía de la Argentina peronista. Barcelona: Ariel.

Holm-Nielsen, L.B./Thorn, K./Brunner, J.J./Balán, J. (no date): Regional and International Challenges to Higher Education in Latin America. http://siteresources.worldbank.org/INTLACREGTOPEDUCATION/ Resources/Overview of HE in LAC.pdf [Date of last access: 31.08.2009]

Chassériaux, J.M. (2005): »Internationalisation of R&D and the European R&D Policy« in Estime, paper presented at ECPR 3rd Conference, Corvinus University of Budapest, September 8th to 10th,

2005. http://www.estime.ird.fr/article154/html. [Date of last access: 7.05.2009]

Kassimir, R. (1997): The Internationalization of African Studies: A View from the SSRC. Africa Today, 44 (2).

Matsuura, K. (no date): Address. portal.unesco.org/education/en /files /31190/10857409691speeches.pdf/speeches.pdf. [Date of last access: 31.08.09]

Neiburg, F./Plotkin, M. (Eds.) (2004): Intelectuales y expertos. La constitución del conocimiento social en Argentina. Buenos Aires: Paidós.

Rhoads, R.A./Torres, C.A. (Eds.) (2006): The University, State, and Market. The Political Economy of Globalization in the Americas. Stanford: Stanford University Press.

Samoff, J. (1992): The Intellectual/Financial Complex of Foreign Aid. Review of African Political Economy, 53.

Torres, C.A./Schugurensky, D. (2001): La economía política de la educación superior en la era de la globalización neoliberal: América Latina desde una perspectiva comparatista, 23 (92), 6-31. Perfiles Educativos, México: Universidad Autónoma de México,

Várnagy, T. (1992): Universitarios: ¿buenos y baratos? Clarín, Buenos Aires, July 3.

Wallerstein, I./Juma, C./Keller, E.F./Kocka, J./Lecourt, D./Mudimbe, V.Y./Mushakoji, K./Prigogine, I./Taylor, P. (1996): Open the Social Sciences. Report of the Gulbenkian Commission on the Restructuring of the Social Sciences. Stanford, CA: Stanford University Press.

Rethinking International Cooperation in the Human Sciences of Brazil

RENATO JANINE RIBEIRO

Internationalization in Research: Diverging Developments of Human and Exact Sciences

This paper will offer me an opportunity to present some theses that may be taken as provocative. I don't mean to be provocative, but these ideas may contribute to think international cooperation as far as human sciences are concerned. If that happens, they will have achieved their goal.

In Brazil more than 95 percent of scientific research takes place within graduate programs, which include master's and well as Ph.D. programs. I am aware that the European tendency is to consider a Ph.D. as research, and a master's degree as a conclusion for professional formation. For example, in the Bologna process some master's degrees, as in engineering, are considered as a final stage of what Europeans themselves could classify before as undergraduate studies.

But in a country like Brazil, with a continental size and deep regional inequalities, the master's programs are a strong element for the incorporation into scientific research (in biological, exact, and human sciences) of less-developed regions, like the Amazonian, the center-western or the northeastern ones. In this paper, following Brazilian custom, we shall consider as human sciences the studies on literature, law, economics, communication and the arts, as well as the »basic human sciences«: philosophy, anthropology, sociology, political science, geography, history, linguistics, and psychology. In such areas there were in

Brazil, in 2007, 377 Ph.D. programs and 889 academic master's programs (besides the professional master's programs, of which there are 50 in the human sciences' field, although the great majority of these—39—belong to business and economics). By January 1, 2007, these programs had 31,968 students at the master's level, and 15,621 at the Ph.D. level. Throughout 2007, in the human sciences' field, 3,218 students got their Ph.D. title, while 12,108 obtained a master's degree. However we must consider also the high regional concentration of these programs. The majority of them belonged to the southern and southeastern states (from Rio Grande do Sul to Minas Gerais and the Federal Capital of Brasilia, with a higher concentration in the states of São Paulo and Rio de Janeiro). Of all programs on human sciences, 97 Ph.D. programs belonged to the three southernmost states, 203 to the southeastern ones, 49 to the northeastern region, and only 17 to the Amazonian and western regions of Brazil.[1] The first Ph.D. program on ethnology in the Amazonian area is only two years old, and that is the region where we find the highest concentration of native ethnic groups in the country. And this program was created due to an initiative of CAPES—Coordenação de Aperfeiçoamento de Pessoal de Nível Superior, the Brazilian Federal Agency for Support and Evaluation of Higher Education. This agency fosters and evaluates the graduate programs in a nationwide range. Every three years CAPES gives each of them a grade that will serve as a criterion for their access to federal financing funds, as well as for their degree of autonomy. A negative CAPES' evaluation can close programs.

When CNPq—Conselho Nacional de Desenvolvimento Científico e Tecnológico, or National Council on Scientific and Technological Development—started an accurate evaluation of individual researchers, and CAPES did the same toward the Ph.D. and master's programs, one of the main criteria was the internationalization of their production. The idea was to assure that the researcher's work be up to date, as well as that his or her work would be evaluated with accuracy. This was seen as necessary to avoid logrolling and political favoritism in the evaluation process.

As far as the exact and the biological sciences were concerned, the outcome was very positive. But in the agrarian sciences, for instance, there is always the risk that a research on wheat may obtain more citations on international journals than one on cassava (or yucca), although Brazil is a major producer of the latter and produces almost no wheat.

---

1 In order to consider regional inequalities we must not include the Brasilia urban area, or Distrito Federal, in the center-western region, otherwise we should have to deal with a great distortion of data.

In the human sciences' field, however, the outcome was not so convincing. On one side, international exchange is highly positive. On the other, it has favored an attitude of subordination that is totally unjustifiable today. At the present time there are no international reference centers in human sciences as strong as the exact and biological sciences. Everyone is able to identify the great research laboratories and centers in the science field. But concerning the human sciences, such references are not beyond dispute. The golden age of »the grand theory« passed into history as far the human sciences are concerned. Coincidently, that was the age of great thinkers like (for instance, in the French case), Lévi-Strauss, Foucault, Barthes, and Derrida. In other words, the human sciences' world scene has changed as long as monographs took the place of the grand theory, and great intellectual leaders left the stage, now occupied by lots of researchers whose works are not so imposing, although they are accurate and precise.

One of the consequences of this movement was that the distance between the best thinkers in North America or Western Europe and those in countries like Brazil became shorter. Unhappily this opened the way to work that, in order to get international relevance, imitated or followed trends that have originated in other countries, but at a moment when these no longer display the extraordinary quality of the great thinkers that dominated the scene during the twenty-five or thirty years after the end of the Second World War. This attitude supposes a world leadership that actually no longer exists in the field of human sciences.

Anyway, we can draw an equation to sum up this problem that affects the human sciences' area. We do have in Brazil a large research and graduate studies realm, generally with a high standard of quality. The main cause of such an achievement is the financing policy of federal agencies such as CAPES, CNPq, and FINEP (Financiadora de Estudos e Projetos—Studies and Projects Financing Federal Agency), or state agencies such as FAPESP (Fundação de Amparo à Pesquisa do Estado de São Paulo—São Paulo State Research Support Foundation, the wealthiest among Brazilian states). We must also consider the policies toward research of the federal public universities and of state public universities from São Paulo, Rio de Janeiro, and Paraná. These universities stimulate scientific research among their faculties. We must mention, among other reasons for this success, the internationalization of scientific research, stimulating people to publish their works abroad, to submit them to the foreign readers, and to maintain contact with their foreign partners. Again, this was a road to success for the exact and biological sciences. But in the human sciences, it opened the way to some very serious problems. The dialogue *inter pares*—between partners—within the country suffered while dialogue between individual researchers or isolated groups and their partners from the wealthiest countries flourished.

The elaboration of common lines of research, which can direct research efforts, researchers, and institutions in the human and social sciences' field also was affected. Suddenly the habit of »franchising« came into vogue: some Brazilian researchers became in a certain way »agents« within their own country, acting on behalf of better-known foreign academic professors. Research synergy faded away. Unlike the exact or biological sciences, human sciences built neither a common agenda planned to gather common areas of knowledge nor transverse ones that could bring different areas together.

Comparing the areas, we may say that the exact and biological sciences accomplished a better internationalization status because of the following reasons.

- They have high-quality international reference centers.
- There is a larger gap between the science research made in Brazil and that made in First World countries; this happens also because the cost of equipment and large research groups is very expensive.
- The concern of the researcher with the culture of the country where he is working is a private matter to him; it has no direct relation with his work as a researcher; it may have some or even a great relevance to him as a citizen.
- It is easier to write in English, as the needed vocabulary and syntax are of a more basic level.
- Last but not least, many scientific communities have successfully established common agendas from and about their own priorities.

In the case of human sciences, we could invert the sense of those five reasons, from plus to minus. But from all of them, only points four and five are actually a real advantage to the exact sciences. We can even say that the fourth one is not exactly something to be proud of. The fifth reason could—and should—be changed in the case of the human sciences through an adequate scientific policy. It is true that there are no decisive international reference centers for the social sciences. There are better centers, but let us say, there are no »central centers«. For this reason, in human sciences the gap between research being done in Brazil and research in Europe is smaller than in other areas. The powerful asset that the human sciences in poorer countries could create would be the setting of common agendas, in order to obtain a better bargaining position vis-à-vis the wealthier countries insofar as themes of common interest appear. Actually what happens is something completely different: countries from the South simply import themes and subjects that had success in the North Atlantic states. This takes the place of a true dialogue between North and South.

The Role of Language in the Human Sciences

If we drew a sketch on the human sciences' knowledge areas, we would see them organized around some basic axes. The first one is the axis of those working with themes and subjects strongly underlined by the place of the world they are framed by. Let us say they are strongly framed by the local culture. Here we find the researches on literature, history, and to some extent in linguistics and the arts. The research corpus of such areas of knowledge changes more from country to country than in other areas. Although researchers from other countries also work with Brazilian literature, history, politics, and our society, the most important production in such areas comes from Brazil, and it is written and debated in Portuguese. Of course this does not mean that theoretical or methodological conclusions produced within such researches could be of no worldwide interest. To make its production reach this potential interest is a real challenge to those areas.

Another cluster of the human sciences is that of anthropology, sociology, political science, and geography. Research in these areas takes place within the country they are in, but it is not so dependent on the nationality of its corpus.

The third case is that of philosophy: it is almost completely independent from the geographic place surrounding the researches.

But in all three cases language plays a crucial role. Even if we distinguish the language of production from the language of communication, we must first acknowledge that for us Portuguese is the major language of production in all cases. Publishing in languages of great international communication in each scientific community is very difficult. When it happens, the publishing house generally is a small one. Besides that, the international references in the case of the human sciences are not necessarily written in English. Very often the Brazilian academic tradition was organized around French references rather than Anglo-Saxon ones. There has been and there is a strong dialogue with German-, Spanish-, and Italian-speaking countries. In the human science area, publishing generally means »books«. But today a great part of online publishing, in reference magazines, sites, or pages, is written in English.

Brazil's language, Portuguese, »a última flor do Lácio, inculta e bela« (»Latinum's sylvan and pretty last flower«), in the words of the Brazilian poet Olavo Bilac, is not an international means of communication as far as the scientific world is concerned. Nevertheless, this is not only a Brazilian problem. It is relevant, although in different levels, to all countries that do not have English as their native language, or at least as second language. Efforts have been made by the Spanish-speaking countries in order to make their language an international means of communication to the scientific world, but without success. And, although there are many Spanish-speaking countries, Portuguese is the official language of only seven countries. Even so it is probably the major spoken language only in two of them, Brazil and Portugal. Lately even the French are publishing their works more and more in English, although French is still an alternative to English in the field of scientific knowledge, above all in the human sciences.

Human sciences have the peculiar feature of employing as their scientific idiom the natural, ordinary language, the one that belongs to everybody. On one hand, researchers in this field do not need such a formalized and specific idiom as other sciences. On the other hand, this makes it more difficult when the language employed as a means of expression is different than the native one. This is a completely different situation than the world of the biological and exact sciences, whose inhabitants, if we may say so, are at ease with a more restricted vocabulary and grammar. Besides that, in Brazil as well in many other countries that are placed outside the ring of Anglo-Saxon influence, the native language is better known and spoken than English, even in the academy. It is very difficult to imagine a Brazilian professor lecturing in English at a Brazilian university, when his or her class has a majority of Brazilian students, even if there are some others who come from foreign countries. It would be a great shock to almost everybody's feelings.

I emphasize that it is not at all a question of nationalism. *It simply would not make sense.* In the Scandinavian countries, or to the people of Benelux, which have smaller populations, to communicate in English with native speakers seems more acceptable. This happens even in Germany, with its big population and strong cultural tradition. But in countries that belong to large language communities, where their national language has an intense daily presence, the English language tends to be collateral and subsidiary to the native one, even if the latter is not recognized for scientific international communication. (This stands for Brazil as well as for the Spanish-speaking countries). In some scientific areas, publishing a work in English may imply the exclusion of much of the national reading community. In Brazil this is true for the agrarian sciences, and it would be the same for the human sciences if researchers published their works primarily in English.[2]

Thanks to all that, there are limits to internationalization as far as the admittance of foreign students is concerned (although there are thousands of foreign students attending classes in Brazilian universities and institutes, many of them coming from the United States and Europe). The foreign student who comes to Brazil to accomplish a human sciences' program must learn Portuguese. The majority do learn this language.

Perspectives for International Collaboration in the Human Sciences

With such an outline in mind, what steps should we propose to foster production and communication of the human sciences in Brazil and in other countries with a similar research model? Obviously internationalization is a fact; but we need to develop strategies in order to amplify its positive aspects and to reduce its less-known or even less-commented-on negative aspects.

Demanda de Balcão

The fundamental point, I think, is that any international relation in any field of knowledge must be the outcome of clearly identified demands, necessities, and capacities in the proposing country. Obviously I am speaking from a Brazilian experience, but I think this should stand for any country. This means that relations with our international partners should not be atomized. Traditionally in Brazil there was and still there is a style of demand that we call »demanda de balcão«, this last noun meaning »counter« in English, as when we say that we buy something on the counter. We could roughly translate, or rather explain, that peculiar expression as meaning a »retail style of demand«. It occurs whenever any researcher presents a financing demand in order to establish some sort of international relationship, but without long-term goals, or without

---

[2] Abel Packer and Rogério Meneghini are responsible for an important online site where people can find almost 200 Brazilian or Hispano-American scientific magazines (www.scielo.br; SciELO stands for Scientific Electronic Library Online). They have suggested that authors publish, if they can, their best texts in two versions, one in English to the international scientific community, and another one in their native language, either Portuguese or Spanish, to the students and to people in general.

the support of a group that could carry on the research not depending on foreign aid. Well, we approve less and less this sort of demand, and it should and must be so. It is true that in human sciences often individuals, instead of groups, conduct research, *and such a feature must be respected*. But when we think of international cooperation, the goals and the achievements that such projects will bring about must be quite clearly stated, even if on each side of the cooperation we find a single researcher.

What We Can Do Together and What We Cannot

There is a second point that is often forgotten. Internationalization works better when there is a previous dialogue among the national researchers (and I am talking about Brazil and many similar countries) on what they want and can do together, on what they cannot do or do not want to do, and also on how internationalization will foster the quality of what they are already doing.

This is a delicate point, because it brings about the costs of and the budget for international relationship. Depending on governmental policies, some countries, like the United Kingdom, charge very high fees for international students. Others do not, like most countries in continental Europe and even the United States. In late 2007 I signed on behalf of CAPES a very positive Memorandum of Understanding with the Royal Academy for the Human Sciences of Britain. They understand perfectly the problem caused by the high fees of UK universities. Anyway, this is a problem that makes clear goals a must for international cooperation.

But, even when the costs of international cooperation are not so expensive, we must keep in mind the possibility and the necessity of clear definitions, which are in our interests in scientific research. This stands for our country as well as for other countries that are in a peripheral position within the capitalistic system. The worst we can do is to assume as ours issues that may be very important within other cultures but that are not so relevant within ours. This could lead us to lose sight of essential questions about the specific trends that build the cultural differences around the world.

An Example: What Is Democracy Good For?

I will give you an example. In the last twenty years or so, almost all Latin American countries have become democratic regimes, after long periods of dictatorship. We have attained historically unprecedented levels of freedom of organization, expression, and voting. People who have lived under some kind of dictatorship will soon be a minority. People who have come to adult life under a democratic regime are now a majority in most Latin American countries.[3] Well, this is hopeful. But at the same time we must recognize that democracy did not put an end to poverty or to misery in the region. This leads to the question: *What is democracy good for*? This question always returns, when democracy does not come along with concrete social benefits. This might be a typical question of poor or underdeveloped countries.

However, at the same time we see that in wealthier countries people who used to follow the vote of *their betters*, of the *aristoi*, of a republican patrician nobility, have now begun to vote on their own, electing as presidents people who seem to be quite inferior to their predecessors. This has happened in the last years in the United States and France. These two countries have opened the modern era of democracy, of human rights. But at the same time that South America was electing several left-wing leaders to rule their countries, and they were inaugurated with no problems at all, in the United States and in France there was a sort of decline in politics. In the United States a president was inaugurated that had not been properly elected and later carried on international policies that made him lose all the world sympathy that the destruction of the Twin Towers had allowed him to gather; and in France a right-wing extremist, almost a fascist, came to the second round of the national election in 2002. In this sense the Latin American experience of the first years of our century may seem more positive than that of these two North Atlantic countries. Often we think the other way around, that is, the poor countries should follow the successful path of the developed countries, the more »civilized« countries, as some people dare to say. But it will be completely irrelevant to compare these two clusters of nations in order to say which one is more successful in democratic terms. The truth is that in all these countries social groups that previously did not have a strong voice in politics are coming to the fore. Quite often the first outcome of this process is not beautiful. Several important heads of state in the world—in Latin America, of course, but also in France and in the United States—are not what we would call literacy-addicted people. They are literate, of course, but they are not especially fond of the literate world. The voice they express may bring xenophobia and other prejudices to the political and social scenes. But the presence of

---

3 There is an apparent contradiction between the two last sentences. But we must consider the great number of young people, nonadults, in countries with high birth rates. Even people who have spent a part of their lives under authoritarian rule may have grown up when the military had already left power.

new actors in politics, even if they do not praise democratic values, is, in a certain way, a democratic feature. It poses a new demand: that democracy open itself to new questions.

I am suggesting that there is something specific about democracy in Latin America, and at the same time that this is not so specific to the subcontinent. We have now a political system endowed with more freedom than in our recent past, but with feeble results on the social front. This specific situation raises many questions about the new actors coming into the democratic scene without having a democratic experience or tradition. But if such questions stand for Latin American societies, they also stand for both American and French societies, because there, as in Latin America, we find large groups (the majority, I would say) who possess incomplete formal education. Strong international cooperation to foster research on such problems could begin to answer several questions of mutual interest. But it is necessary that both sides know exactly what questions they are interested in, otherwise one of them will impose its own views to the other.

The agreement we signed with the British Academy may be a good and successful example of what I mean. The two sides have agreed that it was not simply a matter of scholars from Brazil or Latin America coming together to discuss this region of the world with national and international specialists on this subject. Our goal would rather be to discuss questions of mutual interest, beyond the national frames where they were set. The British example would be a seminar on Plato. At the same time that an agreement was reached with the British, I helped build a European network of specialists on Brazil, like the one that already existed in the United States. This group held their first congress in November 2008 in the Spanish city of Salamanca. One path is complementary to the other.

Internationalization Is an International Theme

In this new world in which we live, internationalization is at the same time a subject and the form of relationship of knowledge production. Internationalization is an international theme. But we must avoid a situation in which a few countries dominate the agenda of internationalization. This is what we were trying to criticize in the previous paragraphs. In the research realm globalization requires a special attention to respecting differences. It is true that it contributes to eliminating differences and destroying identities. But it is also true that, in order to really produce knowledge and to foster real political action, we must avoid assuming that particular features of one culture are universal.

The agendas of UNESCO and some other international organizations and seminars have been dominated in the last twenty years by one important discussion: on the one side of an imaginary table sit those who see Western values as universal ones, while on the other side sit those who deem non-Western values as also positive, and try to find a place for them in the international scene. Maybe the main issue of such a debate is the one concerning human rights. It is true that the idea of human rights was first enunciated in the West during the seventeenth and eighteenth centuries, and since 1948 have been elevated as universal values. A large part of what is understood as human rights, if we imagine them as a ring, presupposes the idea of an individual at the center of the circle. But this character, if I may say so—the individual—is typical of Western cultures. It seldom appears in other cultures, where the social actor par excellence is a collective and not an individual one.

Nothing can justify the suppression of liberty and other fundamental human rights in the name of supposed specific cultural values, but nothing justifies also the *nonchalance* displayed by some Westerners when they simply dismiss values of other cultures. A highly positive African cultural value, for example, is that of »ubuntu«, a word or a concept that could be explained by a sentence like »a person is a person only through other people«. »Ubuntu« presupposes the idea of a collective subject that incorporates, assimilates, and surpasses conflicts through some sort of forgiveness.

»Ubuntu« was one of the keystones of the transition from the apartheid regime to democracy in South Africa (Louw 1997; Teles 2007).[4] Western societies, strongly based as they are on individualization, would not be able to achieve the intended reconciliation in South Africa without going through such concepts like guilt, repentence, and then forgiveness. It would scarcely be an exaggeration to say that the West could not go beyond a sort of Nuremberg trial followed by an amnesty— understood not as one process, but as two successive and separate ones. But these concepts will have a very different meaning in a society where the collective feeling is stronger than in the West.

As Lévi-Strauss wrote at the end of »Tristes Tropiques«, Western societies work with the concept of »anthropoemia«, implying the exclusion or the elimination of the dissident or the delinquent. Some other societies, many of them the so-called savage ones, follow an »anthropophagical« approach, in the sense that they incorporate, assimilate, and

---

4 Ubuntu is today the name of a software of the Linux family, pointing out an interest in appraising non-Western cultures.

include the dissenter (and this does not mean that they necessarily eat him).

The fact that apartheid did not end in bloodshed or with a massive purge of the preceding rulers was the outcome of a society whose culture is not a Western one. Their old rulers were guilty of crimes similar to those of the Nazi in Europe. But their society was able to accomplish something unthinkable in a European society. South Africa has not only achieved a reasonable peace within its society: it was capable even of *building a society* where there was none. The apartheid oppression did not lead to a normal social life: rather, it produced extreme forms of conflict. »Ubuntu« is an excellent tool to help us understand how international cooperation can work in the field of human sciences.

These conditions may be shown if we analyze the ways people who have lived under severe dictatorships were able to deal with the wounds of oppression. »Ubuntu« and its role in South African society may help us understand the problems of reconciliation after intense social traumas, especially those imposed by dictatorships. The South African model seemed initially to follow the script that had been written when Argentina brought its former dictators to trial. Through such a process several Latin American countries—but neither Brazil nor Chile—and South Africa itself made a survey of the cruel crimes against humanity committed in their recent past and were able to see how much of that past remains part of the present. But the South African procedure was also an important innovation, and it was able to open the way to a comprehensive criticism even of the Nuremberg trials, which had inspired the democratic civilian government of Argentina in the 1980s. Later the South African experience encouraged some countries in Latin America to organize their commissions in order to investigate their own past not only as commissions for »the truth« but also for »reconciliation«.

Regaining Democratic Passion

I allow myself to develop some points of divergence between what I would call the North Atlantic approach to political theory and what can be a rich not-so-Western contribution to it. I understand that Latin American countries share a dissident Western culture, since most of our citizens share cultures that have originated in Europe, especially in Spain and Portugal, but at the same time we are quite different from our ancestors: among other factors, the colonial experience, the climate, the *métissage,* poverty, and a century of U.S. interference in our internal affairs have given us some ways of seeing the world that distinguish themselves from the European, and for that matter, also from the African and the Asian ones, even though their blood may flow in our veins.

We could synthesize things saying that North Atlantic countries have been able to develop the rule of law (what we call in romance languages »Estado de direito«, or »l'Etat de droit«), democracy, and capitalism. But the West paid an expensive price for this genuinely progressive revolution in political and social life: it had to exclude passions from the public and especially from the political field. If we refer to Montesquieu's »Persian Letters« and »On the Spirit of Laws«, we can see how the important ideals he contributed to the West were linked to the belief that »Eastern« despots gave an excessive importance to the satisfaction of their desires and passions, that is, were unable to rule their countries since they were moved by lust and not by honor or virtue. Of course, this »East« devised by Montesquieu may mean some real Western practices he was condemning—Louis Althusser has argued, convincingly, that Louis XV with his Parc aux Cerfs, where he confined his mistresses, and his dismissal of the Parlements would be a very proper example of »Oriental despot«. Well, the point I would like to emphasize is that, even though the Western ideals of the rule of law and (later) of democracy have been relevant social achievements, the price that has been paid for them has not been cheap. Once I worked at the Torre do Tombo, the old Lisbon archive, for a number of weeks, reading laws and all sorts of decrees of Portuguese kings. I noticed that when slavery understood as captivity of Africans begins to wane from the collections—maybe due to its abolition in Portugal itself by the Marquis of Pombal in the eighteenth century—it reenters them in the idea of captivity to passions. Blacks were captive »tout court«, but the worst captivity is the one that subjects people to their passions. Actually this »topos« is much older than Pombal and the Oporto revolution of 1820: it has been present for a long time in Christian doctrine.

But the point I want to stress is that with the rule of law and reason, and the equation of both to liberty, the role of passions has been equated to slavery. A person who does not rule his or her own passions becomes their captive. This way it became almost impossible to understand passions that can help people be free. Spinoza could praise positive passions and give them an important role in human life, but most political thinkers have held that passions often create a life devoid of freedom and dignity. We could say that political theory and practice in the West have left aside the possibility that passions could be the allies of liberty. Of course passions have helped the cause of freedom (»Liberté j'écris ton nom«, wrote Eluard), but the emphasis of our thinkers and educators has not been put on them. This has implied that the world of passions has not been »educated«; in those countries where formal education is bad, quite often the politicians who address popular passions will be the ones more connected with corruption, human rights abuse, and authoritarian rule. And this situation has created a divide between the First and the Third Worlds—France, for instance, after the important investment it made in education from the mid-nineteenth century has had a much more democratic policy than, say, countries which have lacked of education. In Latin America itself, countries as Chile and Argentina have attained in both education and politics levels quite superior to the Brazilian ones, at least until the dictatorships of the late twentieth century; and even then, it can be argued that if they have been much more cruel in those countries than in Brazil, it has been because formal education and political consciousness had roots there much deeper than in Brazil, so the military did not face the same hostility in the latter country as they did in the former.

But now the picture has changed. To return to a point we have already raised, it is no longer possible to say that American and French politics exclude demagogues from ruling their countries. Spin doctors, so important in all democratic elections of our days, appeal to the passions of the electorate. This implies that, if we want to keep the democratic gains of the last centuries—and to advance them—we will need to study the authoritarian passions and try to defuse them. We will need to bet on democratic passions. We will have to *build* democratic passions.

This is no longer a task confined, say, to underdeveloped countries. It is a worldwide challenge that must be met by all of us. It involves both research and practice. We will need empirical research, lots of new theory, and also many practical applications. The interesting point is that maybe the Brazilian »Partido dos Trabalhadores«, or PT, notwithstanding the faults it has committed after having its leader Lula elected as president of Brazil, has developed a good record of dealing with popular passions in a democratic way. While its *frère ennemi*, Cardoso's »Partido da Social Democracia Brasileira«, has employed reason, rule of law, and seriousness as its main political assets, PT has challenged the rightist parties that once supported the military on their own field, the one of passions. For several elections a human heart has been the symbol of some right-wing candidates: »I love São Paulo«, »I love Salvador«, they would say. They would draw most of their votes among the uneducated poor. But this has been the same constituency PT has been able to persuade and to incorporate into its electorate.

The interesting phenomenon is that right-wing voters did not go to centrist parties and then to left-wing ones: they went straight from the right to the left, and this happened because both (but not the center) addressed their passions, although in very different ways and giving them different meanings.

To conclude this reflection, we can say that today there is not a big difference between politicians from the United States, Italy, or France that appeal to the prejudices—sometimes even xenophobic—of their voters and right-wing politicians who did the same for several decades in poor countries or in the poorest states of those same countries (e.g., Huey Long in Louisiana). The problem is worldwide; the solution, I repeat, must also be worldwide; we could even say that it fits quite well the original mandate for UNESCO, when this organization was created by the United Nations in order to fight racism. Authoritarian passions are the inheritors of racism. Democratic passions, to be created and/or fostered, will be the cement of fairer societies than those that have existed until now.

How International Cooperation Could Work

In order to achieve high-quality international cooperation, we should start with an accurate survey of themes that could cope with the following conditions.
- They should be interdisciplinary, bringing together researchers from philosophy, anthropology, sociology, political science and theory, history, the arts, and communication.
- They should also be able to develop new questions that have not yet been posed, or that have been approached only with more conventional views.
- They should also open the way to *questions different from the original ones.*

Last but not least, nothing here is meant to suggest that our countries should close doors and think only about themselves. The idea is the opposite. We should advance—and this is not easy—a step forward, or many steps forward, to publishing more and more in languages that are recognized as those of international scientific communication. This means that our scientists should translate their texts into other languages, but this must be done without losing or even loosening their commitment with the several people they are researching about. And it is also important in our case never to abandon the goal of writing in our native languages, so that our own people may have access to academic production. To accomplish all this is a very difficult task, if we consider all the complexity of writing in the fields of the human sciences. This may require in the short-term a good team of translators, and in the long-term a bilingual capacity as part of the skills of our scientists. It is very difficult to publish texts in the wealthy democracies that originate in the underdeveloped world. But we must advance toward this goal. And, most of all, more important than the form are the contents. Publishing either in our languages or in English is not so relevant: what is crucial is to have an agenda that is truly interesting for all involved.

References

Louw, D.J. (1997): Ubuntu: An African Assessment of the Religious Other. In: www.bu.edu/wcp/Papers/Afri/AfriLouw.htm [Date of last access: 03.03.2009]

Teles, E. (2007): Brasil e África do Sul: Los paradoxos da democracia. Memória política em democracias com herança autoritária, São Paulo.

Internationalization of the Humanities and Social Sciences: Realities and Challenges in Jordan

ABDEL HAKIM K. AL HUSBAN, MAHMOUD NA'AMNEH

Introduction

To compensate for its scarce natural resources, Jordan has invested heavily in education at all levels. Jordanian nation-state and academic institutions have come to realize the essential importance of the so-called »global education«. Consequently, education in general and higher education in particular have been given a prominent position in lists of development goals. Since its creation in 1920s, the Jordanian state has invested massively in higher education and scientific research, including social sciences and humanities.

Despite the impressive record Jordan has lately accomplished in higher education and scientific research, it can be easily argued that the scientific research in the fields of the humanities and social sciences suffers from serious problems and challenges. These challenges relate in various ways to the structure of the Jordanian society and the Jordanian nation-state on the one hand and to the structure of institutions dedicated to scientific research on the other.

This paper aims at investigating the evolution of research in social sciences and humanities in Jordan and the multifaceted impact of forces of globalization on the Jordanian higher educational system and scientific research. It seeks primarily to shed light on the realities and challenges of the internationalization of the humanities and social sciences in particular. A great deal of attention is given to the recent wave of Islamization in the humanities and social sciences, which tends to prioritize the religious mode of knowledge over scientific knowledge.

In the midst of academia's current preoccupation with forces of globalization and transnationalism, internationalization of scholarship in the humanities and social sciences in particular has recently surfaced in the context of discussions of the global scientific landscape and the growing international knowledge and information economy. Nonetheless, globalization has raised a number of challenges and concerns, which required academic institutions to adopt and implement several adjustments and initiatives to better respond to the demands of an ever-increasing global interconnectivity. Cross-national collaboration has become a top priority and a major concern for any academic institution around the world.

Generally speaking, the present situation in the humanities and social sciences poses different challenges for the scholarly community. Among the many reasons that contribute to such a situation are the tight job market, which results in a fierce competition for positions, the lack of mobility among scholars due to the fear of losing academic positions, and the absence of efficient international networks in the scholarly community.

Despite the remarkable advancements it has recently achieved, scientific research in Jordan continues to suffer from serious obstacles and problems. Part of this has to do with the fact that Jordan lacks active and productive specialized and independent research institutes in the fields of the humanities and social sciences on one hand, and to the association of research with teaching on the other. Thus scientific research appears to be an inefficient and unproductive practice for reasons that are essentially structural and relate in various ways to the structure of the Jordanian society, the Jordanian nation-state, and the structure of institutions dedicated for scientific research.

The structure of the Jordanian state was historically formed as a product of several interconnected regional and international arrangements. The state has also historically relied on foreign aid rather than on generating wealth through internal, sustainable development based on scientific research. Furthermore, the structure of the Jordanian society does not value or reward scientific research. It also does not recognize the individual through accumulated individual accomplishments but rather through affiliation with a group, clan, or a region. Therefore, the structure of the Jordanian universities tends to be characterized by strict bureaucracy and a close affiliation with the political realm. All of the aforementioned factors play a pivotal role in turning scientific research into a marginal activity in managing the society, state, and individual.

It is important to note that this paper is empirical rather than theoretical in nature. It largely depends on ethnographic data collected through various methods, including structured and unstructured interviews with researchers, scholars, and decision-makers; available records and statistics; and personal observations and experiences.

Historical Background

Intertwined factors of history, geography, and politics had produced a special environment for education and scientific research in Jordan. Thus the development and dynamics of the educational system in Jordan should be addressed within an historical context. Starting from almost nothing in the early 1920s, Jordan has achieved a great success in developing a comprehensive and high-quality educational system.

The creation of the Jordanian state in 1921 was part of the British colonial legacy in the region. The British-Hashemite mutual interests gave birth to this new geopolitical entity. The national struggle against Ottoman rule (1517–1918) and later against the British mandate (1921–46) played a crucial role in shaping the nationalist experience in Jordan. During Ottoman rule, which lasted in the region for almost four hundred years, Jordan, known at the time as Transjordan, was part of the province of Damascus (Wilayat Dimishaq). The northern parts of Jordan were administratively connected to southern Syria, the middle parts to Jerusalem and the southern parts to Hijaz. Therefore, the concept of Transjordan as a location preceded the concept of Transjordan as an independent nation-state.

Given that Jordan was divided into districts whose relations with the center would change in accordance with the will of the Ottoman capital, the people of Jordan formed a society with multiple cultures and multiple political and religious identities. For instance, because it was administratively related to Damascus, the society of northern Jordan continues to share many common features with the society of southern Syria. Residents on both sides are engaged in a close network of kinship and marriage. The same argument can be made about the relation between southern Jordan and northern Saudi Arabia which is manifested in the strong kinship, trade, and economic networks.

Following the Ottomans' departure, Jordan became a British Mandate in accordance with Sykes-Picot agreement. At the time, the social organization in Transjordan remained largely tribal and civil society was almost nonexistent. The historical roots of the early Jordanian society were primarily composed of scattered Bedouin communities and peasant villages given that urban settlement and activity was almost nonexistent. Such activity was actually very limited over the four centuries of the Ottoman rule in Jordan. Due to the absence of urban centers, villages played a significant role in the formation of the Jordanian state and its distinctive social and cultural feature.

To support such a claim, it is sufficient to note that until the 1960s more than 70 percent of the Jordanian population were peasants living in villages whereas city population represented about 25 percent, and a minority living in the Badia and desert areas. It is worth noting here that most of the people living in the cities were originally from neighboring areas, primarily Palestine, Lebanon, and Syria. The tragic historical developments in Palestine and the subsequent massive Palestinian diaspora greatly contributed to the emergence of urban centers in Jordan. For instance, the population of Amman during the establishment of the Jordanian state in 1921 did not exceed 3,200 whereas the number now is more than 1.5 million.

Due to the particularity of the Jordanian state formation process in which regional or international external factors played the crucial role, foreign aid has always been indispensable for the continuity of the state and the Hashemite regime. Another particularity of state formation in Jordan is also revealed by the excessive official attention given to the military and security apparatuses.

Higher Education in Jordan

From the perspective of higher education, Jordan has achieved noticeable progress and distinction at both the Arab and regional levels, despite limited natural and financial resources. Jordan's record in the field of higher education has proven impressive by international standards. Jordan is ranked eighteenth in the world according to UNESCO and first in the Arab world.[1]

Education in Jordan developed steadily over the past two decades in terms of policies, programs, content, and methodology. Jordanian academic institutions have lately began to take positive and rapid steps in order to upgrade their educational system at all levels in attempt to assume a leading role in the expanding global »knowledge economy«. These institutions have come to realize the importance of taking advantage of the recent technological developments and information technologies to meet international standards of education. As a result, Jordan has

---

1 www.unesco.org

adopted various elements of the European system to enhance and modernize its higher education systems.

The adoption of the European experience can be explained by many factors. After long decades of adoption of the American model in education, many Jordanian decisions-makers have come to the conclusion that the American model cannot be fully applied in Jordan since the socio-economic situations in both countries is quite different. In addition, the cooperation between Europe and Jordan economically and educationally has reshaped the Jordanian academic realm. Since 1980s, many Jordanians have begun to pursue their higher education at European universities, especially in England, Germany, France and Spain, after long decades of depending on the American universities.

The Jordanian Council of Higher Education was established in 1982 in order to regulate and develop policies of higher education. It initially formed the core for the Ministry of Higher Education, which was established in 1985. However, the ministry was annulled in 1998, but was reestablished in 2001 in accordance with the instructions by King Abdullah and renamed as The Ministry of Higher Education and Scientific Research. In light of this, a new law (Law no. 41) for higher education was endorsed in 2001. By this law, the Ministry of Higher Education and Scientific Research took over supervising all higher education issues and councils.

The Ministry of Higher Education has constantly sought to raise the level of higher education in Jordan to enable it to reach the best levels of global education. In this regard, it has developed a number of strategies and initiatives to enhance the quality, as well as the quantity, of higher education and scientific research. This includes admission policies of public and private Jordanian universities, education management, quality assurance, developing human resources, and developing curricula.

Postsecondary education in Jordan was not possible until after 1951. Students who sought to continue their higher education had to travel to neighboring countries, especially Egypt, Lebanon, Syria, and Turkey. As Abu Nasir (sixty-five years old and a retired teacher) recalls:

»When someone went to a neighboring country to pursue his higher education, all of his relatives or the whole village would come to say goodbye to him. He was considered an important person. Not many people used to finish their higher education« (interview with Ibrahim abu Nasir on October 10, 2008).

The total number of Jordanians students enrolled in Arab universities in 1948 was estimated to be 223 students.

The establishment of higher education institutions, following the official unity between the West and East Banks in 1950, began with the opening of a class for teachers in al-Hussein College in Amman. After that, the Ministry of Education established an institute for female teachers in Ramallah in the West Bank, which was officially under the Jordanian sovereignty at the time. By 1962, eleven teachers' institutes had been established: eight for men and three for women. The establishment of the so-called *Ma'ahad Mu'alemmeen* (teachers' institutes/ colleges) was intended to satisfy the growing local need for school teachers after the creation of tens of schools in the different parts of Jordan. These colleges were meant to offer specialized, career-oriented training, and prepare students for work in the middle-level professions.

In the last two decades, several public and private universities have been established in different parts of Jordan. At present, there are ten public universities and sixteen private universities. The first Jordanian university, Jordan University, was established in Amman in 1962. During the first decade, Jordan University followed the years' system until 1972 when it adopted the more liberal and advanced credit-hour system. Several public universities were established in the following years: Yarmouk University in 1976 in Irbid, Mu'tah University in 1981 in al-Karak, Jordan University for Science and Technology in 1986 in Irbid, Al al-Bayt University in 1994 in Almafraq, the Hashemite University in 1995 in al-Zarqa, and al-Balqa' Applied University in 1997. The latter was meant to restructure the sector of community colleges.

On the other hand, the first Jordanian private university, Amman Private University, was established in Amman in 1990. In fact, the 1990s can be considered the golden age for private universities as most of them were established during this period. One of these universities, Amman Arab University, specializes in graduate studies (MA and Ph.D.) only. The Higher Education Council has significant power over private universities. The council approves the types of studies and fields of specialization at various levels, sets admissions criteria, reviews performance through examination of budgets and reports, and approves any cultural or technical cooperation agreements. Neither private universities nor their students receive financial support from the government. Additionally, a number of foreign universities, such as the German-Jordanian University and the British University, have opened their doors in Jordan in the last two years.The following table lists the academic staff in public and private Jordanian universities according to academic ranks (data provided by the Ministry of Higher Education in Jordan 2007, www.mohe.gov.jo):

Table 1: Academic Staff in the Jordanian Universities by Academic Rank for the Year 2005/2006

Teach. & Res. Ass.	Lecturer	Instructor	Assistant Prof.	Associate Prof.	Full Prof.	Grand Total		
493	494	810	2572	1172	1101	6542	T	Grand Total*
220	161	303	371	93	48	1196	F	
79	91	54	352	222	331	1129	T	The University of Jordan
35	35	34	73	27	16	220	F	(Pu)
67	0	116	205	176	206	770	T	Yarmouk University (Pu)
30	0	42	18	8	11	109	F	
13	96	22	152	149	106	538	T	Mu'tah University (Pu)
5	25	5	11	4	4	54	F	
15	138	3	282	153	97	688	T	Jordan University of
6	49	0	45	10	1	111	F	Science & Technology (Pu)
88	50	24	200	45	38	446	T	The Hasherrite
41	8	13	35	6	1	104	F	University (Pu)
0	0	55	138	19	17	229	T	AL al-Bayt University (Pu)
0	0	21	10	2	0	33	F	
73	14	51	113	29	22	302	T	AL-Balqa' Applied
25	2	3	4	2	0	36	F	University (Pu)
6	15	9	67	15	6	118	T	AL-Hussein Bin Talal
2	1	1	6	0	0	10	F	University (Pu)
21	20	20	40	14	3	118	T	Tafila Technical
1	0	6	1	0	0	8	F	University (Pu)
0	8	0	5	2	3	18	T	German Jordanian
0	8	0	1	0	0	9	F	University (Pu)
51	5	58	110	49	23	296	T	Al-Ahliyya Amman
36	1	23	7	7	0	74	F	University (Pr)
0	25	35	167	50	27	304	T	Applied Science University
0	16	11	33	2	3	65	F	(Pr)
78	1	81	127	59	22	368	T	Philadelphia University
39	1	33	14	5	3	95	F	(Pr)
0	0	78	115	35	28	256	T	Al-Isra Private
0	0	24	21	2	2	49	F	University (Pr)
0	0	59	76	33	18	186	T	University of Petra (Pr)
0	0	41	24	5	2	72	F	
0	28	38	117	62	38	267	T	Al-Zaytoonah Private
0	15	9	36	12	5	77	F	University of Jordan (Pr)
0	0	33	90	18	8	149	T	Zarqa Private
0	0	9	8	0	0	17	F	University (Pr)

0	0	25	78	16	2	121	T	Irbid National
0	0	11	5	1	0	17	F	University (Pr)
0	0	43	97	15	7	162	T	Jerash Private
0	0	11	10	0	0	21	F	University (Pr)
0	0	8	19	7	7	41	T	Princess Sumaya University
0	0	1	2	0	0	3	F	for Technology (Pr)
2	2	4	3	2	0	13	T	Jordan Academy of
0	0	4	0	0	0	4	F	Music (Pr)
0	1	2	19	2	0	24	T	Educational Sciences
0	0	1	7	0	0	8	F	Faculty (Pr)

* 5663 Jordanian, 797 Arabic, 77 Foreign
Pu = Public University, Pr = Private University

When the Jordanian state was established in the 1920s, the number of schools in Jordan was very limited. For instance, there were only two secondary schools in the whole country: one in Irbid and the second in Salt. As such, the percentage of illiterates exceeded 90 percent of the whole population. So, with the rapid construction of the state bureaucracy, the need to educate citizens to be able to occupy the created jobs and statuses in the newly established bureaucracy seemed urgent.

A case in point here is that planners for higher education in Jordan adopted the philosophy that Jordan could become an important provider of qualified human staff (teachers, doctors, engineers, nurses, bankers, journalists, technicians) to the Gulf countries, which lacked the necessary developed educational infrastructure. Moreover, Jordan provided some Arab countries, especially Morocco and Algeria, with qualified graduates in the fields of Arabic Language and Islamic Sharia after the adoption of Arabization policies in these countries during the 1970s and 1980s of the last centur.

The need of the Gulf countries for qualified personnel has played a significant role in determining the variables of supply and demand in the Jordanian education market. When the Gulf countries were lacking universities to prepare qualified cadres in administration, agriculture, industry, journalism, and services, they turned to Arab markets including the Jordanian market. The Gulf countries sought constantly to attract Jordanian graduates, especially those with MAs and PhDs. By doing so, these countries were able to advance their system of higher education and to be able later to graduate classes of human cadres.

In other words, the Gulf markets and universities have exercised a great pressure on higher education and scientific research in Jordan. A large number of qualified cadre who have received their education in the best

Western universities and have acquired excellent practical experiences within Jordanian academic institutions are attracted to work in the Gulf countries because of the difference in the salary between Jordan and the Gulf countries. Work conditions in the Gulf countries in terms of work hours, availability of facilities, and work social conditions are also attractive to them. Therefore, we can talk about a rising drain in the Jordanian qualified and trained cadre who leave the Jordanian educational system in search for better opportunities.

Scientific Research

Generally speaking, scientific research in Jordan does not have great traditions or a golden history. The roots of the Jordanian experience in this area go back to the 1970s and 1980s when several universities were established in different parts of the country. There is a lack of specialized research organizations, that is, institutes and centers whose main job and mission are totally dedicated for scientific research and scientific development, to support scientific research in Jordan.

Thus, institutes and groups that have historically carried out scientific research and continue to do so in Jordan include:
- Travelers and Orientalists
- Missionary and Biblical expeditions
- Academics from Western universities and research centers
- Western research centers affiliated with Western embassies (IFPO, ACOR, The German Protestant Institute, The British Institute)
- Public/ Governmental Universities
- Private Universities
- Governmental Research Institutes (The Royal Scientific Society, The Higher Council for Sciences and Technology)
- Private and Independent research centers (New Jordanian Center, Jerusalem Center)
- Nongovernmental Organizations (NGOs)

The earliest traditions of scientific research in Jordan can be traced back to the attempts made by some European travelers such as Burkhardt and Schumacher, who conducted scientific journeys in Jordan in the nineteenth century and photographed and documented some aspects of the Jordanian history and life. Prior to this period, it is hard to find writings by Western, or even local, researchers or travelers due to the dominance of oral culture and the absence of written documentation.

Early attempts of social research in Jordan by some travelers were limited to descriptive ethnographic material about areas of settlement and statistical information about numbers of people and houses. Such material was oversimplified and lacked any form of analysis. Travelers' interest back then concentrated on documenting and classifying historical sites in Jordan according to periods of settlement. It is worth mentioning here that this interest in the archaeological heritage of Jordan was triggered by Biblical studies as travelers sought to look for evidence to prove what is mentioned in the Bible about the historical Jewish settlements in this region of the world.

The 1990s, the period when martial laws in Jordan were lifted, witnessed a remarkable improvement in the Jordanian-American and Jordanian-Western relations. In light of this, the number of nongovernmental organizations in Jordan increased noticeably. Research agenda also changed. Topics such as human rights, freedom of speech, democracy, market economy, and privatization became prominent.

It should be noted here that the issue of funding determines in many ways the research types and priorities. Funding for these nongovernmental organizations does not come from local Jordanian agencies but rather from foreign ones such as European Union, the U.S. Agency for International Development (USAID), or the Japanese International Cooperation Agencies (JICA). Thus most of these nongovernmental organizations do not function, in terms of the scientific research, in accordance with the Jordanian national agenda but rather with the international agenda in general, the Western in particular. Thus this research is directed toward an imaginary and a hypothetical society rather than a real society as reflected by empirical evidences.

It is worth mentioning here that most of these NGOs are usually directed or supervised directly by officials or members of the royal family or indirectly through joint projects between them and the Jordanian state, especially the security forces, which exercise a strong authority and surveillance over these organizations. The establishment of any society or organization must meet strict official requirements.

In addition to NGOs that carry out social research in Jordan, there is rapid increase in the number of private study centers that usually register themselves as commercial companies. Therefore, these centers provide research-related services to the local and foreign markets. Several factors have contributed to the establishment of these private centers, including the lifting of martial laws in the 1990s, the increase in the need of the private sector for more research and studies, and the huge influx of Western and foreign capital into Jordan.

By examining the records of the Department of Press and Publications (www.dpp.gov.jo), which is the department that gives authorization to any research or study center, we find a large number of these study centers are inactive or do not even have offices, and so all they have are the name and the permit.

Research Centers

As has just been discussed, Jordan—like many countries in the Middle East region (see Mitchell 2003)—does not have enough specialized research institutions in the fields of the humanities and social sciences. Though several governmental and private research centers have lately been established, the experience of these centers in the fields of the humanities and social sciences is still limited and immature. Among these centers are:

- The New Jordan Center (UJRC) which was established as an independent research center by one of the Jordanian dissents who took advantage of the lifting of the martial laws in Jordan. It is very active and carries out activities that sometimes compete in quality and quantity with those carried out by Jordanian universities. The center's main mission is to provide distinguished research and foster public debate on internationally important issues. The center is currently publishing a newsletter in Arabic on issues such as civil society, democracy, and sustainable development.
- Center for Strategic Studies (CSS), which is part of Jordan University. It is mainly concerned with research in the fields of regional conflicts, international relations and security.
- The Jordan Center for Social Research (JCSR), which was established as an independent, nonprofit center. The main areas of interest to the center are development, social change, social inequality, poverty, human rights, social problems, and social policies.
- Civil Society Development Center (CSDC) which was established as a nonprofit interdisciplinary educational entity based at Jordan University of Science and Technology with the main mission of enhancing the participation of the university community in civil society programs to meet Jordan's development challenges.
- Royal Institute for Inter-Faith Studies (RIIFS), which is dedicated primarily to interdisciplinary study and rational discussion of religion and religious issues, with particular reference to Christianity in Arab and Islamic society.
- Royal Scientific Society (RSS) which is a nonprofit center that conducts scientific and technological research related to the develop-

ment process in Jordan. It also aims at disseminating awareness in the scientific and technological fields and providing specialized technical consultations and services to the public and private sectors.

In addition to these local centers, several international research and study centers and agencies have been established in Jordan to conduct research in the fields of social sciences and the humanities. This includes the American Center of Oriental Research (ACOR) and the French Research Institute.

Flow of Researchers and Students

According to the statistics of Higher Education Ministry, a total of 24,699 foreign students enrolled at Jordanian universities in the 2006–7 academic year compared to 23,053 in 2005–6. Arab students constitute the majority of foreign students, with a total of 6,202 Palestinians, 2,866 Syrians, 2,725 Saudis, 1,720 Omanis and 1,423 Iraqis. This noticeable increase can be attributed to the good reputation enjoyed by Jordanian universities coupled with the security and political stability in the country (Hazaimeh, *Jordan Times,* 8/21/2007). Ayman Adnan, a Saudi graduate student at Yarmouk University, mentions, »I decided to come to Jordan because Jordanian universities are well known for their strong education system, in addition to the kindness and hospitality of the Jordanian people« (interview with Ayman Adnan on July 21, 2008). In addition to Arab students, some international students enroll at Jordanian universities for one semester or a year.

Moreover, several Jordanian universities offer intensive Arabic courses to international students. For instance, the Language Center at Yarmouk University has hosted students from a number of prestigious American, Asian, and European educational institutions, including the University of Virginia, University of Arkansas, and University of Richmond (USA), Du Provence University (France), and Bremen University (Germany).

Several international scholars come to Jordanian universities to teach or to conduct research. For example, many scholars have joined different Jordanian universities through the Jordanian-American Commission for Educational Exchange (Fulbright), which seeks to enhance mutual understanding between the people of the United States and other countries. Similarly, a number of Jordanian scholars have received Fulbright fellowships at some Western universities.

Scientific Research in Social Sciences:
The Case of Social Anthropology

For many reasons, anthropological research can be considered a good example of the ongoing process of the internationalization of social sciences in Jordan. Anthropological research in Jordan was historically practiced by Western pioneers with complete absence of any local tradition in practicing anthropological field studies. Anthropology was and is still viewed of as a Western and colonial discipline used by the Western powers to dominate the Arab world. Until recently, anthropology was not known in most of the Arab countries.

Over time, anthropology has produced a good number of scientific and fieldwork research on various aspects of the Jordanian society and culture. The beginning of this type of research goes back to the 1950s when the anthropologist Richard Antoun conducted fieldwork research in one of the villages in northern Jordan (Antoun 1972). Moreover, the first and only academic Department of Anthropology in Jordan was established at Yarmouk University in 1984 with the primary purpose of conducting scientific research in the Jordanian society.

The Department of Anthropology at Yarmouk University was initially dedicated to graduate studies as it used to offer MA degree in both social and physical anthropology, with a graduation requirement of writing a dissertation on one of the anthropological phenomena in the Jordanian society. The department continued to focus on research in addition to teaching until the year 2002 when a bachelor program was established. As a result, teaching rather than scientific research became the primary concern of the department. The Department of Anthropology was founded at the hand instigation of a number of Western researchers from England, United States, Germany, and Turkey who designed the course plan, taught courses, and supervised MA dissertations. The language of instruction was mainly English. Currently, all employees in the department are Jordanians. Most of the teaching staff have their Ph.D.s from Western universities.

In addition to the anthropological research carried out by instructors and students of the Department of Anthropology since the year 1984, many anthropological studies in Jordan have been conducted by Western researchers from the United States, France, England, and Scandinavian countries. Far more anthropological studies in Jordan are conducted by Western researchers than by their Jordanian counterparts.

It should be noted that some of the anthropological studies on Jordanian society have been conducted by Jordanian students during their graduate studies at Western universities. They usually choose for their graduate requirements a topic of study related to Jordanian society. This is why Edward Said (1978) has called the researcher an »informant« when he argues that the role of the Arab anthropologist who studies in a Western university and chooses a research topic related to her/his own society is one that helps the Western society itself.

Within the same context, some anthropological research has been conducted by researchers affiliated with Western research centers based in Amman such as the French Research Institute and the American Center for Oriental Research (ACOR). In addition to these centers, a small part of the anthropological research has been conducted by NGOs such as Queen Noor Foundation, the Hashemite Fund, or al-Badia Development Fund.

Topics of the anthropological research have changed over time and varied according to the educational institutes of the researchers. During the 1980s and 1990s, most studies focused on the relationship between the state and the tribe. Another topic that was of a major concern for anthropologists was the status and living conditions of the Palestinian refugees in Jordan. Some studies focused also on the peasant economy while others focused on women and family in terms of roles and statuses.

Lately, the research interest has switched to focus on more recent topics such as identity, gender, oral history, Palestinian immigrants, women's rights, child labor, political development, city formation, and poverty.

Analytical frameworks and approaches employed in these studies, especially those conducted by Western researchers, have also changed. Studies conducted in the 1960s, 1970s, and 1980s employed the Marxist and functionalist approaches whereas recent anthropological studies tend to adopt theories of postmodernism, especially the theoretical frameworks developed by Michel Foucault and Pierre Bourdieu.

Regulations followed by Jordanian universities do not distinguish between theoretical or library research and fieldwork research. Thus ethnographic research is treated in the same way a research in linguistics or economic sciences is treated. We all know that an anthropological research requires complicated measures and procedures. These steps require a long time, intensive effort, and a financial support. Therefore, they should not in any way be compared to a research in economics, linguistics, or history.

Current Situation

Islamization of the Humanities and Social Sciences

The relationship between the social sciences and dominant religious and Islamic ideologies in Jordan and other Islamic countries is very complicated and multifaceted. Existing Islamic ideologies seek to impose its perspective and thus shape the topics and methodology of social sciences. To a great extent, the tension between the two represents a tension between two modes of knowledge (Mendelsohn 1993): the scientific mode employed by social sciences and the mythical mode, mixed with metaphysical and philosophical knowledge.

It has become easy for researchers and scholars interested in studying the Jordanian society to notice this pluralism of knowledge is used not only by elites but also by the masses. In seeking to understand the different phenomena, whether natural or social, people tend to employ a unique mixture of inherited mythical, religious metaphysical, philosophical, and scientific modes of knowledge.

Using one mode of knowledge in particular depends largely on the topic under investigation as well as on the social category of people. For instance, issues such as death and illness are usually explained by employing both the medical empirical knowledge and the metaphysical one. The latter uses explanations that include witchcraft and magic.

Most scholars seem to agree that the Arab defeat by Israel in 1967 war proved the failure of the Nasserist, semi-secular, and pan-Arab ideologies. The postwar political and nationalistic situation left the door wide open for Saudi Arabia to adopt and present an alternative ideology. In addition, Saudi Arabia took advantage of the oil boom during the 1970s to assume a leading role in the Arabic and Islamic worlds, especially after the Saudi-American coalition was constructed during the cold war. All these situations helped in presenting Saudi Arabia as an influential financial power that possesses a strong symbolic religious power within the Islamic World.

The Saudi alternative ideology stems from a religious thought that glorifies the past and adopts a *salafi* (fundamental) agenda (Stenberg 2004). It argues that the solutions for all human and social problems were proposed by Islam a long time ago. Thus what the modern society has to do is to go back to its Islamic roots. According to this ideology, Islam is a comprehensive religion that fits all times and places, and one that organizes different aspects of the relationship between God and believers.

It should be recognized that the Saudi campaign, supported by the United States under the banner of resisting the communist bloc, which sought to revive and spread the Islamic feeling in the Arab and Islamic region, achieved great success. This Islamic ideology spread to several social and economic spaces inside the Arab and Western societies. Many of the Arab secular constitutions in the newly decolonized countries were amended to add a religious flavor to them. Prominently emphasized were articles in national constitutions that state that Islam is the official religion of a country.

In addition to legal scripts, the process of Islamization can be noticed in most social and economic structures. For instance, the slogan »Islam is the solution« has been used by various religious ideologies to reshape social, political, economic, cognitive, linguistic, symbolic, and architectural realms.

This ideological shift in the Arab and Islamic world has led to the emergence of a trend that seeks to Islamize knowledge of all kinds, including scientific knowledge. Scientific knowledge is depicted as Western and is thus associated with the imperialistic policies. Moreover, the Arab-Israeli conflict and the Western support for Israel in oppressing the Palestinian and Arab peoples has had a critical role in what can be called »satanization of sciences« and in associating all forms of scientific knowledge with the West and colonialism (e.g. Alatas 2006). This in turn has led many Arab and Islamic scholars to adopt an opposite trend, to Islamize knowledge. As a result, several forms of Islamic sciences have emerged and spread for the purpose of emphasizing the uniqueness and distinctiveness of the Arab and Islamic culture (e.g. Bakar 1998).

Quranic knowledge has lately assumed a great importance among the academic and scientific elites. This doctrine argues that the Quran obtains in it all forms of knowledge and scientific discoveries. For instance, it is believed that recent discoveries in fields such as chemistry, physics, medicine, bacteriology, history, archaeology (Waheeb 2005), and linguistics were revealed by the Quran long time ago. During the 1980s, the topic of Islamic economy was presented as an alternative to the communist and liberal economic theories. Several Jordanian universities established academic programs in the field of Islamic economy. At the same time, Jordanian Islamic banks began to spread throughout the country.

Another example can be noticed in the shift toward what is called »the Islamic medicine«, or »the Quranic medicine«. Some doctors, trained in the West, practice the Quranic treatment in which they ask the patient to recite some verses from the Quran. Thus, the doctor, the patient, and the masses all participate in the Islamization of medicine and medical treatment.

This process of Islamization has also reached the fields of the humanities and social sciences. Islam is conceived of as a primary source for various sociocultural symbols, representations, and explanations in the Jordanian society. What is interesting is that some social, economic, and cognitive networks have largely contributed to de-emphasis and doubt about scientific knowledge while encouraging and promoting the religious one (for example see the work of Khan 2004).

While modern sciences are based on the principle of causality, linking every phenomenon to a certain cause, the dominant collective conscious in Jordan believes in fatalism rather than causality. Thus, there is a strong belief that phenomena, whether natural or social, are all predestined and are of God's creation. The collective social understanding of phenomena promotes such belief.

In the field of archaeology, the Islamization wave has created a balance between archaeological excavations and findings on one hand and the Quranic script on the other. Interestingly enough, some archaeologists educated and trained at Western universities believe in the possibility of achieving this balance. They combine the Quranic stories about ancient peoples and civilizations with recent excavations. In some cases, they try to prove that what is mentioned in the Quran preceded the archaeological evidences and results.[2]

In light of this, it can be easily argued that the humanities and social sciences need to liberate the human mind from this understanding should they seek to develop and prosper. The conviction that God rather than humans is the sole producer of events and phenomena can lead to a latent and obvious conflict in various contexts.

Problems and Obstacles for the Internationalization of SSH

As has been mentioned before, internationalization of social sciences and humanities in Jordan suffers from serious challenges and problems (Bani Hani 1996; on general aspects of internationalization of the social sciences and humanities, see Rampelmann 2005) despite the noticeable progress it has achieved in the last few years. In comparison with the scientific research in the fields of natural sciences which has come to be part of the global scientific research, the scientific research in social sciences and humanities still lags behind.

---

[2] Interview with Dr. Abu Ghanima, professor of archaeology at Yarmouk University, on August 21, 2008.

This situation can be attributed to several factors, which are related to the nature of humanities and social sciences itself, compared with the natural sciences. The latter is not influenced by the ideological debates and enticements that are related to historical, social, and economic conditions special to the Jordanian society. In contrast, social sciences and humanities have had to develop and flourish through these conditions.

Factors that play a crucial role in impeding the integration of scientific research in social sciences and humanities into the global structure of scientific research can be divided into six categories: linguistic, ideological, political, administrative, cognitive, and financial.

The linguistic factor contributes to the formation of a cognitive estrangement between researchers in the social sciences and humanities and between the theories circulating globally in these fields. A vast literature is published in English, French, German, or Spanish. In contrast, the quantity and quality of scientific output in Arabic does not constitute a significant proportion in this regard. As a result, specialists in humanities and social sciences find themselves in a linguistic and cognitive estrangement with what prevails worldwide.

Arab social scientists have traditionally been heavily influenced by Western theoretical paradigms of social science, particularly Marxism and functionalism (Ibrahim 2000). Most of those with postgraduate education in social and human sciences received their degrees from Arab universities—mainly in Syria, Iraq, or Egypt—or Indian universities, and just a few of these researchers obtained their degree from Western universities. The situation is quite different in the natural sciences. Most of the holders of higher degrees in Jordan in the fields of physics, chemistry, medicine, engineering, and mathematics have completed their education at Western universities, specifically American. This gives them the relative advantages in terms of having access to the latest theories circulating globally. Additionally, mastering the English language would enable them to acquaint themselves with recent theories and scientific production in fields related to their majors.

Discussing the linguistic factor leads us to examine the political factor. Most of the natural sciences professionals are graduates of Western universities, and therefore do not face a linguistic barrier to access the new knowledge in these sciences. This can be attributed to the political and administrative factor represented by the historical priorities of the Jordanian political system and the policies adopted by the administrations of the Jordanian universities.

The Jordanian political system has traditionally had close relations with Western countries, particularly with the United States and some Western European countries. These close ties have led to the development of a philosophy that emphasizes the importance of knowledge and technology transfer from these states. The successive administrations of Jordanian universities have put such philosophy into practice by sending thousands of graduate students to these Western countries in order to transfer the Western knowledge and technology.

The preference of the Jordanian government and university administration to devote most of the already limited budgets of scholarship to those who wish to study natural sciences and the allocation of modest budgets to scholarships in the areas of humanities and social sciences has resulted in a big division in the community of researchers and academics in Jordan. Specialists working in the natural sciences are qualified in one of the world languages in general, while the degree of proficiency of researchers in the social sciences and humanities of these languages is at a lower level.

There are also some other barriers related to administrative policies in Jordanian universities that hinder the integration of Jordanian researchers in the humanities and social sciences in scientific research worldwide. The point here is that there are no clear university policies that enable researchers to be in continuous contact with the Western research centers and universities. The budget allocated for conferences or to send academics to Western universities is usually very limited. Additionally, the selection of researchers to participate in global scientific conferences or in other research activities are not objective but subject rewarded on the basis of personal relationships.

Furthermore, researchers in the humanities and social sciences who wish to update their knowledge and information face the obstacle of access to books and references. Due to the high prices of new books, most universities in the Third World, including Jordanian universities, cannot afford the latest issues of publications in various areas of science, including the humanities and social sciences.

In regard to ideological obstacles, the level of scientific empiricism in the humanities and social sciences is much lower than that in the natural sciences. Therefore, social theories are usually subject to many cultural and ideological peculiarities. Given that the production of these theories is controlled by the Western scientific elites, most researchers and students believe that humanities have a Western perspective and flavor. It is ironic that such a view does not prevail in the natural sciences.

Part of the rejection of the social sciences comes from the perception in the social and even academic imagination that the social sciences are of a Western nature. For example, there is still a belief that anthropology is a colonial science by highlighting the link between the beginnings of anthropology in Western universities and research centers which were linked with the colonial administrations, particularly in the nineteenth century and the beginning of the twentieth century.

Obviously, this leads us to talk about the conflict over the issue of identity and its related symbols in the Arab societies, including the Jordanian society. Debate over the question of identity in Jordan is governed by three currents: a liberal current that rests on the belief that the future of Jordan lies in the adoption of the Western socioeconomic model; a pan-Arab current that stresses societies' specificities, though it does not reject cautious ties with the West; and finally an Islamic current that focuses on the cultural specificity of the Muslim communities and often presents itself as opposing the West.

The abovementioned fact evokes an additional obstacle to the integration of the social and human sciences in Jordan with the global research trends. It is the epistemological manner through which the Arab and Islamic societies, including the Jordanian society, approach human and social phenomena. The prevailing view in the Islamic societies is that social phenomena are a product of God rather than human beings. It is certain that such a view does not promote the creation of real development in the humanities and social sciences, which are primarily based on the idea that social and human phenomena are the product of social agents and the groups rather than the product of other forces whether natural or metaphysical.

Furthermore, the practices of Jordanian researchers, especially in sociology in terms of the adaptation of Western concepts without any attempt to adapt their epistemological tools to the particularities of the local communities and local cultures, encourages local people to consider these sciences as western and strange knowledge and consequently leads to further neglect and disregard the social and human sciences.

References

Alatas, S.F. (2006): Alternative Discourses in Asian Social Science: Responses to Eurocentrism. Thousand Oaks: Sage.
Antoun, R. (1972): Arab Village: A Social Structural Study of a Transjordanian Peasant Community. Bloomington: Indiana University Press.
Bakar, O. (1998): Classification of Knowledge in Islam: A Study in Islamic Philosophies of Science. Cambridge: Islamic Texts Society.
Bani Hani, A. (1996): Scientific Research and Economic, Administrative, and Social Impediments: A Descriptive Study of a Jordanian University. Al Mustakbal Al Arabi Publications.

Ibrahim, S.E. (2000): Arab Social-Science Research in the 1990s and Beyond: Issues, Trends, and Priorities. In: E. Rashid/D. Craissati (Eds.), Research for Development in the Middle East and North Africa (pp. 111-140). Canada: International Development Research Centre.

Khan, M. (2004): The Role of Social Scientists in Muslim Societies. www.ijtihad.org/IslamicSocialSciences.htm [Date of last access: 13.08.09].

Mendelsohn, E. (1993): Religious Fundamentalism and the Sciences. In: M. Marty/R. S. Appleby (Eds.), Fundamentalism and the Society: Reclaiming the Sciences, the Family, and Education (pp. 23-41). Chicago: University of Chicago Press.

Mitchell, T. (2003): The Middle East in the Past and Future of Social Science. UCIAS Edited Volumes, 3 (pp. 1-32). The University of California: International and Area Studies Digital Collection.

Rampelmann, K. (2005): What Factors Impact the Internationalization of Scholarship in the Humanities and Social Sciences? Bonn: Humboldt Foundation. [Available at: www.humboldt-foundation. de/pls/web/docs/F1610/tshp.pdf]

Said, E. (1978): Orientalism. New York: Vintage Books.

Stenberg, L. (2004): Islam, Knowledge, and »the West«: The Making of a Global Islam. In: B. Schaebler/L. Stenberg (Eds.), Globalization and the Muslim World: Culture, Religion, and Modernity (pp. 93-110). Syracuse: Syracuse University Press.

Waheeb, M. (2005): The Archaeological Discoveries through the Quranic Verses. Amman: Arabic line printing press publications.

Internationalization of Social Sciences: The Lebanese Experience in Higher Education and Research

JACQUES E. KABBANJI

Introduction

The concept of internationalization of social sciences is relatively new (Smelser 1991) and has competed with other more prominent concepts, such like universalism and modernism, that played a major role in legitimizing the spread of new, essentially Western, scientific knowledge. Since the latter concepts are related to a particular civilization—although it is composed of many cultures—with a particular history of political and economic dominance and hegemony, internationalization therefore seems to be more neutral. It gives the impression that all the involved parties in the production and dissemination of scientific knowledge worldwide, be it hard or soft, are equals. In a recent article, Neil Smelser highlights the fact that »internationalization means that knowledge is broadly applicable without reference to national and other boundaries« (Smelser 2003: 645). It is this very dimension that sociologists and social scientists from the »Third World« in particular have been challenging since at least the 1970s. In the wake of this critique, many other scholars have also criticized internationalization. They essentially belong to the cultural domain. The question of cultural specificity and identity comes first when sociologists have to face the internationalization of social sciences, and most notably sociology. To cite an Indian context, according to one sociologist, »If the cosmopolitans ignored the deep and specific historicity of Indian society thereby creating intellectual alienation, the traditionists with their overemphasis on the uniqueness of Indian civilization created intellectual claustrophobia« (Oommen 1991: 79).

To cite another context in the developing world in which sociologists in the 1970s faced the challenging effects of the social science's internationalization one can rely on what has been reported by some Western sociologists. *Dependency* is the key word used by these sociologists to make an inventory of social sciences in the developing countries. Dependency is also viewed as a consequence of the dominance within the social sciences worldwide of the theories and methods produced in the industrialized world. Therefore such dominant tools constitute the referential frame in teaching as well as in researching in the developing world (Chadwick/Lyons 1974; Saint-Pierre 1980).

In dealing with the same situation, Arab sociologists have adopted the same perspective. In the mid 1980s, they published in Arabic a coauthored book with a very significant title: *Toward an Arabic Sociology* (Hijazi et al. 1986). In this book the major challenge to the hegemony of Western sociology was both epistemological—the question of validity of sociological knowledge for Arab society since it is, conceptually, based on ready-made approaches—and methodological. In fact, this book was intended to introduce the »state of the art« of the discipline of sociology in the Arab world at that time. The book contained articles that had been first published in the Center for Arab Unity Studies monthly review *al-Mustaqbal al-Arabi* (The Arab Future).

The first section of this book, which includes ten contributions, deals with the »nature of the crisis« of sociology in the Arab world. It's obvious that the participants to this book are particularly aware of the uneasy situation in which sociology had to face theoretical, methodological, and practical, mostly institutional, problems. The most important topics gathered under the so-called crisis umbrella are sociology and ideology, theoretical and methodological problems encountered while conceiving an Arabic sociology, sociology and social problems in the Arab world, and Arab sociologists and the study of Arab society. The second section is composed of five sociological case studies that are mainly concerned with the study of North African societies. Finally, the last section includes three articles that are preoccupied with the desirable conditions for the establishment of an Arab sociology. In this section sociology is expected to give answers not simply to social problems but also, and first of all, to problems facing Arab societies while making their transition from underdevelopment to industrialization and modernity. Other issues in this section are additionally taken into account. Hence theoretical dependence on Western sociological literature is viewed as a great obstacle to the development of an independent Arabic sociology. Therefore, Arab sociologists are blamed for not doing enough to coin their own conceptual and methodical tools for the study of their own societies. Propositions were consequentially made to lead Arab sociologists toward the best gateways to map their societies.

Twenty years later, a new essay on the same topic has been published (Kabbanji 2005). Thirty-three articles, published by *al-Musqbal al-Arabi*, have been reviewed. While this is only a sample of Arab sociological production, the essay nonetheless sheds light on the way Arab sociologists have managed, between 1978 and 2003, to deal with the theoretical, methodological, and empirical problems they continued to face. The main subjects tackled by this review are development, woman, Arab social structure, Arab and foreign migration for work to the Arab Gulf states and Saudi Arabia, Arab intellectuals, problems related to the crisis of sociology and the understanding of sociology by Arab sociologists, and globalization. Although some of these issues are the same as those found in the earlier book, new subject matters are prominent. What is more important to note is that treatment of the new issues has also changed. Instead of the dominant figure of the politically and socially engaged sociologist of the 1980s, the new sociologist is more diversified and suggestive. Thus the essay sketches three types of Arab sociologist: the militant, the disillusioned, and the expert. Each type is concerned with a specific kind of approach to social and methodological problems. The militant continues to undertake topics within the perspective of whole societal change. The disillusioned represents the prominent figure of the professional sociologist who is concerned with finding specific solutions to specific social problems. And, finally, the expert is committed to working within the framework of the international organizations, such like the World Bank, the International Monetary Fund, the United Nations Development Program, the European Community, and the like.

From the 1990s on, some scholars, inspired by these changes and other critical comments, have clearly and solemnly claimed the necessity of an Islamic sociology. Whether religious identity suffices by itself to make a new sociological school was not an important issue for them. Although these criticisms, requirements, and changes of perspective are particularly illuminating for the understanding of new tendencies within the broader sociological view, whether it is »Arabic« or international, they nevertheless pointed out a very serious problem in the internationalization process of social sciences. This problem consists in the overwhelming new trends and realities created by globalization and information technology. To choose only one of these trends, one may point to the fact that most important research activities are by now mainly the product of R&D worldwide. Although social sciences are less concerned than applied sciences with such a mechanism, they mirror the trends set by big corporations and organizations—especially those of the UN, North America, and Europe.

Universities represent no exception to this trend. One may even speak of a new phenomenon that is emerging from the globalization process in the domain of academics. Since the beginning of the 1990s, »turnkey« universities have been founded in different regions of the Arabic world. The oldest of this type of university were begun almost two decades ago in the oil-based states in the Arabic-Persian Gulf, in Jordan, and in Lebanon.

Internationalization of Higher Education in Lebanon: the Starting Point

According to Neil Smelser (1991), internationalization of social science knowledge is best understood as a process of multiple-facets dynamism that characterize the whole process of producing, coding, normalizing, validating, and disseminating social scientific knowledge mainly through a specific modern institution, which is the university. Smelser goes on further to propose seven criteria in order to specify this particular process: the development of universal, general principles of knowledge, applicability of general variables without reference to national and other boundaries, the cognitive validity of knowledge across international lines, international institutionalization of the development of specific scientific infrastructures, international scholarly associations, international interaction and networking of scholars through collaboration on research and exchange, and finally the development of formal knowledge among scholars and communities.

To take them together, these criteria imply a de facto acceptance of the recent trajectory taken by mainstream, »Western« social sciences, and one has to wonder if, in assuming the universality of principles of science and their validation, social scientists and their institutions in the developing world have the same opportunities as their colleagues from the developed world, and if they have at their disposal the same scientific infrastructures and professional networks. Finally, one has to ask if they have the same access to the publishing world and the circulation of ideas on a world level. Isn't it therefore more appropriate to speak about an uneven process of production, circulation, and networking of scientific ideas worldwide? Lebanon will serve as a test case for these hypotheses.

Higher education in Lebanon started in the second half of the nineteenth century. The first newcomer was an American-related college that constituted the prototype for the American University of Beirut (AUB, 1920). AUB began as the Protestant College, which was founded by American missionaries in 1866 to teach and train religious (Protestant) and educational staff. The AUB was shortly followed by a Jesuit university (Université St-Joseph, USJ) in 1875. The latter also specialized first in training religious (Catholic) and health care staff (Bashshur 1988).

Even when these two universities acquired their real status as modern universities they restricted their role to being »teaching universities« specialized in shaping new professionals rather than making research one of their main activities. Research as such started, in the field of social sciences, in the 1930s when some teaching staff helped authorities with their skills and know-how, especially in economics.

By the time that the state university was founded in the early 1950s, three other universities were already working in Lebanon, namely AUB, USJ, and the University College for Women (first called American Junior College for Women and later called Beirut University College [BUC], and since mid 1990s Lebanese American University [LAU]),[1] which was founded in 1924.[2] Although the Lebanese University (LU) adopted Arabic and French languages in teaching, the USJ chose to teach in French and the AUB and UCW in English. This linguistic diversity remains a main characteristic of higher education in Lebanon. It is one of the main obstacles that hinder direct communication among the social scientific community, since there is no reference language adopted on the national scale.

In the end of 1950s, Beirut Arab University (BAU) was created as a branch of Egypt's Alexandria University. It is a university that had been

---

[1] According to the internet site Wikipedia, »LAU was founded in 1835 by American Presbyterian Missionaries as the American School for Girls. In 1924, the high school added a two-year junior college program. Three years later, the college was separated as the American Junior College for Women. In 1948–49, the college expanded into a four-year, university-level institution, and changed its name to the Beirut College for Women. In the same year, the college obtained a New York state charter. In 1955, the college was authorized to grant associate's and bachelor's degrees. In the early 1970s, the college accepted a limited number of men into selected programs. In recognition of this change, the college changed its name to Beirut University College (BUC). BUC became fully coeducational in 1975. In 1994, the college changed its name to Lebanese American University«.

[2] To be more precise one should notice that the University of the Holy-Spirit (USEK) was founded in 1949 but it did not start teaching until 1961.

founded mainly because of the language problem—the need to teach in Arabic—especially in the domain of law. The latter field was essentially taught in French, the law's language par excellence. By and large, religious origin and dependency toward foreign institutions are the birthmark of the private Lebanese universities since their very early age. Most of these universities »have a confessional origin or basis and many of them are a systematic extension of a 'mother institution' outside Lebanon or a blueprint of a foreign model with little effort, if any, to adapt the model to the local situation« (Gaillard et al. 2007).

Religiously, by the end of the twentieth century, all of the working universities except the state university were founded alongside a religious affiliation (see table 1 in the chapter appendix). This implies that those universities are in close relationship with their religious-backed authorities, be they local or especially international. The main examples are the AUB, which has solid ties with its religious community in the United States. Later, these same ties were transferred at the academic level via the American university system, based in the United States. The USJ until now has its roots deeply tied to the Jesuit order, which constitutes an international institution. To that one can add that the same university has been sponsored academically for most its existence by the Université de Lyon in France. The University College for Women (UCW/BUC/LAU) followed the same design.

Still, one should keep in mind the fact that the evolution of higher education is a part of the evolution of the whole cultural scene in Lebanon. One of the most important features of the latter, until the beginning of the war in Lebanon in 1975, consisted in a fierce competition between two visions. The French-educated elite, generally formed at the USJ and implanted in the different state agencies, was eager to preserve the hegemonic position of the French language within the education system, especially at the tertiary level, while the rising English-educated professional elite, law professionals excluded, globally trained at the AUB, sought greater social and cultural recognition.[3] English finally gained momentum and strength as the whole Lebanese educational system adjusted to the needs of the new wave of the globalization process.[4] By

---

3   Law teaching was monopolized by French language until the 1950s, when the newly established public Lebanese University started teach it in both French and Arabic languages. In 1959, the new, Alexandria-affiliated university BAU (see Table 1) was authorized to teach law only in Arabic (Kabbanji 2009).
4   S. Khalaf, a Lebanese sociologist who teaches at the AUB, describes this fierce competition in the following words: »The rivalry between the two sister institutions, a relic of the bitter hostility between French Jesuits and

now, twenty private universities out of twenty-seven have adopted English as their language of instruction, either completely or partially, compared to only twelve totally or partially French-based universities, including the LU.

Since the 1980s through 2000, almost all the newly founded universities and colleges in Lebanon managed to keep these two blueprints—the religious affiliation and external ties—as their birthmark. However, the dependency on foreign universities is more obvious in the new generation of universities. In fact, the latest ones are conceived of as extraterritorial bodies of some American and Canadian universities in particular, with a »ready-to-import« American teaching system. »Most of them are institutions for the transmission of »second-hand« knowledge« (Gaillard et al. 2007). As of 2009, there are twenty-seven certified universities in Lebanon (Ministry of Education and Higher Education, 2008).

The whole academic scene in Lebanon is therefore organized alongside a threefold inner division: the communal or confessional origin, language of teaching, and ties to an external academic entity. All of them are, although unevenly, layers or levels of internationalization of higher education. Furthermore, because of the dominance of the private interests in the domain of higher education, especially in the latest two decades, the whole task of the academe was, and still is, closely associated with market constraints.[5] It is striking that the new legalized universities have clearly shifted way from the religious allegiance to the profit-making rationale. The latest tendency is widely and chiefly supported by American and Canadian universities and colleges as it is shown in table 1 (see the chapter appendix).

To be exhaustive, one should add to this list (table 1), another figure. As data provided by the Ministry of Higher Education in Lebanon (MEES) show, there are, in addition to the working universities in Lebanon, eight university colleges and four religious colleges. The grand total of higher education institutions is therefore thirty-nine. It goes without saying that such an amount of higher education institutions is disproportionate to the size of the country and its population. In fact, Lebanon is a country of roughly 4 million inhabitants, including 143,000 stu-

---

New England Protestants, had given this competition a rather creative and vibrant edge« (Khalaf 2001: 178).

5 It is interesting to note that many new universities, in introducing themselves on their websites, emphasize their status of »non-profit institution«. If this is really the case, the question is then: how is it possible for these universities to survive without making a profit?

dents in higher education.⁶ About half of them are usually enrolled in the Lebanese University—about 70,000 for the academic year 2006-7. This figure helps us understand that the majority of universities are of a small size. They are mainly specialized in teaching technical skills in business and accounting, graphic design, computer science, IT, health care, and the like.

Another fact may better explain the proliferation in the number of new universities. Since 2000, demand, especially from the Persian Gulf, oil-based countries, has grown for skilled Lebanese manpower.⁷ According to a new World Bank study on »Migration and Remittances« (World Bank 2007), Lebanon ranks at the top of major Middle East countries in terms of university-educated, diploma-holding emigrants: 29.7 percent of the total of its emigrants in 2005 belong to this category.⁸ Therefore, demand for rapid training may somewhat explain the growth in the number of new universities. Recent trends in internationalization of higher education in Lebanon could henceforth be understood, partially at least, as an adjustment to a deep globalization trend.

Status of Lebanese Social Science Research

Taking into account the foregoing characteristics of the Lebanese university system, it is possible therefore to speak, with regard to internationalization process, of a »one-sided internationalization«. As a consequence, social science research isn't a priority for most Lebanese universities, especially the newest ones.

Moreover, only a handful of working universities in Lebanon teach integrated social sciences curricula. Only two major universities, namely, the LU and the USJ, have specific institutes or departments that are dedicated to teaching and research in social sciences. Other universities, such like the AUB, LAU, USEK, NDU, and Balamand, while teaching social sciences as a part of a social and behavioral sciences (SBS) pro-

---

6   According to the lastest statistical estimations (Kasparian 2008), the size of the resident population in Lebanon is estimated to be 4.05 million inhabitants.

7   Lebanese emigration is not a new sociohistorical phenomenon. In fact, it started in the second half of the late nineteen century and grew after the First War World. It reached its highest peak between 1975 and 1990, during the war in Lebanon.

8   It's interesting to compare these data with other Middle East countries according to the same source: Djibouti (17.8%), Iran (13.1%), Morocco (10.3%), Tunisia (9.6%), Iraq (9.1%), Algeria (6.5%), Jordan (6.4%), Yemen (5.7%), and Syria (5.2%).

gram, do not have specific departments of sociology and anthropology per se. However, one should point out that the AUB used to have such departments until 1976, when the internal and regional war broke out and made it impossible for this university to continue its mainstream curriculum (King/Scheid 2006).

Therefore the question we should ask is whether research activity in SHS has a solid base in, and outside, the Lebanese academic arena.

The historical background shows some strong evidence in matter of research in social sciences. The main research activity within the university is composed of two types: academic staff who are doing research according to a personal agenda, and Ph.D. and master students' theses.[9] In addition, a few social sciences research centers are available on campus, but there are no laboratories. The research unit is still the individual researcher. Funds are subsequently allowed to researchers on the base of individual research projects, with some few exceptions.

Off campus, the leading research center is the National Center for Scientific Research (NCSR), which is a public institution created in 1962 to promote research in the field of hard sciences. The center was first vehemently opposed by private universities and local big business.[10] The center had no interest in exploring and monitoring research in the social sciences since they were not considered »fully scientific domains« until recently. In fact, starting from 2003 the NCSR changed its definition of science and integrated research in social sciences in its scheduled program for projects to be funded. Once all fields of scientific research are taken together, the NCSR should be considered as the leading research body in Lebanon as shown in diagram 1.

The weakness in university social science research has partially been »corrected« by an increasing presence of the private, external-tied research centers. Remarkably, during the 1970s, which was a decade of great political and security turbulence in Lebanon, several social science research centers were founded, including those that are still very active nowadays. In 1974, on the eve of the devastating Lebanese war, the Arab Institute for Development, funded and monitored by Libyan authorities, who sought at that time visibility and influence in the Arab world, was launched in Beirut. This center, whose main activity ceased

---

9   Only the LU and USJ confer a doctorate degree in social sciences in compliance with their respective curriculum. Another university, the BAU, delivers doctorate degree too but it does not hold an autonomous Department of Social Sciences.

10 For instance, clause 2 of article 5 of the decree that created the NCSR stipulates that the center has no right to engage in an »unfair« competition by undertaking activities that exist within already working institutions.

in the early 1990s, played a major role in attracting many Lebanese social researchers and, for the first time, offered specialized units for research in many social fields. The latter were composed of Lebanese and Arab researchers. As such, it was an unparalleled opportunity for Arab social researchers to work together on a regular basis. Many published books and research reports resulted from this limited research experience. In addition, the institute regularly published many periodicals, including one, *Arab Thought*, dedicated to social sciences issues.

Diagram 1: structure of research system in Lebanon

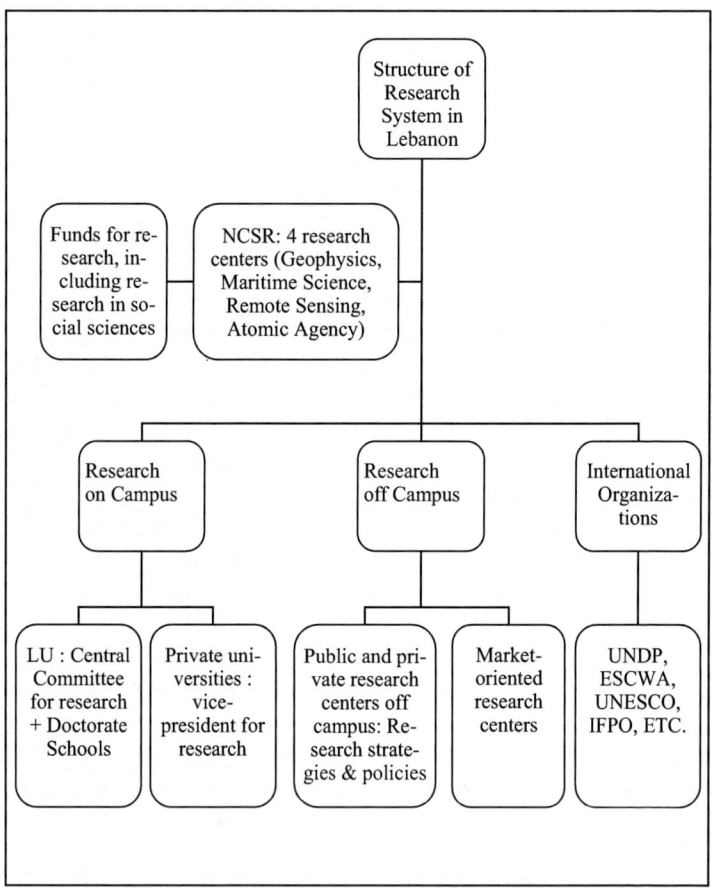

Another research center, the Center for Arab Unity Studies, based on a vision of Arab nationalism, was launched in Beirut in 1978. Unlike the Arab Institute for Development, the Studies Center for Arab Unity is still active and growing but curiously has attracted fewer Lebanese researchers since it appeals to Arab researchers who are partisans of the conception of Arab unity. The bulk of its research work was, and still is, the outcome of a network of Arab researchers laid down by the center itself. Consequently, its main focus is put on Arab issues and affairs. Hundreds of books, materials from seminars, colloquia, and conferences in different fields of social sciences have been published by the center through three decades. In addition, an Arabic social sciences monthly review, *al-Mustakbal al-Arabi*,[11] was launched when the center was founded in 1978. It should also be noted that in January 2008 the center started publishing a new English-language journal to introduce Arab social scientists to a wider range of Western readers.[12]

One of the salient features of the aforementioned centers is their regional nature. Both of them have affected a sort of »gravitational pull« for many Arab researchers from different Arab countries, so that the two centers have served as a »substitute« for internationalization. Lebanon, in hosting the two centers, has established its ability to be a host country for academic and research activities with strong external ties. But, unexpectedly, the two centers have had a weak impact on the Lebanese system of research because of their lack of institutional connection with it on all levels. Except for some circulation of ideas, through publications in particular, there has been no effective cooperation with universities and local research centers, public or private. Hence, the most important effect they created was on individual researchers.

The extended war, with its devastating human casualties and the destruction of physical structures and equipment, has directed attention toward obvious local problems and, to some extent, some methodological issues. Since the implementation of a relatively stable political and civil situation in Lebanon since 1990, new research centers in social sciences have been set up in the late 1980s and 1990s mainly to deal with the social, political, institutional, and economic problems aggravated by the lasting war.[13] As diagram 1 shows, the present system of social research in Lebanon is mostly dominated by internationalized universities and centers rather than regionalized ones. With multiple resources—direct cooperation on the institutional level between major local universities on the one hand and French and North American univer-

---

11 In English this means The Arab Future.
12 The journal's title is Contemporary Arab Affairs, and it is jointly published by the Centre for Arab Unity Studies and Routledge (Taylor & Francis Group).
13 The official version says that this war started in April 13, 1975, and terminated in October 13, 1989.

sities principally on the other hand, direct funds and grants from external universities and institutions, foreign institutions' interest in some areas of the social field, the international networking—the internationalization process, with an unequal effects, is playing a major role in social research as well as in university-level teaching.

Thus, in the early 1990s, a challenging period for social sciences and social scientists began in Lebanon. It was a period when the »reconstruction« of the country was at stake. But no single university in the beginning of the early nineties was prepared for, or became, a real partner in creating the reconstruction plan. Only few individual researchers have had this opportunity. In fact, being a real local partner is new to Lebanese social sciences. The main representatives of social science research during that time consisted of international bodies, such like the United Nations Development Program (UNDP) as well as the United Nations Economic and Social Commission for West Asia (ESCWA), in addition to the World Bank (WB). Those international bodies deeply changed the whole scoop of social sciences in Lebanon. In fact, too many social scientists as well as groups within some universities worked during the 1990s under the auspices and directives of these international bodies. Add to that the conceptual changes these bodies were able to introduce into the perception of the »social« and society. According to the UNDP and the WB, society is no longer seen as a unique, cohesive body. Rather, it is taken as a composite one, in which every component is autonomous. To take an example, poverty, in their approach, is isolated from structural background. No explanation of its causes and effects should be searched beyond people who are affected by it. Many studies and reports have been produced to propagate this approach, which has been generalized to make it applicable in understanding and treating other social phenomena and problems. It goes without saying that the bulk of Lebanese social researchers who worked, and still work, with these international bodies were generally considered as researchers of the »second zone«.[14] And the results of their fieldwork were expected to match these bodies' theoretical and methodological framework.

This overwhelming fact, along with the new sort of cooperation between local and international universities and formal and informal networks, is the main internationalization tool in the domain of social

---

14 It is no exaggeration to use the term »second zone researchers« in this particular context. Keep in mind the famous »terms of reference«, which assign to the local researcher her/his specific tasks and then her/his limits and which are at the same time the basic institutional link between each one of these international bodies on the one hand, and the individual local researcher on the other hand.

sciences in Lebanon. Our inquiry among social scientists themselves, composed of sociologists, anthropologists, law scholars, and economists, carried out in 2006 and ended in 2007, explores, among other things, the extent to which they have been concerned with, and affected by, this type of internationalization.[15] The starting point that lies behind our inquiry is that components of the »Lebanese system« of higher education and research are best defined from the research community's situation and viewpoint. Hence, results and findings we have at our disposal show that most social researchers are at the same time teachers at universities. This is mostly true for sociologists and anthropologists. Economists, political scientists, and researchers in law are more attracted to market and government positions.

In the following section findings from the aforementioned field research will be provided in order to see the way internationalization of social sciences is working in Lebanon in particular domains.

Internationalization of the Social Sciences: Current Issues

Languages and Heritage

One of the main tools used in internationalizing the research as well as the teaching process is language. In the Lebanese context languages have always played a determinant role in opening up scientific activities to foreign institutions. In fact, since the establishment of modern higher education in this country the question of teaching language was always at stake. The American University of Beirut, which is the first institution of higher education to be founded in Lebanon, adopted English as the teaching language, after a few years of trial and tergiversations (Bashshur 1997). The Jesuit university, shortly established after the American

---

15 The ESTIME project (Evaluation of Scientific and Technological capabilities in Mediterranean countries) 2004–7 described the scientific and technological capabilities in eight research partner countries of the Mediterranean (Morocco, Tunisia, Algeria, Egypt, Lebanon, Syria, Jordan, and Palestinian Territories). The project attempted to provide precise indications on research, technological development, and innovation in the Mediterranean countries, supported by empirical investigations and a thorough revision of sources of information. The activities were coordinated by the French Institut de Recherche pour le Développement (IRD). In Lebanon, the ESTIME project included both the natural and social sciences: 45 social scientists and 27 natural scientists were interviewed. In addition, research on innovation has also been done.

one, chose French as its main teaching language. The other universities, with the partial exception of the state university (LU), BAU, and Islamic-oriented universities, have approved the same alternative in choosing either English or French, or a mixture of both of them (Balamand, AUST, and partially the LU) as their teaching languages.

The main consequence of this choice is that a deep gap has been created between two worlds within universities: the self-centered one with the choice of Arabic language and the other-centered one by choosing either the English or French language. It is a cleavage that is more palpable and hence more concrete in the field of social sciences. The LU, where more than 70 percent of students and about 50 percent of the whole Lebanese faculty staff in the domain of social sciences are concentrated, teaches social sciences in Arabic principally,[16] while the prestigious private universities (AUB, USJ, LAU, etc.) teach social sciences in foreign languages. Therefore the choice of language becomes a very important tool in the new era of internationalization. Choosing French, or especially, English, implies a symbolic dimension, that is, the symbolic and cultural capital, that enhances at the same time both personal skills and social status.

Language also plays an essential role in creating two processes of internationalization: the »poor« one on the one hand and the more rewarded one on the other. By »poor internationalization« we mean that teaching and researching by only using the Arabic language makes the impact of international interaction weak: no direct communication with international scientific communities; translation, regardless of its quality, is one-sided: from English and French essentially into Arabic, and when it is available, it replaces genuine sources; no or meager academic and intellectual exchange, etc.

This is a fact that is best represented in the domain of research. In taking into account our own partial findings at the conclusion of the Estime survey (Kabbanji 2009) we can assert that a majority of researchers in social sciences are publishing in Arabic: 60.5 percent of the total publications, including articles, research reports, and books, are in Arabic, while 27,5 percent of them are in French and a meager 12 percent in English. Although these data do not represent the whole production of researchers in social sciences, they give an accurate idea of the fragmentation of the Lebanese community of researchers in social sciences with regard to the language frontiers. Finally, researchers who publish in

---

16 The Institute of Social Sciences, one the main faculties of the LU, has five sections in different Lebanese regions. One of them is totally dedicated to teaching in French and, to a lesser extent, in English.

Arabic do not usually publish in another language, and they are barely translated into foreign languages.

Funds for Research

Many resources for funding scientific research are available in Lebanon. External resources, mainly from UN agencies and other international bodies, as well as internal ones, especially from universities themselves and government (NCSR), are unequally promoting research activities in all fields. Research in social sciences is mainly underwritten by external finances and university resources. Public funds—LU not included—sustain basic and natural sciences. Available data show the nature and the extent of the latest source in supporting research activities.

»Over the last seven years (2000–6), 614 projects have been approved [by the NCSR] for a total budget of US$3,274,050, or an average of US$5,332 per project. On average, and according to statistics for the last four years, half the projects submitted (48.5 percent) have been approved. According to international standards, this is a rather soft selection rate« (Gaillard et al. 2007).

In addition, the distribution of the NCSR's public funds to different fields of science gives more evidence of the marginalized status of social sciences. At the same time, it may indicate that social scientists, all universities included, are reluctant to propose their own research projects for public funding according to international norms.

Table 2: Grant research program percent allocation of funds by discipline (2000–6)

Field	% of Funds
Medical sc. & public health	40.6
Env. sc. & nat. res.	16.4
Agricultural sciences	14
Basic sciences	12.1
Tec. eng. & applied sc.	14.0
Social & behavioral sc.	2.9

Gaillard et al. 2007: 12

The basic unit of distribution of these funds is essentially the individual researcher him/herself. In other words, there is no global research strategy based on groups of researchers or on research centers in the field of social sciences. The UL, an important source for funding research in social sciences, constitutes no exception in this context. Furthermore, the same university, which is the host of the essential bulk of staff and students in social sciences, does not seem to be succeeding in supporting research in the field of these sciences.

»Although the research grant scheme is open to all disciplines, 60% of the grants and an even higher percentage of the budget have been allocated to the basic sciences. The other disciplines, except for economics (19 grants) and media/communications (13 grants), received fewer than 10 grants between 2002 and 2006 (socio-anthropology 9, psychology 5, philosophy 1).« (Gaillard et al. 2007: 21)

Leading private universities, especially because of their international standard and ties, make a consistent effort in backing research. The AUB, to cite as an example, dedicated for the year 2005–6 an estimated research budget of US$2,500,000 in addition to the amount of US$6,794,88 for R&D in the same academic year (Gaillard et al. 2007). However, no detailed information is available in order to detect the social science's part of this budget.

The most striking difference between these two types of managing funds for research lies, however, in the fact that private universities embark on international cooperation, which helps them out through mutual agreements and accords so that their research activities are not confined to local parameters. Evidence is provided with the AUB support to »faculty by providing short- and long-term development grants primarily for short-term travel to conferences and workshops to present research and long-term visits to research facilities« (Gaillard et al. 2007: 19).[17]

However, the main supply for research in social sciences in the case of Lebanon rests on the shoulders of international and regional agencies. The most significant of the later agencies are the World Bank, the European community's agencies, and the UN bodies that work in different domains of socioeconomic and cultural development, in addition to the domain of health and childhood.[18] Consequently, the most important research activities in the social field have been funded, since the early 1990s, by the same international agencies. In this kind of on-demand research, as noted earlier, researchers are asked to fulfill specific tasks that

---

17 External funding sources are far greater than funding from the AUB budget. Medical research is the main recipient area. The NGOs include a number of North American NGOs. The foundations include the Arab Science and Technology Foundation, Welcome Trust, and Ford, among others. The private sector includes many pharmaceutical companies (e.g. Novartis).
18 Agencies like: UNDP, UNICEF, ESCWA, UNESCO, etc.

are relevant to the agencies' priorities. Such a situation may be called, according to our findings, remote-directed research. In this later case, funding may be described as a tool that deepens the uneven internationalization process in the domain of research. In addition, the growing part of international agencies in financing social research may partially explain the limited funding from public sources.

It is worth noting that the government itself is encouraging such a growing weight of international bodies in the social research. Some indicators on this trend are the major surveys and field studies that have been conducted since the mid 1990s with funding from international agencies and in compliance with their own standards embedded in the famous »terms of reference«.[19]

Publishing

One of the standing mechanisms for internationalization is supposed to be a universal normalization of scientific publishing. Two criteria follow in the wake of enhancing such a mechanism: the mode of scientific evaluation and tools of publishing. The former is only partially applied in the Lebanese academic arena. New universities do not seek research activities nor publishing materials from their staff since their unique concern is limited to teaching technical skills. To the contrary, high-profile universities are more or less committed to evaluating the scientific production of their staff. If academic promotion depends therefore on the outcome of the evaluation process, the »faculty«'s job security is unevenly affected by the same process. In the case of the AUB, the motto »publish or perish« is fully working according to »universal norms«. In the case of the LU, to take a counterexample, while promotion does follow the verdict of evaluation, publishing in itself is not a requirement for a faculty member to keep his/her position. Concerning the latter issue, because the LU is a state university, it is different from the private universities. Once tenure track is acquired, the job security is then guaranteed regardless of the »publish or perish« rule.

Not all published material is accepted for evaluation in the AUB. Only articles published in scientific journals recognized by the university are acceptable. They also should be English language–based journals. Hence, the AUB is promoting English language at the same time it pro-

---

19 The most important surveys are: Health of Mother and Child (UNICEF); the Family (League of Arab States); Lebanon Family Health (League of Arab States); many inquiries and surveys on Poverty (UNDP & World Bank); and many inquiries and surveys on local development (UNDP and ESCWA in particular), etc.

motes international standards in publishing. This is a reality that has caused some dissent in the community of social scientists.[20] The LU, which I will use once again as a counterexample, applies another strategy in the matter of publishing. It accepts materials written in three languages, Arabic as well as French and English. Furthermore, it has no required list of scientific journals. All kind of periodicals, scientific as well as unscientific, with a reading committee or not, are accepted. Thus international publishing norms are not compulsory for it. To be more precise, one should emphasize the fact that academic and social scientific journals and periodicals in Lebanon are few, and most of them do not have a reading committee to select and evaluate materials to be published. The same situation mainly prevails in the publishing houses.[21] In fact, the latter have no specialized editors and no strategies for editing. Finally, in the French-oriented universities the publishing policy is less coercive. Although using the French language is the rule, publishing in English or even in Arabic, under certain conditions, is also accepted.[22]

The main consequence of this language division consists in a mutual ignorance, especially about subjects published in Arabic in the case of Lebanese researchers who have adopted English as their publishing language.[23] Thus internationalization in this case seems to have deepened the original cleavage between the two Lebanese academic worlds.

Relations with International Institutions

As a hallmark of the process of internationalization of social sciences, relationships with international and regional scientific associations are crucial to determine the extent to which Lebanese social scientists are being »internationalized«. Along with publishing, being a member of one or more of social sciences' international associations could be a use-

---

20 One of Lebanon's best-known sociologists, in an interview during our field research, had this comment on the AUB's attitude : »Professors of the AUB do research and at the same time they have their eyes on the promotion issue. Their research is carried out right here [in Lebanon] but they are published there [in the United States]. They are researchers who have ignored their society [...] and we know nothing about what they are publishing« (Kabbanji 2009).
21 For a broader view on publishing policies, marketing, and problems in the specific case of the Lebanese capital Beirut, see Mermier (2007).
22 The main condition consists in the kind of publishing support. The latter should respond to academic criteria.
23 It is interesting to note that most social scientists from the AUB and LAU, who publish in English, in general ignore the material of their colleagues published in Arabic.

ful indicator of the level of global cooperation—scientific networks, professional ties—and involvement in the field of practice.

In general, Lebanese social scientists are not so keen to be involved in international professional and scientific organizations and associations. Except for some who work for major private universities, the bulk of Lebanese social scientists have no regular ties to such kind of organizations. Their major host associations are home-based ones. Once more the language barrier seems to be a major handicap that hinders them from being associated with international associations. Language is either a facilitator—by being published and read in English or French—or an impediment—by being published and read in Arabic—to the opening up to international connections.

In addition, the university for which a researcher works plays, in this case, a major role. By its rules and regulations it may encourage or inhibit such international links. Major universities, public and private, are the most important channels of institutionalized internationalization. They promote close ties with other international universities, according to lingual and cultural affinities. American-oriented universities as well as French-oriented ones have, as previously noted, a long history of internationalized attachments to other American and French universities. Lebanese as well as foreign scholars travel to their each others' countries, and programs of mutual cooperation are set up.

The public Lebanese University (LU) also has its own network with international partners. They are mainly French-based universities that seek to preserve and enhance the important role played by French language in teaching and researching in Lebanon. One of the best-known programs generated by such strategy is called CEDRE.[24] It allocates grants to researchers from both countries who are working together. Also agreements with French partners allow faculty and students from the LU to be hosted and trained in France.

Off-campus international institutions such like Ford Foundation, Fulbright program, and similar European initiatives play an important role in the internationalization of the Lebanese social science community. The major tools used by these institutions are hosting grants and other privileges for Lebanese visitor scholars.

---

24 This program was established 12 years ago between France and Lebanon to support partnership in matters of mutual scientific research. Thus, 4 million Euros have been spent between 1996 and 2006 in funding 126 research projects and exchange activities in the domain of higher education.

Conclusion

Higher education and research activity in Lebanon have already become heavily internationalized. Much evidence has been presented in this chapter to affirm this main result. However, as we have seen, internationalization in this particular case is unequal in its effects and significance. Although all universities are more or less affected by the same process, some of them are more involved in trying to fully adapt to its logic and functioning. The AUB and LAU, among others, are well-advanced in internationalization. In their case, it seems that internationalization is a matter of cultural affinity—hegemony of American academic traditions and of the English language—in addition to their endeavor to adapt to globalized market conditions. The new wave of »technical universities« seem to be no more than an extension of the market needs. Therefore, they are »internationalized« by obligation, not by choice. French-based universities, and especially the Lebanese State University (LU), are struggling to create a genuine path in their international integration. As for the LU, evidence on its hesitant approach to internationalization are given by its shaky experience in applying the international higher educational system's norms. In trying to break away from its old teaching system, the LU in the beginning of the third millennium has decided to adopt the system called »License (Bachelor Degree), Master, and Doctorate« (LMD).[25] Resistance to this decision from faculty in social sciences and humanities in particular as well as from internal institutional barriers have resulted in an odd situation: faculties of sciences and applied sciences have already started functioning according to the new system, while faculties of social sciences and humanities are still stuck with the old system.

The main significance that could be inferred from such a situation is obvious. The internationalization process is not as homogenous as it could be. Social and cultural restrictions have deprived some within the scientific community of institutional, logistical, and linguistic support. Add to the fact that internationalization is seen as a major challenge by some national institutions. When social sciences have no solid base and full recognition from national institutions, it is therefore hard for social scientists who are affiliated to LU in particular to see themselves as possessing the main condition of a fully constituted scientific community. Consequently, they feel that internationalization of their own fields

---

25 This system is the French European version of the North American one. L stands for license (bachelor degree), M for master's degree, and D for doctorate (Ph.D. degree).

would represent a major challenge they are not »yet« able to face. But this is only one side of the whole process. The other side is that internationalization is embedded into another overwhelming kind of internationalization processes. If globalization now provides the whole world with the best material and communicational conditions for universalizing norms and criteria, the international division of scientific labor constitutes the hub for a very subtle discrimination among science institutions as well as among social scientists worldwide. Changing the rules of such an uneven division is a key for fostering better cooperation among scientists and better institutional integration on a world scale.

References

Bashshur, M. (1988): The Role of Education: A Mirror of a Fractured National Image. In: H. Barakat (Ed.), Toward a Viable Lebanon (pp. 42-67). London & Sydney: Croom Helm & Center for Contemporary Arab Studies, Washington.

Bashshur, M. (1997): Higher Education in a Historical Perspective. In: E.-A. Adnan (Ed.), Higher Education in Lebanon (pp. 15-93). Beirut: LAES. [In Arabic]

Chadwick, F.A./Lyons, G.M. (1974): Social Science as a Transnational System. International Social Science Journal, 26 (1), 137-49.

Gaillard, J. (with J. Kabbanji, J. Bechara & M. Assaf) (2007): LEBANON: Evaluation of Scientific, Technology and Innovation Capabilities in Lebanon. Availabe at: http://www.estime.ird.fr/IMG/pdf/JGLebanonFinal28Sept_ar7.pdf [Date of last access: 10.12.2008].

Kabbanji, J.E. (2009): Rechercher au Liban. Communauté scientifique, chercheurs et innovation en sciences sociales. (État des lieux). ISS-Lebanese Univ. Press

Kabbanji, J. (2005): How to Read the Arabic Sociological Production. In: J. Kabbanji (Ed.), Knowledge and the Process of Social Change in Arab Countries: A Sociological Approach. Beirut: Arab Sociological Association & Friedrich Ebert Stiftung.

Kasparian, C. (2008): L'émigration des jeunes libanais et leurs projets d'avenir, Premiers résultats, Conférence de presse, Université Saint-Joseph. Beirut: OURSE/USJ.

Khalaf, S. (2001): Civil and Uncivil Violence in Lebanon. A History of the Internationalization of Communal Conflict. New York: Columbia University Press.

King, D.E./Scheid, K. (2006): Anthropology in Beirut. Anthropology News 47, 44.

Hijazi, M.I. et al. (1986): Towards an Arabic Sociology: Sociology and Current Arabic Problems. Beirut: Center for Arab Unity Studies [in Arabic].

Mermier, F. (2007): Beirut: Public Sphere of the Arab World? In: B. Drieskens/F. Mermier/H. Wimmen (Eds.), Cities of the South, Citizenship and Exclusion in the 21st Century (pp. 280-304). London: Saqi.

Ministry of Education & Higher Education in Lebanon, Directorate General of Higher Education (DGHE). http://www.higher-edu.gov.lb/ English/default.htm.

Oommen, T. K. (1991): Internationalization of Sociology. A View from Developing Countries. Current Sociology, 39 (1), 67-84.

Ratha, D./Mohapatra, S./Vijazalakshmi, K.M./Xu, Z. (2007): Remittance Trends, Washington: World Bank, In: http://siteresources.worldbank.org/EXTDECPROSPECTS/Resources/476882-11571335 80628/BriefingNote3.pdf [Date of last access: 10.03.2009].

Saint-Pierre, C. (1980): Internationalisation de la sociologie ou résurgence des sociologies nationales? Sociologie et Société 12 (2), 7-20.

Smelser, N.S. (1991): Internationalization of Social Science Knowledge. American Behavioral Scientist 35 (1), 65-91.

Smelser, N.S. (2003): On Comparative Analysis, Interdisciplinarity and Internationalization in Sociology. International Sociology, 18 (4), 643-657.

Appendix

Table 1: Lebanon's 27 universities by chronological order of creation

University	Year of foundation	Affiliation at foundation	Total Faculties
American University of Beirut (AUB)	1866	American Protestant Missionary	7
Saint Joseph University (USJ)	1875	Jesuit Congregation & University of Lyon, France.	12
La Sagesse University	1875*	Lebanese Maronite congregation	13
Lebanese American University (LAU)	1924	American Presbyterian Missionaries	4
Saint Esprit de Kaslik University (USEK)	1949	Lebanese Maronite Patriarchate	13
Lebanese University (LU)	1953	Lebanese state	15
Haigazian University	1955	Union of the Evangelical Churches in the Near East and the Armenian Missionary Association of America	3
Beirut Arab University (BAU)	1960	Alexandria University in Egypt and ›el-br & el-Hissan‹ Sunnite Assoc.	9
Notre Dame de Louaizeh University NDU)	1986	Maronite order of the Virgin Mary	6
Balamand University (BU)	1988	Greek Orthodox Church	6
Al-Manar University (MUT)	1990	Founded by a former prime minister.	3
Islamic University of Beirut	1996	Islamic Sunnite High Council (Dar el-Ifta'a)	1
Islamic University of Lebanon	1996	Islamic Shiite High Council	5
Antonine University (UPA)	1996	Antonine Maronite Congregation.	4
Global University	1999	Al-Azhar University in Egypt	3
Al-Jinan University	1999	Islamic Brotherhood	4
Makassed University	2000	Islamic Sunnite Association of Education	3
Arab Open University	2000	Saudi Arabia	–
Middle East University	2001	Protestant Mission	3

Lebanese International University (LIU)	2001	Former minister of education	6
Hariri Canadian University – HCU	2006	Former Lebanese prime minister & Canadian Institution	3
American University of Science & Technology – AUST	2007	Arrangements with American and African—mainly Nigerian—Institutions	3
Arts, Sciences & Technology University in Lebanon – AUL	2007	Undetermined	3
Modern University for Business & Sciences – MUBS	2007	Undetermined	3
Lebanese Canadian University – LCU	2007	University of Quebec at Montreal (UQAM), Canada	3
Lebanese German University – LGU	2007	Arrangements with a German Institution	3
Lebanese-French University of Technology & Applied Sciences – ULFTSA	2007	Undetermined	3

* closed in 1913 and reopened in 1961

The Internationalization of South African Social Science

JOHANN MOUTON

Introduction

The policies of the apartheid government between 1948 and 1994 have had many deleterious and tragic consequences for South Africans. Intellectually and academically the apartheid policies of the nationalist government resulted in the gradual and widespread isolation of South African science and scientists over this period. This isolation took many forms. Most South African scientists, and perhaps even more so the social scientists, in one way or the other were affected by these policies. This included lack of scientific contact with fellow scholars overseas, bans from attending scientific conferences, rejection of scientific publications by international publishers, and a general lack of international scientific collaboration.

The demise of apartheid and advent of a democratic political dispensation in 1994 would also have its impact on science and scientific institutions in the country and—as one might expect—lead to the increasing liberalization and opening up of science. These trends were undoubtedly reinforced and further accelerated by worldwide globalization dynamics as well as the internationalization of academic institutions and subsequent increased mobility of scientists and students across the globe. This chapter presents a broad overview and exposition of these trends as well as proposing some explanations for these.

In order to understand the context and position of the social sciences during the apartheid and more recent post-apartheid periods our discussion commences with a very brief history of the establishment and institutionalization of social science in South Africa during the 1920's and 1930's. More specifically we show how these early years of the formation of the social sciences in the country were embedded in a South Africanist ideology that emerged in the wake of the Anglo-Boer wars of the late nineteenth and early twentieth century. The new ideology of South Africanism, driven by its biggest exponent—Jan Frederick Smuts—provided a powerful rationale for locating South African science within a new world order and in fact presents a peculiar understanding of the internationalization of science in the period between the two world wars.

The second section of the chapter will then discuss key features of the social sciences during the apartheid years (1948–1994) which witnessed both a growing international but also domestic insular and isolationist science system. We argue that the isolationism of South African science over this period was both geographical (isolated from international trends and events) as well as ideological (increasing internal insularities between Afrikaans, English and African scholars and institutions within the country).

The third main section of the chapter is devoted to a presentation and preliminary discussion of a series of quantitative and bibliometric analyses of research output data. This section presents data on co-authorship patterns as well as trends in the international visibility of South African science as measured by papers in the ISI Web of Science.

In the concluding section I present some general and more qualitative observations on internationalization and its current and potential future consequences for the social sciences in South Africa. I argue that recent and continuing trends towards the internationalization of South African social science must be understood within the context of prevalent policy and intellectual contexts. In this respect I make reference to South Africa's attempts to re-establish itself as an African country and nation and the effects of these on regional and continent-wide initiatives such as Nepad, as well domestic science and research policies and most notably a research funding policy that continues to privilege local and domestic journal publication and support.

South Africanism and Internationalism in the Preapartheid Years

It is widely accepted that South African social science (especially in such disciplines as sociology, anthropology, social work, education, and psychology) emerged as academic disciplines during the 1920s and 1930s when these disciplines found institutional homes at South African universities. The country's first sociology course was taught at the University of South Africa (UNISA) in 1918. The first professor of sociology, appointed at Stellenbosch University in 1932, was Hendrik Verwoerd, renowned as the architect of apartheid. But it is also worth reminding ourselves that the earliest universities in the country (UNISA, Cape Town, Stellenbosch, Pretoria, Natal and Rhodes) were initially mainly devoted to teaching and the reproduction of knowledge. Original and organized academic research was not the first priority: the education and training of teachers, social workers, lawyers, and other highly skilled professionals was seen to be the primary responsibility of these institutions.

This was equally true of the natural sciences where research did not emerge within the universities but mainly in government-based departments and laboratories, which were driven by the mining engineering, agricultural, veterinary, and health demands of the newly formed South African union in 1910. The first organized scientific endeavors in South Africa came about as a direct result of increasing industrialization and subsequent urbanization in the wake of the discovery of gold and diamonds between 1867 and 1875. The pastoral era was over. Highly concentrated populations agglomerated in search of these precious materials. In order to cope with such masses, rail and road communications had to be developed rapidly, enclosed or isolated farms had to be opened up, mass food production had to be ensured, and unprecedented shortages had to be coped with (for instance a lack of timber for pit props). This situation necessitated industrial as well as scientific capacities which had hitherto been nonexistent.

The mining enterprises found that they needed engineers, geologists, and later on geophysicists, chemists, and even doctors of occupational medicine or parasitologists. The colony could not supply such professionals, so qualified people had to be brought over from Europe. Government, confronted by a series of recurring disasters (plant diseases, animal parasite attacks, linked to the transformation of agriculture for mass production or the opening up of frontiers and increased circulation of people), began to expect science to come up with solutions.

A good example of this is veterinary research. There was certainly no shortage of diseases among livestock, which were recorded a long time before. Some of these are legendary (1719: massive mortality in horses; 1780: all herds were hit; 1854: loss of half of the horse population; 1882–86: anthrax decimated both domestic and wild animals and was transmitted to humans through eating meat). At the same time, the

Transvaal called on its own experts to combat rinderpest. This province put its faith in the Pasteur Institute, whose delegated scientists (J. Bordet and T. Danysz in 1897) were to help develop a serum. These missions by expert scientists forged the durable links necessary for cooperation. They strengthened the position of the local scientists who recommended such interventions by outside specialists. A good example of one such expert was Arnold Theiler, a young Swiss immigrant, who would eventually become the founding director of the world famous Onderstepoort Institute for Veterinary Science established in 1908.

The establishment of the South African Institute of Medical Research (SAIMR) in 1912 is another example in which the establishment of a scientific institution had its root in the material realities of the emerging industrial state. At the turn of the twentieth century the Chamber of Mines was worried about the terribly high incidence of illness among the many workers who flocked to their sites. In the face of neglect on the part of government agencies, the mining houses pressed the authorities to enter into an agreement which led to the establishment of a Research Institute (largely at their expense and directed by them). This institute was to use the resources of fundamental biology to elucidate the causes of infections or occupational illnesses, which were then poorly known and disrupted production.

The first major and organized social science study was the Carnegie-funded investigation (1929–32) into the plight of the poor white Afrikaner. In the wake of the First World War and due also to increasing employment of cheap black labor on the mines, the position of many white Afrikaners deteriorated rapidly. This was further aggravated by the world depression of the early 1930s when poverty became endemic. The Carnegie study of the poor white problem (as it would subsequently become known) is recognized to be the first major interdisciplinary, applied and policy study in the social sciences in South Africa that involved both academics and government policymakers.

It is generally recognized that the person who identified the poor white problem as a major object for social investigation was E. G. Malherbe, although the role of Loram in getting Carnegie support for the study was equally decisive (Bell 2000). Malherbe had studied at Columbia University in the 1920s and on his return became lecturer in education at the University of Cape Town. In 1927 the Carnegie Corporation of New York undertook a study tour to Africa. During this visit, Malherbe met with Fred Keppel, the president of Carnegie, and convinced him to provide the much-needed funds for a study of the poor white Afrikaner. As Saul Dubow observes, the involvement of the Carnegie Corporation in this study was not entirely unexpected as this commitment resonated with their overall philanthropic ideals as well as local experiences back home.

»The Corporation's espousal of progressivist ideas was marked by a commitment to the preservation of Anglo-Saxon Protestant institutions which, in the case of several of its influential trustees during the 1920's, easily translated into an interest in white poverty and social degradation, issues with obvious parallels in America, as were the similarities between racial segregation in South Africa and the American South.« (Dubow 2006: 225)

But the role that the Carnegie Corporation and other foundations such as Rockefeller and Ford played during this period should be understood against the background of broader international developments. I again quote from Dubow:

»The Carnegie Corporation exuded a progressive and strikingly modern developmentalist ethos that reflected the rise of the New World over the Old and presented the United States as an alternative international force to Britain. Unencumbered by accusations of imperialism, the Corporation travelled abroad with light ideological baggage; it was eagerly embraced by ambitious educationalists like Malherbe, Loram and Cook, who all sought scientific solutions to South Africa's urgent social and political problems.« (ibidem: 225)

And as Morag Bell (2000) has shown, the ideals of the Carnegie Corporation were also quite consistent with Woodrow Wilson's postwar interests and with the developmentalist ideology that was prevalent in the United States at the time:

»Woodrow Wilson's inclusionary and idealist vision of international citizenship based on a partnership of peoples not merely governments, provided a persuasive political and cultural context for the study and also a validation of its goals. It chimed effectively with a commitment within America to the »international mind«, a concept actively supported both by the Corporation and the Peace Endowment.
Based on the principle that nationalism and internationalism could be mutually reinforcing identities, it represented a particular variant of the liberal developmentalist ideology which, from the late-nineteenth century to the mid-twentieth century, couched American imperial interests in terms of altruism, evolution and world progress. The focus of the South African Study on Poverty gave it further public credibility.« (Bell 2000: 457)

During the first three decades of the twentieth century not only did major shifts in international geopolitical arrangements occur but also scientism and positivism—the belief in the progressive power of science to solve all societal and natural problems—were ascendant. Science was seen as a neutral site of universal truths that transcends all political, cultural, and geographical divides. This new epistemology (even ideology) was equally influential in the early days of the establishment of South African science and perhaps not surprisingly also the South African state.

Dubow shows convincingly how the advent of institutionalized and state-funded science in South Africa in the 1920s and 1930s occurred against the backdrop of a new political ideology—South Africanism. It was always to be expected that the unification of the South African colonies, which was brought about by the establishment of the Union of South Africa in 1910, would trigger new attempts to forge a truly South African state and nation that would overcome the colonial legacies. Political leaders such as General Botha (the first prime minister of the Union) and Field Marshall J. F. Smuts believed that science could and should be mobilized to help forge this new identity.

In the wake of the South African War of 1899–1902 and the subsequent establishment of the Union of South Africa, it was recognized that the new state needed various new institutions that would be able to build and nurture a new national culture. In this regard, as Dubow points out, scientific and technical agencies, as well as professional bodies, made a significant contribution to South Africanist ideology. »The notion of science as transcendent truth rendered it possible to cast the language of progress and universality within the imperial ›chain of civilization‹« (Dubow 2006: 6). It is therefore not surprising that one of the first institutions to be established after 1920 was the South African Association for the Advancement of Science—a body that was seen as being instrumental in establishing a »closer union« within the country.

Our brief review has revealed how these early years of organized social inquiry in the country were influenced by a variety of interrelated factors—the concern for the upliftment of the poor white Afrikaner (already to the exclusion of the poor black in the country), the role of international philanthropic agencies in advancing a new world order ultimately to be dominated by the United States as well as the growing commitment to a scientific positivism and universalism that viewed science as a progressive and unifying force worldwide. The internationalization of South African social science in this period was clearly not simply a matter of local scientists trying to establish themselves within the »commonwealth« of knowledge. But it was certainly one of the primary motivations, and the evidence provided by Dubow clearly shows that prominent social scientists of the time were intent on demonstrating the universality of the social problems they faced. In the biggest social science meeting to be held to date, Malherbe in 1934 organized an international conference to »showcase« the emerging capacity and scholarship in the country. The conference was attended by more than four thousand delegates and included such eminent scholars as John Dewey, Bronislaw Malinowski, Beatrice Ensor, and Pierre Bovet (Dubow 2006: 229).

But this was also a case where the interests of scholars and the political leadership coincided. Leaders such as Botha, Smuts, and Hofmeyr were all eager to reaffirm South Africa's status in the international league of nations, and science was seen as a credible and nondivisive vehicle to attain this. An investment in science was seen as an investment both in nation-building as well as improving South Africa's international standing. In his address to the conference Hofmeyr argued that South Africa was changing from an isolated society to a country conscious of its integration in the world family of nations. Using a metaphor that would become very popular later on, he also suggested that South African society—with all its racial and cultural diversity—should be seen as a *microcosm* in which social problems can be studied with unusual clarity (Dubow 2006: 231).

But the rise to power of the National Party in 1948 and the enforcement of the apartheid ideology would turn all of this on its head and the emerging internationalization of the prewar period would soon turn into an all-encompassing and debilitating isolationism.

The Growing Isolation of the Social Sciences During the Apartheid Years

The history of the social sciences during the apartheid years have been well documented (Dubow 2006; Jansen 1991; Rex 1981; Sharp 1981). Our aim is not to revisit this history. What is relevant to our discussion is how the rise of the apartheid ideology and state would gradually lead to the external isolation and internal insularity of South African social science.

Although the National Party came into power in 1948 South Africa's political isolation can be traced to various critical events that occurred in the late 1950's and early 1960s only—the banning of the ANC in 1960, Verwoerd's decision to leave the Commonwealth in 1961, the Sharpeville massacre in 1962 and Nelson Mandela's incarceration in the same year, and subsequent United Nations arms embargoes in 1963. The latter was soon followed by comprehensive cultural sanctions (which were only enacted for the most part in the 1970s and 1980s), which included severing academic links and contacts with South African scholars and scientists. The results are well-known: South African scientists could not for the most part attend international conferences and meetings and visits by foreign scholars to South Africa dwindled and scholarly exchanges became negligible. International scientific collaboration became impossible as increasing numbers of South African societies and professional associations were banned from being members of international bodies (such as UN Educational, Scientific and Cultural Organization [UNESCO], International Council of Science [ICSU], and many others). The South African government through its censorship laws further contributed to academic isolation by banning books by authors (mostly Marxist and neo-Marxist) that it saw as a threat to the civil order. The pariah status that was attributed to the government spilled over its citizens and also its scientists.

But the ideology of apartheid also had major negative effects on the state of South African science itself: on the one hand, it led to increasingly polarization within the so-called »white« academic community at the historically white universities; on the other hand, the creation of the historically black universities, or so-called »bush colleges«, led to huge inequalities within the higher education system with very little or no contact between white and black academics.

As to the former, the relations between Afrikaans and English-speaking, or between the more conservative and more liberal academics within universities and science councils, became extremely ideologized and polarized. A significant number of Afrikaans scientists and academics had sided with the government and supported state institutions such as the Broederbond (a secret society of government supporters who exerted major influence in all spheres of society and government) and the Akademie vir Wetenskap en Kuns (an exclusively Afrikaans Academy of the Sciences and Arts). Some Afrikaans scholars in such fields as education, anthropology, history, and sociology not only publicly supported the apartheid ideology, but provided scientific and academic justifications for it.

Most English-speaking academics on the other hand dissociated themselves from the apartheid state and engaged in varying degrees of critique, dissension, and protest. These divisions would soon spill over into a form of internal academic isolation, which was most clearly manifested in the social sciences. Afrikaans and English academics in such fields as sociology, anthropology, psychology, and education soon split along ideological lines. Liberal and progressive English-speaking scientists refused to apply for funding from the government or to collaborate with Afrikaans-speaking academics in conservative institutions. Professional societies split over issues of membership of black academics. This would—during the 1960s and 1970s—soon lead to the establishment of two professional societies for anthropology, sociology, education, and psychology respectively. In each case one society (the conservative and Afrikaans-dominated one) would not allow black academics to join and would often be closely aligned to the political leadership of the day; the other society (more liberal and critical and English-dominated) would be a society for open members and would encourage critique and dissension of the apartheid state.

This also meant that separate journals would be established in most of these fields, which coincided with these ideological divisions. So, for example, the *South African Journal for Sociology*, the *South African Journal of Ethnology,* and the *South African Journal of Psychology* were seen as the mouthpieces of the conservative state-supported and predominantly Afrikaans-speaking societies whereas alternative journals (*Transformation*, *Social Dynamics,* and *Psychology in Society*) were established in the 1970s and 1980s by liberal English-speaking and some black academics.

The creation of the historically black universities (HBU's) and their underfunding by the government led to a different kind of polarization, that is, between privileged white universities and disadvantaged black universities. The HBU's integrated with the apartheid governments policy of establishing homelands for each ethnic group. Hence the University of Zululand was established to serve the Zulu community, the University of Bophuthatswana to serve the Tswana ethnic group. They were also viewed as predominantly teaching institutions with little investment and encouragement to engage in research and scholarship—lest these endeavors bring them in confrontation with the state!

In summary then: because of international bans and boycotts, many South African scientists had little scientific contact with their international colleagues during the seventies and eighties. More seriously, however, was the lack of contact within the scientific community in South Africa. Collaboration with colleagues across political and racial divides was minimal to nonexistent, leading to an isolationist scientific culture that produced a system that was compartmentalized in the extreme.

It is worth pointing out, however, that despite the hegemony of the apartheid state, the late 1980s and early 1990s were a time during which a powerful, critical social science tradition arose within South Africa. Some scholars refer to this—with regard to sociology specifically—as the phase of public sociology. For many of today's leading South African sociologists, these were indeed the »golden years« of social science in the country. ASSA (Association of the Sociology in South Africa) conferences linked academics—who came from various disciplines, not just sociology—with radical students and organic intellectuals. Eddie Webster, professor of sociology at the University of the Witwatersrand (Wits) and one of the key figures in this movement, recalled:

»ASSA became an academic forum for a rich and vibrant sociological community in close dialogue with the new social movements struggling against apartheid [...] instead of limiting the possibilities of genuine scholarship, this [...] seems to have provided the impetus for a flowering of original sociological studies. Furthermore, the engaged nature of these studies inspired a generation of graduate students to work in these new social movements and to establish developmental NGO and alternative publications.« (Webster 2004: 30)

The inspiring memory of a public sociology that was intellectually invigorating and politically influential was one legacy of the last years of apartheid. But there was another, flip side. As a consequence of the intensity of the struggle and the impact of academic boycotts, South African sociology had been weakened by its lack of participation in international debate. So, when Immanuel Wallerstein visited South Africa in 1996, he noted »a certain parochialism and South African exceptionalism« (Webster 2004). Social scientists liked to believe that the problems of race, ethnicity, class, power relations, and so on that were pervasive to the apartheid society were somehow unique and required specific South African solutions. This cultivated a belief—ironically amongst conservative and progressive scholars alike—in the extraordinary nature of the South African case. Given more than twenty years of scientific and intellectual isolation, this was perhaps not entirely unexpected or surprising.

In a study on patterns of scientific collaboration that was undertaken in 1996 (only four years after the new political dispensation) and published in 2000, the author mapped the extent of academic isolation at that stage. This study was based on a comprehensive postal survey of South African scientists (more than four thousand completed questionnaires) that asked respondents to comment on the extent of their collaboration at that point, including collaboration across fields and institutions. In the final paper we concluded as follows:

»The results presented in this (paper) support two general conclusions: Levels of collaboration across scientific fields and institutional boundaries in South Africa are low. Inter-field collaboration averages at 8 percent for ›strong‹ and 19 percent for ›weak‹ collaboration. Inter-sectoral collaboration constitutes only 13 percent of all research activities for the total sample. These data would tend to support the more general observations made at the beginning of the paper, i.e. that academic science in South Africa is conducted within rather confined disciplinary and institutional enclaves. Even if one allows for the fact that ›inter-field collaboration‹ is a more stringent requirement than interdisciplinary collaboration, the overall averages remain low.« (Mouton 2000)

In our final conclusions we described the South African science system at that stage (mid-nineties) as »an isolationist system where many of the barriers to collaboration that developed during the apartheid years« were still in place (Mouton 2000). Against this background, it is not surprising that the new democratic government in 1996 produced a new white paper on science and technology that made reference to various mechanisms and incentives for increased collaboration: collaboration across institutional and disciplinary boundaries to address the socioeconomic challenges facing the country; regional collaboration between institutions that were formerly divided by ideology; and collaboration between historically advantaged and disadvantaged institutions in order to promote the transfer of knowledge and expertise especially to black scholars.

Have these measures been effective and has South African social science in particular broken through the isolationism of the apartheid era? We turn to these questions in the next section.

Social Science in Postapartheid South Africa: Some Quantitative Evidence

This section is devoted to the presentation and preliminary discussion of three sets of quantitative/bibliometric data. The data address three different dimensions of international collaboration and cooperation:
- Trends in output in local versus foreign journals
- Trends in foreign coauthorship
- Trends in the international visibility and impact of South African (co-)authored papers

Trends in Output in Peer-Reviewed Journals

Our first quantitative analysis focuses on absolute volume of output as measured by peer-reviewed journal articles. When bibliometric analyses of journal output are undertaken, it is commonplace to use the Web of Science (ISI-Thompson) database as the main authoritative point of reference. One of the downsides to such an approach is that it invariably underestimates publication in local journals, which are not indexed by the ISI. This is particularly true of developing countries (where local journals are often very poorly represented in international indexes) and—for our purposes—particularly negative as far as the social sciences are concerned as the latter often utilize local journals more than the natural and health sciences.

Any analysis of South African (social) science must take these factors into consideration as the country has a fairly long and established tradition of journal publication with more than 250 scientific journals recognized by the state for subsidy purposes. In addition, through a fairly unique system of state subsidization, South African universities are directly rewarded for the number of publications in journals that are »accredited« by the South African Department of Education. This system was established in 1985 as a way of incentivizing South African science amidst its growing isolation. The system basically entails that articles published by South African academics (with an address at a university) in any of the accredited journals as stipulated by the department would qualify for a subsidy to be determined each year as part of the »block subsidy« granted to each of the public higher education institutions. The original list of accredited journals was compiled by including all ISI-journals and adding South African journals on an »Accreditation List«.

Due to regular submissions of successful proposals for additions of new journals, the list of accredited South African journals grew to 210 by the end of 1997. Between 1998 and 2003, however, the Department of Education did not augment the list; its »freezing« was justified in anticipation of a planned review of the whole program. In September 2003, the Department of Education published a revised policy on SA research output—»Policy and Procedures for the Measurement of Research Output for Public Higher Education Institutions«, which came into effect on the 1 January 2005 for the 2004 research outputs. The policy stipulated that journals listed the following journal categories qualified for subsidy purposes:

- The Sciences Citation Index of the Institute of Scientific Information (ISI)
- The Social Sciences Citation Index of the ISI
- The Arts and Humanities Citation Index of the ISI
- The International Bibliography of Social Sciences (IBSS)
- The Department of Education (DoE) List of Approved South African Journals

The list of approved South African journals (excluding the ISI-listed titles) was appended to this new policy, numbered 197. A supplementarylist, containing the names of a further 23 South African journal titles, was circulated in 2004 (although one of these titles was duplicated in the ISI list). The heading of this list indicated that these 23 journals included one journal that had been removed from one of the ISI lists (SA Statistical Journal). This brought the total of South African journals titles (still excluding those on the ISI-list) to 219 journal titles. At the time, 23 South African journals were listed in one of the ISI indexes. In addition there were 14 social science journals included in the International Bibliography of the Social Sciences, of which 2 were also included in the ISI (South African Archaeological Bulletin and the South African Journal of Economics), which means that the total number of SA journals that are recognized in one way or the other as being of acceptable quality by the DoE numbers 254.

The effects of this system on scientific publishing in the country have been far reaching (Mouton et al. 2006) and even more so for the social sciences. A breakdown by broad scientific field shows that approximately 44 journals can be classified as being social science journals and a further 76 as being humanities journals. That means that although the social sciences and humanities have consistently produced about 40 percent of total national output, they have access to nearly half of the local journals. Simply stated—scholars in the social sciences and humanities in the country have an abundance of opportunities to publish in these journals. Coupled with the fact that the average acceptance rate of articles in many of these journals is more than 70 percent it is easy to see why many social scientists continue to publish in local journals rather than attempting to get accepted by internationally indexed journals which have much more rigorous acceptance rates. In fact, we have shown (Mouton et al. 2006) that in many cases these journals are published by a single university department or faculty and that large proportions of articles published in such journals (especially in fields such as law and theology) are authored by members of the same faculty. This form of »protectionist publishing« not only raises serious questions about quality and the enforcement of proper peer review practices, but also serves to maintain a local publishing system that furthers parochialism rather than international visibility (more below).

Analysis of trends over the past fifteen years reveals some interesting results. First, as table 1 below shows, the social sciences (with 19 percent) and humanities (21 percent) constitute a significant share of total scientific output for the country.

Table 1: Total South African article output by broad field, 1990–2004

Natural & agricultural sciences	34 %
Humanities	21 %
Social sciences	19 %
Health sciences	19 %
Engineering & applied technologies	7 %

When we look more closely at trends over time, these proportions have remained virtually unchanged. What is interesting though, is how these trends differ for the social sciences and humanities (compared to the other science cultures), when we control for South African (local) and ISI journals. Tables 2 and 3 respectively presents an overview of publication patterns for these two sets of journals. The breakdown by journal shows that the majority of articles in the natural sciences, health sciences, and engineering sciences are published in ISI journals, whereas the majority of articles in the social sciences and humanities are published in non-ISI journals. There has been a small shift in the social sciences toward publishing in ISI-journals over this period.

Table 2: Broad scientific field distribution of South African article output in ISI journals

	Total period N=45917	1990 – 1992 N=8470	1993 – 1995 N=9153	1996 – 1998 N=9111	1999 – 2001 N=9914	2002 – 2004 N=9269
Natural & agricultural sciences	53 %	53 %	53 %	52 %	52 %	53 %
Humanities	4 %	5 %	4 %	5 %	4 %	4 %
Social sciences	9 %	7 %	8 %	10 %	9 %	11 %
Health sciences	26 %	27 %	26 %	25 %	26 %	23 %
Engineering & applied technologies	8 %	7 %	8 %	8 %	9 %	8 %

Table 3: Broad scientific field distribution of South African article output in non-ISI journals

	Total period N=32688	1990 – 1992 N=6590	1993 – 1995 N=6733	1996 – 1998 N=6569	1999 – 2001 N=6631	2002 – 2004 N=6164
Natural & agricultural sciences	11 %	14 %	11 %	10 %	10 %	8 %
Humanities	45 %	44 %	47 %	46 %	44 %	45 %
Social sciences	31 %	30 %	31 %	31 %	29 %	32 %
Health sciences	10 %	9 %	8 %	10 %	13 %	11 %
Engineering & applied technologies	3 %	2 %	3 %	3 %	3 %	4 %

Standard responses (by social scientists and humanities scholars) when presented with these results are predictable, as they emphasize the fact that social science scholarship is typically more embedded in the local social and cultural context of a specific country and hence the predominance (more than 80 percent) of social science and humanities articles in local journals. However, as we have indicated above, even if one accepts this argument, the fact remains that the social sciences (and humanities) in South Africa continue to benefit hugely—somewhat ironically—from a state subsidy system established during the heyday of the apartheid regime. Although one can certainly make a persuasive case for the protection and even advancement of local social science scholarship in many countries (and more so perhaps in developing countries), it does also mean that much of this scholarship is not readily accessible to an international audience. The vast majority of the social science journals published in the country have very small subscription and circulation lists. Inclusion of South African social science journals in the ISI Web of Science is also quite limited (only 6 in the ISI Web of Science and 14 in the International Bibliography of the Social Science and unless a scholar has access to the Index of South African Periodicals (an Index produced by the South African National Library), it is unlikely that he or she will find it easy to identify South African social science scholarship. It is therefore one of the rather interesting and somewhat ironic consequences of the research funding policy of the South African state that although it protects local publication especially in the social sciences, it also continues to isolate much of this scholarship from the international scientific community!

Trends in Foreign Coauthorship

The results presented in the previous section do however still point to increasing scientific collaboration over the past ten years when measured in terms of coauthorship of scientific papers. A comparison of the proportion of coauthored papers (with at least on foreign author) between the early 1990s and the period between 2000 and 2004 shows a doubling and even trebling of international collaboration (table 4 below). Although coauthorship remains the exception, this is not surprising as single authored papers remain the norm for most of the social sciences and humanities (Moed 2005).

Table 4: Trends in foreign coauthorship for the social sciences between 1990 and 2004

Social Science field	Share of foreign co-authorship in the period 1990–1992	Share of foreign co-authorship in the period 2000–2004
Economics and management sciences	3.1 percent	8.7 percent
Sociology and related disciplines (such as anthropology, social work)	2.8 percent	10.0 percent
Psychology	3.0 percent	17.1 percent
Education	1.4 percent	.3 percent
Other social sciences (Geography, Political Studies, etc.)	3.0 percent	10.8 percent

Trends in International Visibility

Our final analysis focuses on the »visibility« of South African social science. This analysis is confined to South African social science papers in ISI-citation indexes, as it allows (1) for comparison with other countries, and (2) more important, for the possibility of citation analyses. Utilizing data produced by the CWTS at the University of Leiden of all South African authored papers published between 1995 and 2007 in ISI journals, we are able not only to look at overall trends in coauthorship but also at trends in visibility or impact as measured by the field-normalized citation rate.[1]

The dataset for these analyses consisted of 5,907 unique papers (with at least one South African author). The breakdown by subfield together with the overall field-normalized citation score for each field is presented in table 5 below.

Table 5: Output and international impact of social science papers by subfield: 1995–2007

	Number of Papers	CPP/FCS
Economics & management sciences	848	0.41
Education	737	0.51
Psychology	1,525	0.60
Sociology, anthropology & related studies	675	0.96
Other social sciences (including Political Studies)	2,122	0.90
	5907	

A detailed breakdown by subfield and by three categories of authorship is presented in table 6. For each subfield we list (by year) the number of papers according to whether the author(s) is/are from a single South African institution (SI), whether the authors of the paper are from two South African institutions (NC) or whether at least one of the authors of the paper is from an overseas institution (IC).

---

1 The field-normalized citation score (CPP/FCS) is represented by the mean citation rate of the fields in which an institute or—in this case—a country is active. The CWTS definition of fields is based on a classification of scientific journals into categories developed by Thomson Scientific. Although not perfect, it is at present the only comprehensive classification system that can be automated and updated consistently in our journals-based bibliometric information system. In summary, CPP/FCS indicates the impact of an institute/group's articles, compared to the world citation average in the (sub-)fields in which the institute/group is active. Self-citations are excluded.

Table 6: Trends in scientific collaboration as measured by coauthorship: 1995–2000

		95	96	97	98	99	00
Economics & management sciences	SI	30	25	28	31	26	42
	NC	6	2	9	4	5	5
	IC	5	9	12	14	14	21
Subtotal		41	36	49	49	45	68
Education	SI	24	30	26	30	35	22
	NC	0	6	4	3	6	2
	IC	8	7	8	9	15	7
Subtotal		32	43	38	42	46	31
Psychology	SI	59	61	73	63	79	68
	NC	16	16	19	21	19	14
	IC	20	16	20	29	17	28
Subtotal		95	93	112	113	115	110
Sociology, anthropology, and related studies	SI	26	17	23	25	23	18
	NC	8	7	5	2	5	1
	IC	6	12	14	17	22	17
Subtotal		40	36	42	44	50	45
Other social sciences	SI	98	97	69	67	69	82
	NC	20	19	24	20	15	13
	IC	12	14	27	24	25	34
Subtotal		130	130	120	111	109	129
Grand Total Social Sciences		338	338	361	359	385	374

Table 7: Trends in scientific collaboration as measured by coauthorship: 2001–2007

		01	02	03	04	05	06	07
Economics & management sciences	SI	36	41	32	33	49	62	70
	NC	8	10	8	4	9	13	18
	IC	17	17	25	26	17	34	31
Subtotal		61	68	65	63	75	109	119
Education	SI	30	19	29	45	57	68	61
	NC	5	5	9	18	20	15	14
	IC	11	7	17	8	17	22	18
Subtotal		46	31	55	71	94	105	93

Psychology	SI	44	52	52	33	32	93	69
	NC	20	17	17	14	13	44	25
	IC	28	42	50	45	57	72	68
Subtotal		92	111	119	92	102	209	162
Sociology, anthropology, and related studies	SI	10	29	31	26	34	32	36
	NC	1	7	5	1	6	4	12
	IC	20	24	28	29	26	39	27
Subtotal		31	60	64	56	66	75	75
Other social sciences	SI	80	97	95	118	133	143	127
	NC	19	21	25	31	39	30	22
	IC	43	44	54	54	57	84	77
Subtotal		142	162	174	203	229	257	226
Grand Total Social Sciences		372	412	480	485	566	755	675

The general trends for all of five subfields are similar: a clear increase in international coauthorship (even where this is often from a very small base in 1995). For fields like Psychology and Other Social Sciences (which include Political Studies and International Studies) the proportion of IC papers as share of overall output is consistently higher than for the other fields. In order to gain a better understanding of the overall trends, we have combined the fields in table 8 below, which presents the trends for this period for the three categories of authorship. This shows even more clearly that the share of internationally coauthored papers for all the social sciences has steadily increased since 1995 (when it constituted 15 percent of all papers) to 2007 (when it had more than doubled to 33 percent of overall output). This constitutes a significant shift in scientific collaboration behavior and is one of the strongest indicators of an evidently more internationalized social science corpus.

Table 8: Trends in scientific coauthorship; 1995–2007

	95	96	97	98	99	00	01	02	03	04	05	06	07
SI	237	230	219	216	232	232	200	238	239	255	305	398	363
NC	50	50	61	50	50	35	53	60	67	68	87	106	91
IC	51	58	81	93	103	107	119	114	174	162	174	251	221

Our final bibliometric analysis looks at the impact of these papers as measured by arguably the most robust citation measure available—the »field-normalized citation score« (CPP/FCS). This measure normalizes for differences in citation behavior across journals within the same field. It also excludes author self-citations, which makes it a very credible indicator of the international visibility of a set of papers—in this case social science papers produced by South African authors over a certain period of time. The detailed citation analyses—by subfield and by rolling three-year window for the period 1995–2007 are presented in table 9 below.

Table 9: Visibility of social science papers (as measured by field-normalized citation scores) for the period 1995–2007

Field of social science		95-98	96-99	97-00	98-01	99-02	00-03	01-04	02-05	03-06	04-07
Economics & Management sciences	SI	0.33	0.33	0.27	0.30	0.44	0.44	0.48	0.49	0.55	0.39
	NC	0.06	0.63	0.50	0.39	0.72	0.67	0.32	0.20	0.23	0.40
	IC	0.34	0.49	0.86	0.69	0.71	0.77	0.59	0.64	0.89	0.75
Education	SI	0.29	0.27	0.25	0.21	0.35	0.54	0.46	0.38	0.27	0.34
	NC	0.22	0.32	0.15	0.23	0.39	0.39	0.35	0.40	0.49	0.44
	IC	0.97	0.76	0.88	1.02	0.60	0.86	0.61	0.56	0.48	0.49
Environmental studies	SI	0.22	0.40	0.64	0.99	0.77	0.67	0.56	0.88	1.07	0.98
	NC	0.29	0.55	0.75	0.78	0.56	1.68	1.37	0.29	0.49	0.78
	IC	0.49	1.06	2.66	2.23	1.49	1.65	1.44	2.20	1.00	1.23
Psychology	SI	0.39	0.44	0.42	0.31	0.35	0.50	0.44	0.51	0.50	0.59
	NC	0.68	0.45	0.26	0.40	0.45	0.56	0.45	0.52	0.71	0.66
	IC	0.73	0.79	0.96	0.93	0.97	1.27	1.35	0.96	0.85	0.98
Sociology, anthropology, related studies	SI	0.77	0.58	0.51	0.44	0.51	0.73	0.87	0.92	0.98	1.03
	NC	0.54	0.23	0.56	0.69	0.92	0.91	0.98	0.94	1.00	1.31
	IC	1.07	1.11	1.45	1.20	1.33	1.44	1.28	1.40	1.63	1.31
Other social sciences	SI	0.65	0.63	0.81	0.74	0.69	0.62	0.74	0.87	1.01	0.99
	NC	0.52	0.60	0.73	0.97	0.90	1.55	1.31	0.97	0.80	0.82
	IC	0.69	0.90	1.05	1.27	1.37	1.29	1.25	1.18	1.22	1.43

SI = Single institution authored papers; NC = National coauthored papers (at least two authors from two different institutions within South Africa); IC = International coauthored papers (at least one author from another country than South Africa)

The most salient point to emerge from the data presented in table 9 is the clear increase in the values of CPP/FCS over the period 1995 to 2007 for most of the subfields. This is especially the case for papers which involved some international cooperation (IC). So, for example, the field-normalized citation impact of papers published in sociology and related disciplines for this category (IC) increased from 1.07 in the first period to 1.31 in the most recent reporting period. For »other social sciences« the values for the same period increased from 0.69 to 1.43. The only exception is Education, where the citation score has declined for IC papers over the same period.

It is also worth emphasizing that three subfields (environmental studies, sociology and related disciplines and »other social sciences«) recorded above average citation scores (more than 1.00) for the most recent three-years period. This is true for all internationally coauthored papers in these fields and is an indication of significant international recognition and impact for papers in these fields.

In conclusion, the quantitative data presented in this section presents strong evidence for the growing internationalization of South African social sciences over the past ten years. Although our measure of »internationalization« is restricted to publication-related indicators (output in overseas journals, extent of coauthorship, and international visibility), and therefore do not attempt to measure anything related to the content of these papers, the evidence still points to an increasingly internationalized body of scholarship. Clearly, at least as measured by these indicators, the isolationism of the apartheid era has been overcome and new collaborative relations and networks have been forged.

Concluding Observations and Speculations

The need for the transformation of South African science in the post-1994 era was driven—amongst other things—by the realization that the research interests of the majority of South African citizens had not been served by apartheid science. One of the most damning critiques of apartheid science was that it had developed one of the most sophisticated nuclear and defense R&D industries in the world, a vibrant energy research industry as well as other internationally competitive research niches, but could not provide shelter, clean water, power or basic health and social services to the majority of its population.

Hence the call—in many official documents of the new government (including the White Paper on Science and Technology of 1996)—to the scientific community to mobilize their resources in the service of the new national social and economic goals. Under the rubrics of »strategic science« and »new modes of collaborative and transdisciplinary knowledge«, scientists of all persuasions were urged to make a contribution to the reconstruction and development of the new society (Mouton 2001; 2003). These calls for a transformed science in the service of the new South African nation are not dissimilar to the sentiments expressed by the political leadership in the 1920s and 1930s when they urged science to serve the emergence of new fledgling South African state! But, of course, circumstances were not identical. The demise of apartheid coincided with other major global events such as the collapse of the Soviet system and the communist bloc and perhaps, most noteworthy, the rise of globalization fueled by the Information and communication technology (ICT) revolution and free trade arrangements.

In addition, therefore, to demands for the transformation of the science system through the inclusion of more black and female scholars and scientists in scientific endeavors and for the production of science that would serve the development and reconstruction needs of the whole society, the new South African government urged scientists in all disciplines to reach out to the international community and reestablish South African science as a significant partner in global knowledge production. But the question of course is whether this was more than mere rhetoric. Did the government in fact put in place policies and programs that would enable and support such an internationalization? We have seen in the previous section that South African social sciences has become more collaborative and internationally more visible. But can these trends be attributed to any specific government initiatives, or do they merely reflect the impact of broader globalization trends and the normalization of a postapartheid isolationism?

In our assessment, it is not obvious that any of the new policy and funding regimes that have been put in place over the past fourteen years have actively contributed to the greater internationalization as far as the social sciences are concerned. In fact, one could argue that some of the existing policies work contrary to such trends!

I conclude this paper with a brief review of some of the major policy initiatives and national interventions that the new South African state have put in place and their possible impact on the internationalization of the social sciences.

Reestablishing international contacts at the government level through signing new bilateral and multilateral agreements and cooperations have received much attention and effort. A review of these agreements will show that the Department of Science and Technology has been successful in this domain (estimate of more than a hundred new agreements entered into since 1994). However, a closer inspection of many of these agreements will also show that the natural and health sciences are almost always emphasized to the exclusion of the social sciences. Collaborations in fields such as infectious diseases, space science, and the environment are prioritized with little mention of social and economic agendas.

Reestablishing South Africa's links with international bodies such as the EU, UNESCO, ICSU, and especially the huge effort to become a significant partner in African science through the New Partnership for Africa's Development (NEPAD) initiative has been another major policy imperative. The latter received great impetus and government support in the late 1990s and early twenty-first century as it also coincided with President Mbeki's new ideology of an African Renaissance. But again, if one looks, for example, at the priorities identified in the NEPAD Program of Action (www.nepad.org), the social sciences and humanities do not feature in any of the twelve main thematic areas! In fact, NEPAD's approach to the social sciences is not unlike that which is found in many documents of the South African Department of Science and Technology where the social sciences are often mentioned as an afterthought or adjunct to other policy priorities or at best as an instrument to serve broader scientific and technological priorities. The social sciences are most often understood to be valuable only to the extent that it serves to sheds light on scientific and technical issues (for example, as in the social consequences of innovation, or the ethical implications of genetic engineering). There is very little evidence in any official documentation that the social sciences and humanities are appreciated for their intrinsic scientific or cultural value and essential contribution to knowledge production. Given the continuing social, economic, and cultural challenges that the country face, this is not only surprising but quite incomprehensible!

At the level of research funding and support, two national policies have been most instrumental in shaping the direction of science and especially social science scholarship over the past decade. These are: (1) the policy on funding journal article output in accredited journals, which has been continued pretty much unchanged since its inception in 1986; and (2) a policy and program on rating individual scientists by the National Research Foundation (NRF) as a means of recognizing and rewarding international excellence. What have the impact of these policies been on strengthening South Africa's international position? The evidence does not suggest a clear positive impact—even the contrary.

We have already discussed above the consequences of the journal subsidy policy and shown that it has served to support a large number of local social science and humanities journals. The current system is particularly advantageous to humanities scholars as numerous journals in many disciplines are financially supported through the funding system. So, for example, we find that out of the total number of 255 local journals supported under the system, 25 are in the field of theology and 19 in the field of law. The situation as far as the social sciences are concerned, is less »generous« but the system still supports 9 education journals, 4 psychology journals and 3 sociology journals. The way in which the funding formula works also means that an article published in one of these local journals earn exactly the same amount of subsidy for a particular university than an article in an ISI- or IBSS-indexed journal. There is hence no monetary incentive for the average social science scholar to publish in the best overseas journals. Whether one publishes in a high-impact journal such as the *American Sociological Review*, which has a huge membership network, or in a local South African social science journal that might at best reach a few hundred scholars makes no difference to the way in which subsidies are allocated. But there is another more interesting disincentive to coauthor. Because the subsidy formula is based on the institutional affiliation of the author, a single-authored article earns the full subsidy amount (approx. $8 000 in 2008) for the author's university. If one coauthors with a non-South African author, the article would only earn half of that amount for the institution concerned! The evidence produced in the previous section shows that international coauthorship in most social science fields has in fact increased over the past ten years. However, it is fair to say, that this has occurred despite and contrary to the existing research funding policies of the South African government. One could only speculate what the impact of a system would have been that would actively encourage and reward international scientific collaboration in high-impact journals.

The other research program that has had some impact on the publication and collaboration behavior of local scholars and scientists is the NRF rating system. This system, which was established in 1985 and recently revised, involves peer-based rating of individual scientists in the country. On the basis of these ratings, the individual is subsequently »categorized« as an international leader in his or her field (A category), a leader within the national scientific context (B category) or as simply a productive and solid scholar (C category). On the basis of these ratings, the individual scientist then qualifies for an automatic research grant (that is commensurate with the rating) for a period of three years.

Many criticisms have been leveled against this system (Auf der Heyde/Mouton 2007) and whether such a fairly »elitist« reward system is in fact the most appropriate research reward and support system for South Africa at this stage in its history. I would argue that the major imperatives and challenges that the South African science system faces (such as broadening the knowledge base because of continuing gender and race disparities, the ageing of the white and male-dominated research workforce, the need for science that addresses local developmental needs) are not in fact addressed at all by this rating system. It is also not obvious that the system, despite its best intentions, has had any impact on the overall production and growth in knowledge production in the country. Has it made any contribution to the internationalization of the sciences? It is certainly the case that some of the criteria that are used in the rating of individual scientists would seem to encourage international collaboration and networking. For social science and humanities scholars, publication of an international monograph or papers in the top journals in their fields would usually generate a higher rating. In fact, based on anecdotal and personal experience, and in an ironic comment on the journal subsidy system discussed above, it is clear that the review panels do not actually place a high premium on articles published in many local journals. It is known that many scholars who submitted themselves for ratings received quite low ratings (mostly in the C categories) because the majority of their publications have appeared in local accredited journals.

But even though the rating system recognizes the value of good quality international publications, is still does not actively encourage other forms of internationalization—establishment of international networks of excellence, capacity building initiatives that involve foreign experts, or collaborative and interdisciplinary ventures to address local issues. The problem might simply be one of »measurement« since these achievements are not easily quantified or rated. But it might also be that—as in the case of the research funding system—no specific thought has been given by policymakers about how to encourage and strengthen the internationalization of science in the country.

In summary, out paper has provided convincing evidence that South African social science is on an internationalization trajectory. We have confined our supporting arguments to the available quantitative and bibliometric evidence. More qualitative indicators would, we would argue, also support such an interpretation. The flow of international scholars to the country and exchanges of scholars and students in the social sciences have noticeably increased over the past ten years—more cynical commentators might point out that some of this is due to the »curiosity value« of studying an African society that has made a peaceful transition to a democratic dispensation. International conferences and meetings (most notably the International Sociological Association conference held in Durban in 2006) have proliferated. South African scholars are again taking their rightful places in international bodies; in fact, many of these bodies (such as UNESCO, the Organization for Economic Cooperation and Development [OECD], ICSU) have invested in research and scholarship that have opened up new opportunities for scholars from the country.

But we have also argued that all of these trends have occurred mostly despite the best intentions and policies of the South African government. In fact, as our analysis of the national journal subsidy system has shown, some of these policies actually work counter to more international collaboration and cooperation. It is clearly necessary for the policymakers at the national level to reconsider ways and means of encouraging even greater internationalization and foreign collaboration. The fact remains that there are still huge social, cultural, and political challenges that the fledgling democracy faces. These are not exclusive or necessarily unique to South Africa: issues of multiculturalism, xenophobia, religious fundamentalism and tolerance, social cohesion, inclusive education, and many more are common to many countries in the North and the South. We do not need a new universalist epistemology that advocates single (usually Western or American) solutions to these problems, but we certainly do need the best minds of the world to come together and address these. The local, particular, and culturally specific parts of any country's social and cultural problems and challenges still need to be subjected to the most rigorous and theoretically robust scholarship that the international community can mobilize. This remains the challenge for the social science community in South Africa as well. It is perhaps appropriate to close with comments by two prominent South African sociologists: Arie Sitas and Eddie Webster:

»Globalization has opened opportunities for cross-national links between sociological communities with nodes of scholars linked to each other in cyberspace. But there is a grave danger in the global age of a kind of »pseudo-universalism«. It is only through understanding of our different histories that we can arrive at an understanding of the many voices in our discipline. We need to remind ourselves that part-breaking cultural creativity in world history has often come, not from the centre, but from the periphery of cultural worlds. This is the challenge, Sitas argues, facing the South African social sciences: to find a critical space and a voice that is at once particular, unique, and at the same time universal.« (Webster 2004: 40)

References

Ally, S./Mooney, K./Stewart, P. (2003): The State-Sponsored and Centralised Institutionalisation of an Academic Discipline: Sociology in South Africa, 1920–1970. Society in Transition, 34 (1).

Auf der Heyde, T.F. /Mouton, J. (2007): Review of the NRF Rating System: Synthesis Report. In: http://evaluation.nrf.ac.za/ Content/Documents/synthesis_report.pdf

Bell, M. (2000): American Philanthropy, the Carnegie Corporation, and Poverty in South Africa. Journal of Southern African Studies, 26 (3), 481-504.

Dubow, S. (2006): A Commonwealth of Knowledge, Cape Town: Double Storey.

Jansen, J.D. (Ed.) (1991): Knowledge and Power in South Africa, Johannesburg: Skotaville.

Moed, H.F. (2005): Citation Analysis in Research Evaluation, Dordrecht: Springer.

Mouton, J. (2000): Patterns of Research Collaboration in Academic Science in South Africa. SA Journal of Science, 96 (9/10), 458-462.

Mouton, J. (2001): Between Adversaries and Allies: The Call for Strategic Science in Post-apartheid South Africa. Society in Transition 32 (2), 155-172.

Mouton, J. (2003): South African Science in Transition. Science, technology and society, 8 (2), 235-260.

Mouton, J./Tijssen, R.J.W./van Leeuwen, T.N./Boshoff, N. (2006): How Relevant Are Local Scholarly Journals in Global Science? A Case Study of South Africa. Research Evaluation 15 (3), 163-175.

Rex, J. (Ed.) (1981): Apartheid and Social Research, Paris: Unesco Press.

Sharp, J. (1981): The Roots and Development of ›Volkekunde‹ in South Africa. Journal of Southern African Studies, 8 (1), 16-36.

Uys, T. (2005): Presidential Address Tradition, Ambition and Imagination: Challenges and Choices for Post-apartheid Sociology. Society in Transition, 36(1), 113-120.

Webster, E. (2004): Sociology in South Africa: Its Past, Present and Future. Society in Transition 35 (1), 27-41.

Internationalization of Social Sciences and Humanities in Turkey

SENCER AYATA, AYKAN ERDEMIR

Introduction

Development and internationalization of sciences in Turkey has lately drawn increasing scholarly attention (Altınok et al. 2006; Glänzel 2008; Yurtsever and Gülgöz 1999).[1] Most scholars, however, have chosen to concentrate on the quantitative aspects of change. As Glänzel notes, »Recent literature on the rise of Turkish science embraces more than 15 papers alone in the journal Scientometrics,« a journal focusing on quantitative features and characteristics of science (2008:10). One of the leading reasons for the burgeoning quantitative interest in the Turkish case is the country's remarkable boom in international academic publications in the last few decades. As Önder et al. state, the »total number of articles published in journals indexed in Science Citation Index (SCI), Social Science Citation Index (SSCI) and Arts and Humanities Citation Index (AHCI) have soared from 148 in 1973 to 13,830 in 2005« (2008:544). For the period of 1991–2007, the average annual growth rate of the share of Turkish publications in the world total was 14.4 per-

---

1 Earlier versions of this paper were presented in GlobalSSH Working Conference on »The Internationalization of Social Sciences« in Beijing, China (14–15 November 2007) and the ESSHRA Working Conference on »Internationalization of Social Sciences and Humanities: How about the Developing Countries?« in Middle East Technical University, Ankara, Turkey (9–10 May 2008). A summary of the conference presentations was reproduced in Kuhn and Okamoto (2008:77-80). The authors would like to thank Esra Can for her research assistance.

cent, which moved Turkey from thirty-eighth to nineteenth in the world's publication output rank (Glänzel 2008:11–12).

In attempting to make sense of the apparent success of Turkish scientific performance as it is reflected in publication output, scholars have turned to economic development and the accompanying increase in R&D spending. Glänzel, for example, points out that the increase in Turkey's scientific publications »mirrors the growth of R&D expenditure on GDP which rose from 0.32 percent in 1992 to 0.67 percent in 2004, but does not yet reach the EU standard of about 1.9 percent« (2008:11–12). İnönü meanwhile explains this increase by stating that for developing countries like Turkey, »scientific research is something to be carried out as a sign of development, not as one of its causes«, and therefore, »the effort which is devoted to this area is approximately proportional to national income« (2003:145). Uzun, however, emphasizes that the »Turkish R&D system is performing better than expected, despite its rather disadvantaged position as reflected by the R&D input indicators« (2006:556).

Another aspect of the development and internationalization of sciences in Turkey that has drawn attention is »how this development relates to co-operation with and integration into the European science system« (Glänzel 2008:10). Glänzel, for example, argues that »the accession and integration process is accompanied by increasing co-publication activity with the European Union and, to a certain extent, by convergence of the corresponding national science systems« (2008:10–11). This perspective is yet another example of the wide range of studies that examine the process of Europeanization in different sectors in Turkey by assessing the success of the country's convergence to European standards. This, however, is a limited perspective since Europeanization—which is indeed the strongest source of internationalization in contemporary Turkey—works alongside increasing cooperation with other areas of the world such as North America, the Middle East, and Central Asia.

Overall, the study of the development, Europeanization, and internationalization of Turkish sciences, particularly from a quantitative perspective, is significant. Such a study, however, should not be limited to a quantitative analysis of indicators and how they measure with respect to European and global benchmarks. As Kuhn and Okamoto (2008) show, the Turkish case offers a rich array of issues to explore that goes far beyond an accession country's incorporation into and convergence with the European Research Area. It is, therefore, important to note that although European Union accession and Europeanization is currently the strongest factor for internationalization in Turkey, there are also other parallel processes at work. Turkey has strengthening ties with its wider neighborhood with which it shares historical, political, cultural, and linguistic affinities, namely Central Asia, the Middle East, the Caucasus, the Balkans, and the Eastern Mediterranean. There is also interaction with the Islamic countries beyond the neighborhood and increasing opportunities for cooperation with the emerging global powers such as Brazil, Russia, India, and China. Thus this study aims to present a more holistic analysis of the recent developments in Turkish social sciences and humanities from multiple vantage points and by incorporating both quantitative and qualitative assessment. Although Europeanization, as the single most effective source of transformation in Turkey, is still the main focus of the analysis, Europeanization will be situated as part of a wider set of internationalizing influences at work within the country.

Research Questions

The recent developments in the field of social sciences and humanities in Turkey, as will be demonstrated below, do not show much divergence from the overall picture of the internationalization and development of sciences in Turkey. In an attempt to present the complexity of the internationalization of Turkish social sciences and humanities and to analyze the scholarly and policy implications for broader audiences, the following questions will be tackled in this study.
- What has been the nature of internationalization of social sciences and humanities in Turkey as one can observe from indicators?
- What has been the outcome of Europeanization of the Turkish scientific community?
- How has the Turkish political and institutional environment changed as part of Europeanization and internationalization?
- For whom and for what reasons do international research projects matter in Turkey?
- What can be achieved through internationalization in Turkey?
- What are the drawbacks of internationalization in Turkey?
- What are the scientific challenges of internationalization?

Research Methods

The term internationalization of social sciences and humanities is used in this study to denote a process of contemplating, practicing, delivering, and institutionalizing social sciences and humanities beyond the nation-state. International mobility, collaboration, exchange, publication, education, and institution-building are different faces of the phenomenon at hand. »Europeanization«, meanwhile, a process by which Turkish accession process to the European Union leads to greater incorporation of Turkish higher education and research in pan-European transnational science practices,[2] is one specific example of internationalization that is again beyond the nation state but limited to a region. In that regard, Europeanization, for the most part, is a contradictory process that simultaneously promotes and hinders internationalization of Turkish social sciences and humanities. While encouraging Turkish scholars to be further incorporated into pan-European collaboration, it concentrates the bulk of Turkish energy and resources on one region of the world often at the expense of other regions of the world. Europeanization, therefore, provides opportunities for Turkish scholars to transcend national isolation while at the same time leading to the risk of regional parochialism. It is, nevertheless, possible that Europeanization could develop synergies and lead to spillovers for Turkish social sciences and humanities to become more global in the long run. When the terms internationalization and Europeanization are used together in this study as parallel processes, their ambiguous, contradictory, and open-ended nature is assumed.

For this study, a survey of relevant literature and secondary analysis of databases were carried out. Three sets of data compilations were of particular use:
- Social Sciences Projection Study 2003–2023 (Turkish Academy of Sciences 2007).
- Turkish Scientific Publication Indicators I (1981–2006) (Demirel/Saraç/Gürses 2007).
- Perspectives for Future Research Collaborations between the EU and Turkey in the Social Sciences (Kuhn/Okamoto 2008).

Besides presenting significant datasets and/or analysis, these compilations are also indicators of the growing interest of the leading public institutions in the development and internationalization of sciences in Turkey. Moreover, in 2007, five semistructured in-depth interviews were carried out with senior scholars who have been actively involved in internationalization efforts at Middle East Technical University, Ankara. The scholars were selected from the departments of economics, international relations, political science, psychology, and sociology. They were asked to present their personal experiences of internationalization as

---

2 The authors are aware that »transnationality cannot be regarded a uniform phenomenon« and that »transnational research can be conceptualised and practiced in distinctly different ways« (Kuhn/Weidemann 2005:79). For a critical reinterpretation of transnationality in the context of European research collaboration, see Kuhn and Weidemann 2005.

well as their critical assessment of the process as observed in different academic environments.

The authors can also claim to be participant observers in the internationalization of social sciences and humanities at Middle East Technical University by nature of their work as the dean and deputy dean of the Graduate School of Social Sciences. They spend a significant portion of their administrative effort to promote internationalization and have been involved in a wide range of relevant activities such as the signing of bilateral exchange and cooperation agreements, signing of Erasmus agreements, as well as the design and launching of joint international programs and international research projects. The authors will reflect on their personal experiences where relevant.

Indicators of Internationalization and Europeanization in Turkey

In this study, three different sets of indicators are used to assess the level and nature of internationalization and Europeanization at work in Turkey:
- International Publications (indexed in SSCI and A&HCI)
- FP 6 applications
- ERASMUS mobility

Statistics used reflect the scientific activity of scholars and students based in Turkey regardless of citizenship. It does not include the output of Turkish citizens or people of Turkish-background settled outside Turkey. While the output of a Turkish scholar employed at a Germany university is not counted, the output of a German scholar employed at a Turkish university in included.

International Publications

The boom in international academic publications that one can observe for natural and applied sciences in Turkey is also the case for social sciences and humanities. Scientific articles that were published by scholars based in Turkey in journals indexed by Social Science Citation Index and Arts and Humanities Citation Index indicate a twelvefold increase over the course of the last twenty-six years. While Turkey had 21 publications per million citizens in 1990 (seventy-ninth in the world), it had 252 publications per million citizens in 2006 (forty-fourth in the world). This shows that Turkey has not only increased its total scientific output as measured by the sum total of ISI-indexed publications but also the number of publications per million citizens.

International partnerships of scholars based in Turkey also increased drastically in the same period. The number of articles written by scholars based in Turkey jointly with scholars based abroad in 1981 was as follows:
- USA (47 articles)
- UK (20 articles)
- Germany (12 articles)

By 2006, there was almost twenty-two-fold increase in joint articles with scholars based in United States. During the same period, collaboration with scholars based in Germany increased twenty-four-fold as German partners overtook British ones as the second most popular coauthors.
- USA (1,026 articles)
- Germany (289 articles)
- UK (273 articles)

The German-Turkish migration nexus, as well as the fact that Germany is Turkey's biggest import and export partner, can explain the strengthening of Turkish-German academic collaborations. As for the dominance of partnerships with the United States over the years, one needs to consider the large numbers of Turkish academics with American Ph.D.s and networks.

Although skeptics often assume that international collaboration increases at the expense of national collaboration, the case of Turkey demonstrates that collaboration at international and national levels do not function as a zero sum game. So, for Turkey, internationalization has also led to an increase in collaboration within the country. The number of social science and humanities articles produced through the collaboration of scholars based in Turkey published in Social Science Citation Index and Arts and Humanities Citation Index showed a significant rise within the last twenty-four years.

Table 1: Social science and humanities articles by authors based in Turkey collaborating at national level

1982	1 article
1998	40 articles
2003	81 articles
2006	183 articles

Another indicator of the rapid internationalization of social sciences and humanities in Turkey is the overall increase in the number of publications by scholars based in Turkey, citations to publications written by scholars based in Turkey, and their impact (calculated by dividing total number of citations by the total number of publications) in the Social Science Citation Index.

Table 2: Number of publications by scholars based in Turkey in the Social Science Citation Index

1981	27 articles	5 citations	0.19 impact
1994	96 articles	225 citations	2.34 impact
2006	897 articles	3,507 citations	3.91 impact

A similar increase can also be observed in the Arts and Humanities Citation Index:

Table 3: Number of publications by scholars based in Turkey in the Arts and Humanities Citation Index

1981	4 articles	0 citation	0 citation
1994	16 articles	12 citations	0.75 impact
2006	92 articles	78 citations	0.85 impact

Turkey's ranking in the Social Science Citation Index (in terms of the number of publications) improved significantly as its rank went up from thirty-third to twenty-third within the course of seven years.

FP6 applications

Turkey's involvement with the EU's Framework programs began with the FP5. Turkey initially held the status of a »third state« under the category of »Other European« countries that can participate in projects generally without Community funding and exceptionally with Community funding when duly justified as being essential for achieving the objectives of the project. Turkey was also under the category of »Mediterranean Partnership« where countries could participate in projects if in conformity with the interests of the Community and on a self-financing basis. In FP5, there were seventy-three contracts with at least one partner based in Turkey.

Turkey officially joined FP6 in 2002. Applications by scholars based in Turkey to calls under FP6's »Citizens and governance in a knowledge-based society« show that there is not only a growing interest in applying for FP6 funding but also increasing numbers and percentages of successful applicants. Overall, there were 133 applications 21 of which were successful (16 percent success rate).

Erasmus Mobility

Another important feature of the internationalization of social sciences and humanities in Turkey is the European Region Action Scheme for the Mobility of University Students (ERASMUS) and the Bologna processes. These include:
- ERASMUS student mobility
- ERASMUS teaching staff mobility
- European Credit Transfer System (ECTS)
- ERASMUS-MUNDUS

Although the ERASMUS program was launched in 1987, Turkey joined it only in 2004. The exchange program had a successful start with dramatic increases in both incoming and outgoing students and teaching staff (Avrupa Birliği Eğitim ve Gençlik Programları Merkezi Başkanlığı 2008).

There was 389 percent increase in outgoing students (1142 to 4438) and 442 percent increase in incoming students (299 to 1321) during the first three years. In 2005–6, the favorite destinations were Germany, Netherlands, and France.

Table 4: Erasmus student mobility

Favorite Destinations of Outgoing ERASMUS Students (2005–6):	Leading Senders of Incoming ERASMUS Students (2005–6):
Germany (691)	Germany (210)
The Netherlands (293)	Poland (103)
France (239)	Netherlands (100)

In 2006–7, the outgoing/ingoing ratio of students for Turkey was 0.30, that is, only three students arrived in Turkish institutions for every ten students leaving Turkish institutions. Turkey's outgoing/ingoing ratio standing was similar to that of Romania (0.24), Bulgaria (0.32), and Poland (0.33), Lithuania (0.39), and Latvia (0.46), countries with a similar imbalance.

A similar increase is also seen in Erasmus teaching staff mobility. During the first three years, there was 407 percent increase in outgoing teaching staff (1142 to 4438) and 306 percent increase in incoming teaching staff (299 to 1321). In 2005–6, the favorite destinations were Germany, Czech Republic, and Italy.

Table 5: Erasmus teaching staff mobility

Favorite Destinations of Outgoing Teaching Staff (2005–6):	Leading Senders of Incoming Teaching Staff (2005–6):
Germany (99)	Germany (63)
Czech Rep. (35)	Czech Rep. (28)
Italy (25)	The Netherlands (25)

In 2005–6, the outgoing/incoming ratio of teaching staff for Turkey was 0.75, a much more balanced ratio than the outgoing/incoming student ratio. Turkey's outgoing/incoming ratio standing was similar to that of Lithuania (0.68), Poland (0.74), Bulgaria (0.75), and Romania (0.78). In 2006–7, however, the outgoing/incoming ratio of teaching staff for Turkey dropped to 0.48, the lowest ratio in Europe (with the exception of Luxemburg).

Internationalization and Europeanization in the Turkish Context

Although Turkish social sciences and humanities have demonstrated an unprecedented track record of internationalization within the last few decades, both in relative and absolute terms, internationalization has been part and parcel of the development of modern sciences in the country since the late Ottoman era. While the Ottoman reforms of the nineteenth century brought European ideas, institutions, and practices, to a great extent in the French language, the early Republican leap forward of the Interbellum years involved a significant contribution of German scholars and scholarship. The following period of the Cold War brought with it a lasting American imprint and the subsequent predominance of the English language in Turkish higher education that still characterizes the contemporary situation. The fourth and final phase could be described as Turkey's accession process to the European Union, which plays a significant role in incorporating Turkish higher education and research into the European Research Area. It should also be noted that Americanization and Europeanization of Turkish social sciences and humanities quite frequently work as parallel processes reinforcing each other, and in certain cases, promote, and in others, hinder the gradual involvement of Turkish scholars and institutions in regions of the world with which Turkey shares historical, political, cultural, and linguistic affinities such as the Central Asia, the Middle East, the Caucasus, the Balkans, and the Eastern Mediterranean. One could argue that the earlier American influence on the internationalization of Turkish social sciences and humanities is one of the reasons why Turkish incorporation in the European Research Area has developed so rapidly.

The current state of Turkish higher education and sciences, stemming from and situated at the intersection of the abovementioned internationalization processes, presents a significant and quite unique case to explore. The first issue to consider in making better sense of Turkey is the remarkable expansion of higher education since the proclamation of the republic in 1923. While the Turkish Republic had only one university and 2,900 students at its outset, there are currently 94 state universities and 45 privately endowed universities with 85,000 academic staff and 2.4 million students in tertiary education. The second issue that sets Turkey apart from many other developing countries is that some of the most prestigious state and private universities have their curricula and instruction in English (in German or French in a few cases).[3] Most academics in these universities have at least one of their university degrees from universities abroad (mostly the United States), sustain significant international links and collaborative projects, and publish predominantly in English. The third peculiarity of Turkey, particularly among OECD countries, is the high percentage of international students that Turkey attracts from non-OECD countries. In 2002, for example, the stock of foreign students in Turkey was 16,300, and 89.6 percent of these students were from non-OECD countries (OECD 2004:37). This is the highest rate of non-OECD students pursuing university education in an OECD country, attesting to the strength of Turkey's historical, political, cultural, and linguistic affinities in attracting students of non-OECD countries.

Finally, Turkey is also a key example of aggressive policies of internationalization through radical state incentives and disincentives. While academics that publish internationally or carry out international research projects are rewarded financially and are promoted, those who fail to do so find it quite impossible to be promoted beyond the level of assistant professor. Overall, Turkey presents a very dynamic and vibrant portrait in internationalization of higher education and research, and is therefore

---

3   Although a small group of Turkish students attending elite universities have very strong language skills, the vast majority of university students have poor language competence. Overall, Turkish students receive the lowest average scores in TOEFL-IBT exams in continental Europe with the exception of Albania and the Republic of Kosovo (ETS 2009:10-11).

a relevant case to study to make sense of diverse aspects of internationalization in different regions of the world.

The Outcomes of the Internationalization and Europeanization of the Turkish Scientific Community

The internationalization and Europeanization of the Turkish scientific community through international publications, collaborations, projects, and student and teaching staff mobility have various consequences for Turkish higher education. The first impact is observed in the transformation of human resources. Through the development of language, managerial, methodological, and social competences, a new class of internationalized academics has emerged. Kuhn and Okamoto show that international research experience, rather than the higher education system, is the main source of the development of managerial and social competences in the Turkish case (2008:44). Over the years, this emerging class of internationalized academics has attained significant levels of experience and competence in international research, publication, and practice. Not only are they highly mobile and partake in international dialogue via Erasmus mobility, conferences, exchanges, and research, but they also tend to publish in international journals and edited volumes often at the expense of national ones. They seem to be unhindered by any language barriers and function comfortably in the lingua franca of their respective disciplines. This group of scholars comprises the new base of internationalized social sciences and humanities in Turkey.

The emergence of this new class, however, brings with itself major cleavages within the academia concerning values, institutions, and capacities. This seems to be the second major outcome of internationalization and Europeanization. Since internationalized academics are a small subset of all academics in Turkey, and since they constitute a small and exclusive class, there is an inevitable reaction on the part of those who feel excluded or left out. The resentment of the majority of Turkish academics who lack language, managerial, methodological, and social competences are manifested in different ways and contexts. There are frequent criticisms and expressions of mistrust of international projects and funding. These scholars not only lack necessary competences but often underestimate »the manifold challenges for international research with regard to these competencies« (Kuhn and Okamoto 2008:45). Moreover, since natural scientists seem to benefit more from internationalization so far, this also leads to a tension between social sciences and humanities, and natural sciences, further fragmenting the community of scholars in a divided society with already high levels of sociopolitical conflict.

The segmented nature of the social sciences and humanities, in fact, is not unique to these fields. Yurtsever and Gülgöz (1999) also found a similar pattern among chemistry professors in Turkey. They state that »a rather small portion of the studied group is responsible both for high number of publications and for higher quality« (1999:321). Their study has shown that there are three different strata of chemistry professors with distinct publication records: »Approximately 40% of the sample are consistently not publishing. There is a group of 25%, who publishes consistently. The remaining 35% publish about once every two or three years but they also contribute consistently« (1999:336). Studies on professors of social sciences and humanities can be expected to come up with a similar segmentation in terms of international publication and research track records.

The third outcome of the internationalization and Europeanization of the Turkish scientific community concerns funding opportunities and practices. Being part of the EU funding schemes, and particularly the Framework programs, has led to a drastic increase in the availability of funds for international projects, collaboration, and mobility. The recent decision of TÜBITAK to fund social sciences alongside natural sciences, as part of Turkey's Europeanization reforms, has also made available significant funds for social scientists. Although most of the funds are allocated for national research projects, there are also certain opportunities such as seed money and international mobility.

The Changing Political and Institutional Environment in Turkey

Internationalization and Europeanization of Turkey and the Turkish scientific community have led to a transformation of the political and institutional environment concerning Turkish higher education and research. Önder et al. refer to three distinct institutional forces that have influenced Turkish sciences since the 1990s: 1) Policies of semiautonomous governmental agencies such as the Higher Education Council (YÖK) and the Scientific and Technological Research Council of Turkey (TÜBİTAK); 2) Recruitment and promotion policies of newly established and privately endowed universities; 3) Changes in the recruitment and promotion policies of public universities (2008:550). It seems that the interplay of recent political developments and the transformation of academic institutional structures summarized below has been responsi-

ble for the ways in which internationalization and Europeanization were experienced in Turkey.
- The Research atmosphere is getting freer. Sensitive political and social issues that many academics used to avoid, particularly in the aftermath of the 1980 coup d'état, are now discussed widely and freely in public. Younger scholars feel that they have the autonomy and freedom to set research priorities and pursue their topics of interest. There are increased opportunities to disseminate findings and exchange opinions with wider audiences and in public without fear of political backlash.
- With the Turkish government's commitment to European Union accession, particularly epitomized by Turkey's enthusiastic embrace of Erasmus and Framework programs, there is increased research and mobility funding through the EU, TÜBITAK, and universities. It is getting relatively easy to fund micro-scale research. Junior faculty and faculty members of small universities also have reasonable opportunities to receive funding for their research programs. Although it is still quite difficult to find funding for large-scale comparative research directed by Turkish scholars, the amount of funding provided for social sciences and humanities in the last few years seem to be more than the sum total amount of funding that was available in the earlier decades.
- Lately, strategic plans of universities have become increasingly research oriented around the country. As institutions emulate best practices or feel compelled to emulate what is generally believed to be best practices, there are increasing levels of institutional isomorphism, especially at the discursive level. The values, institutions, structures, and organization of research activity, however, are still limited. There seems to be a significant lag between discursive change and actual institutional change. Desire for a wholesale convergence to a »best practice«, which is for the most part set by a handful of elite research universities, and the increasing efforts at attaining institutional isomorphism, hinders institutional diversity, tapping of comparative advantages, and development of local innovation strategies fit for specific niches.
- The Turkish higher education system has lately provided very significant incentives and disincentives for international publications and research. In a country where academics have very low salaries, there are now not only generous financial rewards for international publications and projects, but also strong barriers to promotion unless one passes foreign language thresholds and publishes internationally. While the incentives have a significant effect in encouraging scho-

lars with international competences to further increase their international research and publication output, the disincentives are particularly effective on junior faculty and/or faculty that lack international competences and force them to acquire at least one foreign language and publish internationally at the risk of not being appointed or promoted.
- EU policies and funding affect the research environment in Turkey. Turkish academics pay strong attention to European opportunities and incentives, and seem to be enthusiastic and brave in going beyond their existing networks. Junior scholars, who are socialized in an era of European funding and cooperation, rapidly develop skills and networks that were quite scarce in Turkey until recently. Taking into consideration that Turkey is a special case in the enlargement process, EU policies and funding priorities should be adapted to the needs of Turkey, recognizing that Turkish social science agendas and infrastructure have significantly differing needs and priorities in comparison to the EU.

International Research Projects: For Whom and for What Do They Matter?

For scholars based in Turkey, international organizations and international nongovernmental organizations used to be the main source of funding and providers of opportunities for international collaboration. These opportunities, for the most part, were only accessible to a small group of elite academics with required competences and networks. Beginning with the Sixth Framework (FP6 hereafter), however, Framework programs became the major source and impetus for internationalization of research efforts. Framework programs not only increased the overall amount of funding going into research in social sciences and humanities, but they also integrated a wider body of scholars, both senior and junior, into international collaborative projects.

Although there is great prestige associated with taking part or leading international projects in general, one must admit that not all scholars expect to have access to these opportunities. These projects are still a central concern for a relatively small subset of Turkish scholars. These scholars tend to have the following qualities:
- Highly qualified: Often a holder of a Ph.D. degree from a prestigious national or international university, and with prior national and/or international collaborative research experience.
- Open-minded attitude to internationalization and/or Europeanization: These scholars do not have any ideological or cultural barriers to re-

ceiving international/European support or collaborating with international/European scholars. Moreover, many of them perceive the simple fact of transnational collaboration as a value in itself.
- Language skills: They are not only fluent in foreign languages (particularly English, German, or French), but have a preference to function and publish in a multilingual environment.

International research projects matter for various different reasons:
- International networks established through international research projects bring Turkish scholars prestige and status within Turkey and abroad. There is not only increasing recognition, publicity, and rewarding of international scholarly achievements, but these achievements have become a de facto precondition of academic success, status, and fame. Academics with strong international networks, therefore, continue to build up their networks, prestige, and status, thereby further widening the cleavages in Turkish academia.
- Projects enable Turkish scholars to travel and engage in international dialogue through conferences and workshops. These often lead to international publication opportunities and strengthening of international networks. Moreover, through academic mobility, local or national benchmarks for academic success and achievement end up being replaced by European or global benchmarks as academics find opportunities to have a firsthand experience of different higher education systems and performance criteria.
- Projects help scholars' promotion within their institutions since the Turkish higher education system values international activities highly as opposed to national activities. Besides promotion, scholars also have increased influence within the country.
- International research projects contribute to the identity formation of scholars since they become markers of identity—mainly European and Western. Evaluation of academics by peers and students increasingly depend on their status resulting from their international connectedness.
- International research projects lead to the emergence of a new ethos in Turkish higher education. This new ethos is highly competitive, self-centered, and low-trust based. It leads to competition among and within universities. New kinds of performance measures that value international research and publication highly are established in this process. The stratum of scholars who find themselves left out of new opportunities and feel growing resentment towards their peers further contributes to status competition and the ensuing frustration.

What Can Be Achieved through Internationalization?

Through internationalization, Turkish scientists and higher education can achieve various results. The positive accomplishments listed in the following list were already achieved by some of the leading Turkish scholars but the ongoing process of internationalization opens these opportunities to a much wider group of academics.

- Increasing numbers of academics are integrated with the international scientific community as they find opportunities to partake in conferences, teaching mobility, collaborative research projects, and joint publications.
- Higher standards are being established in the academia. Turkish scholars feel the need to update their grasp of the international literature, methodologies, and theories since this seems to be a precondition of effective interaction and collaboration. Peer review becomes the central evaluation mechanism.
- Scholars gain positive experience in teamwork and project organization as they learn by doing. This allows a vast majority of Turkish scholars, who have no formal training in managerial and social competences, to develop necessary skills for further international collaboration.
- Scholars gain invaluable experience in collaborative and multisited project design. Many researchers, for the first time in their careers, have the opportunity and funding to be involved in large-scale comparative projects.
- A new generation of young scholars with organizational skills comes into existence. They have a chance to develop crucial competences and accumulate experiences at an early point in their career. This in turn raises the possibility that they can successfully compete with international scholars for international funding and institutional support.

What Are the Drawbacks of Internationalization?

- Although scientific content is updated, the global and EU agendas are often borrowed as they are. The agendas that are imported from the outside do not necessarily correspond to social and political priorities of Turkey. This leads to social sciences and humanities that are at times irrelevant for Turkish audiences.
- The processes that lead to the exclusion of a large body of scholars from internationalized academia leads to the widening of the gap between included and excluded. There is, nevertheless, trickle down

and spillover effects, as can be seen from METU's Faculty Development Project whereby approximately a thousand Ph.D. students are trained in an internationalized academic environment to be later employed at state universities around Turkey. These efforts, however, are not sufficient to ensure inclusion of significant numbers of social scientists in internationalization.
- Segmented internationalization leads to ideological reaction within the academia. There are cleavages even in most successful Turkish universities. There seems to be little exchange among faculty members.

What Are the Scientific Challenges of Internationalization?

The main challenges of internationalization are generally about matters of scientific organization and academic culture. Research centers tend to be the academic infrastructure for collaborative social science projects around the world. Functioning and internationally oriented centers, however, are few in number in Turkey (Erdemli 2000). The existing ones are inexperienced, financially weak, and organizationally underdeveloped. There is a dire need to build capacities, provide incubation services, and to establish centers of excellence. The problems pertaining to organizational infrastructure are deepened by the weak culture of cooperation dominant in the Turkish higher education.

Until quite recently, resources (mainly funding) available for social sciences and humanities were extremely limited in Turkey. In the second half of the 2000s, TÜBITAK, EU, and university funding increased dramatically. This increase might be expected to stimulate research centers and academic entrepreneurs. More direct EU-funding of centers, quotas, and preferential treatment for countries like Turkey could be particularly beneficial for the development of centers. Funding, however, is not the major bottleneck since there are other significant problems with the Turkish universities. They are not yet research oriented in terms of values and institutional structure. The main bottleneck seems to be the lack of affiliation system in the universities. Faculty members have to fulfill all the academic and administrative requirements of their faculties. Their involvement with centers is often something they have to carry out in addition to their regular work load. The departments on their own are very diverse and are not a suitable setting for research. Collaboration within and among departments happens on an ad hoc basis, but one cannot always expect this to happen spontaneously on its own. In this organizational and cultural environment, centers are administrative-heavy organizations, proactive in forming teams, and aggressive in searching for funds.

Overall, the scientific challenges can be summarized as follows:
- The main scientific challenge concerns transforming teaching universities into research universities particularly by strengthening research centers. The institutional re-organization of research universities is the key task.
- Autonomous dynamics of Turkish social sciences and humanities are weak and getting weaker. Research agendas and areas of academic interest are to a large extent determined by external or transnational scientific communities and their academic priorities. Turkish scholars immediately follow the set international and/or European research agendas. They have little contribution to setting research agendas.
- Most social science research is dominated by themes, subjects, and agendas of developed, postindustrial, and knowledge-based economies such as advanced consumer culture, surveillance, discipline, the hyper-real, global sectors, and media-centered culture. These do not neatly correspond to the priorities of Turkish society, economy, politics, and culture.
- The alternative to importing global research agendas is reducing Turkish social science priorities to a few spectacles, namely, Islam, ethnic tensions, identity, traditionalism, refugees, and status of women, among other topics. There seems to be an expectation for Turkish scholars to study these »authentic« issues. Study of Turkey-EU relations, in a narrow politico-historical sense, also emerge as a priority of social scientists. There is, meanwhile, a great need to study issues such as the breakdown of traditional rural society, urbanization, informal sector, family, forms of sociability, political parties, and power structures in Turkey.

Overall, one can observe that the research agenda is heavily concentrated in certain areas while crucial areas are either ignored or left out.

Concluding Remarks

Turkey's integration to the European Union and process of internationalization are unique. Turkey is not only an interesting case in Europe but it is also a relevant model for non-EU countries. Turkey's links and access to the neighborhood, particularly the non-EU neighborhood such as the Middle East, the Caucasus, Central Asia, the Balkans, and the

Eastern Mediterranean provide it with certain advantages for comparative and collaborative research beyond Europe. This, however, would require opportunities for Turkish scholars to set research agendas relevant for national and developing country contexts in collaboration with their developing country peers. If internationalization ends up, in Kuhn's words, as a de facto »colonization« of research agenda and practices of Turkish scholars by European and North American priorities, social sciences and humanities could be further estranged from the people that it aims to study and serve. The future development and success of Turkish social sciences and humanities in an increasingly globalizing academic world will be determined by whether the pitfall of colonization can be avoided and emerging opportunities can be utilized by Turkish scholars and their peers around the world.

References

Altınok, T./Mehmet, S./Önder, C. (2006): ISI Atıf Endeksleri ve Türkiye'nin 2001-2005 Yılları Arasındaki Durumunun Değerlendiril-mesi. Savunma Bilimleri Dergisi, 5 (2), 1-23.

Avrupa Birliği Eğitim ve Gençlik Programları Merkezi Başkanlığı (2008): 2005/2006-2006/2007 Erasmus Programı: Değişim İstatistikleri, Öğrenci Değişimi, Öğretim Elemanı Değişimi, Ankara.

Önder C./Mehmet S./Altınok, T./Tavukçuoğlu, C. (2008): Institutional change and scientific research: A preliminary bibliometric analysis of institutional influences on Turkey's recent social science publications. Scientometrics, 76 (3), 543-560.

Demirel, İ.H./Saraç, C./Gürses, E.A. (2007): Türkiye Bilimsel Yayın Göstergeleri (1) 1981-2006. Ankara: TÜBİTAK-ULAKBİM.

Erdemli, Ö. (2000): A Guide to Research Centers in Turkey. Turkish Studies, 1(2), 135-145.

ETS (2009): Test and Score Data Summary for TOEFL Internet-based and Paper-based Tests: January 2008–December 2008 Test Data. Princeton: Educational Testing Service. http://www.ets.org/Media/Tests/TOEFL/pdf/test_score_data_summary_2008.pdf [Date of last access: 31.08.2009]

Glänzel, W. (2008): Turkey on the Way to the European Union? On a Scientific Power Rising Next Door. ISSI Newsletter, 4 (1), 10-17.

İnönü, E. (2003): The influence of cultural factors on scientific production. Scientometrics, 56(1), 137-146.

Kuhn, M./Okamoto, K. (2008): Perspectives for Future Research Collaborations between the EU and Turkey in the Social Sciences. Towards FP7/Enlarging the SSH Research Agenda (ESSHRA) Report.

Kuhn, M./Weidemann, D. (2005): Reinterpreting Transnationality – European Transnational Socio-economic Research in Practice. In M. Kuhn/S.O. Remoe (Eds), Building the European Research Area: Socio-Economic Research in Practice (pp. 53-83). New York: Peter Lang.

OECD (2004): Trends in International Migration, OECD Publishing.

Turkish Academy of Sciences (2007): Sosyal Bilimlerde Öngörü Çalışması 2003–2023. Ankara: Türkiye Bilimler Akademisi.

Uzun, A. (2006): Science and Technology Policy in Turkey. National Strategies for Innovation and Change During the 1983–2003 Period and Beyond. Scientometrics, 66 (3), 551-559.

Yurtsever, E./Gülgöz S. (1999): The Increase in the Rate of Publications Originating from Turkey. Scientometrics, 46 (2), 321-336.

Internationalization of the Social Sciences and Humanities in Russia

IRINA SOSUNOVA, LARISSA TITARENKO, OLGA MAMONOVA

Introduction

For almost all of the twentieth century the Soviet Union was a closed country, and Soviet science was not integrated into world science. The fact is that this situation has not improved radically since 1991, when the Soviet superpower was dissolved. The Russian Federation, a successor-state to the USSR, has social sciences that are still far from the high road of world-recognized scientific achievements. The general historical cyclical attitude of Russia to the West, according to Nikolai Rozov (2006)—either copying Western models of development (including science) as a dominant one or, on the contrary, declaring a special »Russian way« and critically rejecting Western models (i.e., isolating Russian science from the West)—has been repeated regularly in the long Russian history. In case of our topic, internationalization of social sciences and humanities in Russia, we can also indicate a short cycle of Western domination in the 1990s and the beginning of a new, opposite cycle under Putin's presidency. In what follows we'll show in more detail how these cycles work in Russia in relation to science in general and to social sciences in particular.

Russian Social Sciences in Historical Context

The inclusion of Russia into the European community of social sciences and humanities began during the eighteenth century. According to the above-mentioned concept of cycles, this period was a cycle of »European domination«. By the time of reforms of Peter the Great (the beginning of eighteenth century) the need to establish an effective national system of higher education and develop the national science has become very sharp. In Western Europe this need was satisfied within the framework of a quite autonomous »university model« of scientific development. Therefore, the organization of scientific institutions and the regulation of their activity has become one of the principal functions of the government in Russia. The fundamental achievement was the foundation of the Academy of Sciences in St. Petersburg in 1724, which combined the academy, the university, the academy of arts, and gymnasia. Peter the Great, a founder of the academy, understood the significance of science and the specific functions of the scientists as the members of professional community; he saw this unification of science and education as a temporary pragmatic measure of the historical moment (Kinelev 1995).

The activity of the Academy of Sciences (from 1836 to 1917 called the Imperial St. Petersburg Academy of Sciences) in the eighteenth and nineteenth centuries was focused in the fields of mathematics and natural sciences, on the one hand, and Russian population and geography, in search for new natural resources, on the other hand (Kulyabko 1962). Nevertheless, the societal needs and the logic of scientific development were reflected in the growing attention to social sciences and humanities: since 1841 the structure of the academy included the Department of Physics and Mathematics, Department of Russian Language, and the Department of History and Philology.

For quite a long time, the universities were the leading centers in social sciences and humanities in Russia. The first Russian university was founded in 1755 in Moscow; it was organized and later named by the great Russian scientist Mikhail Lomonosov. Moscow University (MSU) was addressed to European demands. It was primarily a humanitarian university till the middle of the nineteenth century. Only then did natural sciences appear in research and education. Since then, a special attention to humanities and social science is the traditional feature at MSU.

By 1917 there were twelve universities in the Russian Empire (six of them are in the present borders of Russia), the leading role belonging to MSU and St. Petersburg University (Ivanov 1991). Up to 1917 the significant feature of Russian scientific activity was the inevitable conflict between the universities and the political regime (between the academic and university autonomies, on the one hand, and governmental status of the Academy of Sciences and universities, on the other). In general, the level of development of social sciences and humanities in Russia up to 1917 was rather high, with work by Russian scientists influencing European scientific process. In particularly, there were internationally known sociologists such as P. Lilienfeld, (1829–1903), A. Chuprov (1842–1908), Y. Novikov (1830–1912), E. De Roberty (1843–1915), and different scientific schools were founded.

During the Soviet period (1917–91), or the cycle of rejection of Western dominance, the conditions for SSH development in Russia changed totally. This period was characterized by the dominance of the Marxist-Leninist paradigm in social sciences, in its simplest dogmatic form (Yudin 1993). From 1920 to 1940, the non-Marxist social scholars were the objects of the state repressions (for example, a significant group of philosophers was expelled in the early 1920s, and many others ended up in the Gulag in the 1930s). For these reasons the traditions of Russian SSH were broken. In 1924 all independent scientific societies, such as the Sociological Society and Philosophical Society, were closed.

By 1930 the dramatic restructuring of science in the USSR was complete. The Academy of Sciences was reformed, the Highest Examination Board of candidate and doctorate dissertations of the Russian Federation was established, and many new scientific institutions and universities were open. The system of scientific habilitation was reestablished, the level of salaries in science and education increased. According to the new Soviet »academic« model of science, the Academy of Sciences played the dominant role, while the other institutions of science and education played a secondary role. This model differed greatly from the Western »university« model of science. In 1927, the Department of Humanities within the Academy of Science was open and included the disciplines of law, economy, history, philosophy, and ethics. In the 1950s and 1960s this structure was slightly reformed, so that independent sections of philosophy, law, and economics appeared.

The following aspects were the basic features of social sciences and research in the USSR: (1) specialized research institutions in each particular field of SSH within the Academy of Sciences of the USSR (Institute of Philosophy, Institute of Sociology, etc.); (2) network of scientific research centers to run applied research; (3) an appropriate nomenclature of scientific degrees (candidate of science, doctor of science). Dominance of Marxist ideology and the ruling role of the Communist Party existed up to the end of the Soviet period and was an obstacle for independent development of SSH and the open discussion of weaknesses of the system of education (in particular, lack of staff and students exchange). Therefore, knowledge of foreign achievements in the field of SSH was limited: only selected foreign books were translated into Russian, while many others were kept unknown for the majority of Soviet scholars. Accordingly, only a few scholars were allowed to represent Soviet science abroad (Soviet scholars started to participate in world congresses only in 1956). As a result, participation of Soviet scholars in international forums was limited and strictly controlled; visits abroad by Soviet scholars were extremely limited, and distribution of Western knowledge was curtailed. Actually, before 1991, ties between Soviet social scholars and their foreign colleagues were considered as »class competition« rather than scientific cooperation (Barabanov/Lebedeva 2002).

However, the Soviet period was not only the »cycle of rejection« for SSH. In spite of the difficulties, the scientific infrastructure in SSH was developed, scientific schools were established, generations of scholars were educated and employed nationally, and some significant theoretical and practical results were achieved. But real internationalization of social sciences was not in place. However, some scholars became famous worldwide—for example, philosophers Alexander Zinovyev and Ivan Frolov, anthropologist and sociologist Igor Kohn, and philologists Mikhail Bahtin and Sergey Averintsev.

The Russian Research Agenda

Current Transformation Processes

Since the beginning of the 1990s the Russian scientific community has been living under the situation of permanent radical transition. The first change received a positive approval of the majority of Russian scholars: it was an elimination of ideological pressure to SSH. According to this change, the nomenclature of scientific degrees has been changed: instead of ideologically biased disciplines such as »History of the Communist Party of the Soviet Union«, »History of Marxism and Leninism«, »Organization of the Communist Party« (in 1990 more than 40 percent of Ph.D.s in history were within these fields), new disciplines were introduced: »Ontology and theory of cognition«, »Social philosophy«, etc. (Osipov 1999:158): The contemporary field of SSH in Russia consists of the following disciplines: history, economics, philosophy, philology, law, pedagogy, psychology, sociology, politics, cultural studies, and arts.

These first radical changes helped to clear the ground for some future international comparative research because they made Russian SSH more or less similar to Western SSH (at least formally) and motivated scholars for international cooperation.

Other radical changes in science have not been completely finished, as the scholarly community has ambiguous attitudes toward them. It is not a surprise that many scholars were not satisfied with the situation: some left the field of SSH, others found extra employment to keep the soul and body together (Guseva 2006). Overall, the number of scholars decreased, and a younger generation is not motivated to enter the scientific field: they demand better salaries and work conditions. According to an official survey of MSU's graduates (Yurevich 2004), major concerns about future jobs were the following:

- low salary in science (92.5 percent of the respondents);
- shortage of up-to-date equipment and devices (47.5 percent);
- small opportunity for professional and managerial growth (41.4 percent);
- poor conditions for realization of their scientific ambitions (37.8 percent).

During the 1990s, many more scholars from the natural sciences than from social sciences left their field or emigrated: there was much less foreign demand for Russian experts in SSH after 1991 than before. Also, the average level of Russian social scholars was not enough for getting employment abroad: language skill and basic knowledge of a discipline do not fit the Western standards Barabanov/Lebedeva 2002). So, instead of learning more and contributing in the process of internalization of SSH, they simply kept their positions at home. However, the number of Ph.D. theses in SSH has increased 2.5 times during 1995–2002 (at least two to three times faster than in natural sciences), as well as the number of publications in the local journals. A few of these are published in the world peer-review journals, and only a small part of those scholars who received their Ph.D. abroad returned to Russia.

According to the existing rules, Ph.D. candidates are not required to publish abroad. Therefore, the majority of Russian scholars in SSH do not have such publications, and a few have papers published in prestigious world journals. Even among the high-ranking scholars in Russia it is not common or necessary to publish abroad (especially now—within the current cycle of rejection of foreign models and search for national ones). As a result, those Russian scholars who are famous abroad and publish their articles in the world journals are not so well-known in Russia, and vice versa. For example, according to calculations of index citations for sociologists in Russian and foreign journals, only four to five names in two lists of citations coincide (for example, the sociologist Vadim Volkov from St. Petersburg is among the most famous Russian scholars in the West while in Russia his rating is much lower) (Sokolov 2009). It was also confirmed by this calculation that the most famous social scholars in Russia are very old (aged seventy or more); although they are still active. In the field of sociology every student knows Vladimir Yadov, who was a leader of Russian sociology and a vice-president of the International Sociological Association in early 1990s. Yadov is still active in research and teaching (including international projects). As for the scholars of middle age (thirty-five to fifty), they are a tiny minority and not famous either at home or abroad (Radayev 2008b). The younger generation is even less known. In order to improve the situation and prepare the new generation of scholars who will be competitive on the international arena, Russian authorities approved a federal program for 2009–2013 to increase the financial support for S&T as well as for higher education staff. However, the current world financial crisis made problematic the implementation of this program and providing enough funds to attract the young scholars in SSH. Among the five hundred scientific projects financed annually by the state almost 10 percent belong to SSH, and the same 10 percent for high technologies in economics. This governmental decision reflects its view on the role of SSH in Russia.

International Orientations

The attitudes of scientists to the crucial issues in SSH are deeply connected with their values, interests, and goals. According to some views, the dominant orientation among Russian scholars is the provision of scientific knowledge in closed elite corporations of scientists (Yurevich 2004), which means that social science is an elitist field in contemporary Russia. Only those with an excellent level of knowledge, employed at prestigious institutions, and receiving appropriate salary can contribute in the process of internationalization of social science. Current practice confirms this fact: only scholars from some prestigious centers in Moscow and St. Petersburg can compete for scientific projects, especially on the international level. The reality can hardly become different, as there are no special government support schemes for Russian social scholars to be included in the European and world networks of social research.

The results of the international cooperation with the Western scholars under the umbrella of the EU Framework Programs are presented in table 1:

Table 1: Participation of Russian social scholars in Framework Program projects (SSH)

Framework program	FP4 - FP5	FP6	FP7 (in progress)
Number of successful projects with Russian scholars	3	11	3
Geography of success	Moscow: 2 St. Petersburg: 1	Moscow: 8 St. Petersburg: 3	Moscow: 3

The next problem is the relevance of scientific results to the public demands. As the results surveys of the Institute of Social-Political Research showed, the pessimistic evaluations of the role of science in Russian society and its perspectives dominate among scholars, while they consider the past period as quite a positive one (Global SSH 2008: 17). The experts think that the role of science has decreased in the countries of the former Soviet Union after 1991. This pessimistic view on Russian science is rather common, even if it is not the absolute truth, and some scholars are productively involved in the internationalization of social science.

Officially, the internationality in SSH is considered important in contemporary Russia: it is addressed in some government documents and in the agenda of the national research foundations. The Ministry of Science determines the major research topics for state-funded cooperation and develops some special joint research programs for cooperation with CIS scholars. The necessity of such cooperation is taken for granted. The problem is always its implementation: funds, motivation, conditions, among other things. When there is no state funds and backing for internationalization, it is a matter of a particular scholar or the institution to be involved. Within the country the funds are distributed not according to international success but to other criteria. Therefore, non-government research firms and private universities are more involved in the international programs than the state-funded centers (however, foreign scholars often prefer to cooperate with prestigious Russian centers; therefore, such institutions as the Academy of Sciences, and major universities in Moscow and St. Petersburg are among those in demand).

What are the most popular forms of inclusion of Russian social scholars in the process of internationalization? According to the result of research presented by Pipiya (2008, for the original study see Kuhn/Okamoto 2009), out of 142 respondents (social scholars) 77 percent (109 persons) were engaged in the international scientific activities. However, the majority participated in conferences (see table 2):

Table 2: Forms of international activities of Russian social scholars, in percent to number of respondents

Forms of international activity	Per cent of participants
International conferences	99
Transnational research activities	70
Publishing in foreign languages	64
Study visits in foreign countries	49
International scholarship	39
International exchange programs	38
Others	15

Russian scholars have partners from many countries, including the following preferences for transnational research (Kuhn/Okamoto2009): Germany (35 percent), USA (28 percent), France (24 percent), Switzerland, Italy, and UK (13 percent each), Ukraine (11 percent), and Poland (9 percent). As one can see, Russian scholars prefer to cooperate with scholars from the developed countries. This can be easily explained by the desire of Russian scholars to cooperate with wealthy partners, because Russia itself does not provide enough funds for research. For this reason, Russian social scholars are ready to participate in international projects regardless of the topics: this is an extra source of money and a chance to learn for those with such ambitions.

Research Topics and Themes of the Russian Academy of Sciences

The main fundamental topics and themes of research in social science and humanity are elaborated in deep collaboration between the Russian Academy of Sciences and different institutions. Since 1990, the Russian Foundation for Basic Research (RFBR) and the Russian Foundation for Humanity (RFH) have also become active participants in this field.

Below we have listed the main themes recently approved by the Presidium of the Russian Academy of Science for both departments related to SSH:

Department of Social Sciences (sections: foreign affairs; philosophy, sociology, psychology, law; economics):
1. Civilization changes in modern Russia: inner world, processes, and cultural wealth;
2. Political relations in Russian society: power, democracy, personality. Problems and ways of consolidation;
3. Transformation of social structure of Russian society;
4. Federative relations and strengthening of Russian state system;
5. Individual as a subject of social changes: social, human and psychological problems. Problems of mass consciousness development;
6. Methodological problems of economics and establishment of »knowledge based economy«;
7. Theory and methods of economic and mathematical simulation of scenarios in socioeconomic and innovative development of Russia;
8. Complex socioeconomic prognosis of Russia;
9. Problems and mechanisms of economic, social, and ecological protection in Russia;
10. Scientific foundations of regional policy and sustainable development of regions and cities;
11. The formation of foundations of contemporary systems of foreign affairs;
12. Complex research of economic and political development of foreign states and world regions in correlation with Russia's national interests. The experience of reforms in foreign countries;
13. Russia in the world economy. Peculiarities of Russia's integration into the world economic community;
14. International terrorism. Problems of national protection and security in Russia.

Almost all of these topics can be included in international research projects; so, on the official level, the goals of internationalization are the following:

Department of Historical-Philological Sciences:
1. Complex research on ethnogeny, ethnic aspect of population, contemporary ethical processes, historical and cultural cooperation in Eurasia;
2. Protection and study of cultural, archeological and scientific heritage: revelation, systematization, scientific description, restoration and conservation;
3. Studies of historical aspects of terrorism; monitoring of xenophobia and extremism in Russian society; anthropology of extremist groups

and subcultures; analysis of ethnical and religious factors in local and global processes of the past and the present;
4. Theories of historical processes; experience of social transformations and social potential of history;
5. Studies of evolution: humans, societies, states. The individual in history and the history of everyday life. Retrospective analysis of forms and contents of relations between the state and the society;
6. Research of public development in Russia and its place in world historical and cultural processes;
7. Studies of aesthetics and inner world of Russian literature and world literature and folklore;
8. Issues of theory, structure, and historical development of world languages; studies of evolution, grammar, and lexical order of Russian language.

The important role in defining the themes of fundamental research in Russia belongs to the major scientific foundations, especially RFBR and RFH: they organize scientific competition of the research proposals, invite the independent experts, give advice about how to adjust a topic or make it closer to the declared preferences and the needs of a society.

In general, the conditions of social sciences and humanities in modern Russia are defined by three basic issues:
- the amount of money and mechanism of financing;
- organizational transformations in the system of scientific institutions;
- the practical effectiveness of the research and application of the achievements of social science to societal needs (production, public sphere, culture, being the most important).

In case of collaborative research with foreign countries these conditions also matter as soon as any state finances are involved. The most painful for the scientific community is the issue of research financing: lack of financing under the increasing cost of surveys, unfair decisions, the opacity in its distribution, shortage of independence of small-scale research groups and laboratories; poor selection of the most perceptive proposals; and lack of a high-level quality scientific expertise (Guseva 2006).

In 2008 the whole amount of financing for SSH from the Russian federal budget consisted 120 billion rubles (near U.S. $4.7 billion); this was 14 percent more in comparison with 2007, but this was only 0.36 percent of the Russian GDP. According to Decree No. 47 from February

12, 2008, approved by the Presidium of Russian Academy of Sciences, the amount of funds for social sciences and historical-philological sciences (two departments within the academy structure that fit the term »SSH «), was about 4 percent each. It is hardly possible to expect any increase of international contacts with such funds. As for the current crisis situation in 2009, it is much worse because of inflation and an economic slowdown, and nobody can predict when the economy will pick up.

Therefore, despite of the sincere desire of many Russian scholars in SSH to participate in international research and their enormous scientific potential, the real situation in the field of internationalization of SSH is far from perfect. Russian scholars need much more support from the government and the international community of scholars—both can help bring Russian SSH to the international arena and make Russian SSH more visible.

Institutional Aspects

The reforms in the field of higher education in Russia are deeply connected with transformations of organizational basics of social-humanitarian research. These transformations reflect the governmental efforts to activate the potential of Russian society by all means, including the rapprochement of the Russian system of scientific centers and the system of universities (Panfilova/Ashin 2006). The adaptation of the so-called »European university model« is a part of these reforms. The official addition of Russia to the Bologna process in 2003 also stimulated the reformation of the Russian system of higher education.

The current federal legislation considers the scientific and research activities of universities as »their necessary condition«. For example, the federal law »About Higher and Post-doctoral Professional Education« declares the basic aim of Russian universities: the development of science and arts through scientific research and creative activity of the students and professors. The principles of autonomies and liberty are extended to the sphere of research. In particular, the universities have to realize fundamental and applied scientific research in different fields of science. Furthermore, the real scientific activity of an institution of higher education is controlled through the state accreditation (without the state accreditation the institutions can't award the valuable degrees to the students). The functioning document »Condition about the State Accreditation of Higher Educational Institution«, approved by the Russian government, established the strict qualitative and quantitative criteria of accreditation. In particular, it requires the implementation of scientific research in more than five fields of science; at the same time, the total amount of scientific and research activities and amount of financing from external sources for one unit of professor and lecturer staff is controlled.

The potential of Russian higher education institutions requires special attention. According to data provided by the rector of Lomonosov Moscow State University, Victor Sadovnichy, at present there are more than seventy classical universities in Russia. The number of humanitarian faculties expanded from six to seven in the beginning of 1990 to ninety. Among seventy rectors of these universities, thirty-three are specialists in social science and humanities, including eleven with Ph.D.s in economics, four in philology, four in sociology, and others. According to the view of ex-minister of science and technical policy, Boris Saltykov, MSU is a good the example of the reformed research university, an example of »island of excellence« in Russia.

One of the strategic aims of higher school reform is the integration of science and higher education. This aim is embodied in the federal special program, The Integration of Science and High Education in Russia in 2002–2006. The establishment of educational-scientific centers conducted by the institutions of the Russian Academy of Sciences and other research organizations and universities is seen as a tool of such integration. The optimal conditions for implementation of new technologies and new specialists with high innovative potential have to be achieved. In this case the new centers will become a part of Russian technopolis.

A significant problem is a job placement of graduating students, especially from the humanitarian faculties (Radayev 2008a). During the government session of May 14, 2008, the chairman of Russian government, V. Zubkov, characterized the situation this way: «The problem of an unbalanced labor market and educational service is rather important. Out of 1.2 million graduated of 2008, near 40 percent are the lawyers and the economists«. From the government point of view, Russia does not need so many SSH specialists. So, humanitarian faculties can be reduced, and the financing level can be decreased for them. As a result, less talented young people will select SSH as their future career, and less talented young scholars will join this field in a five-year period.

Overall, the situation with the international research within the universities is similar to the situation with the research institutions: limited funds, little institutional support, and lack of correlation between an academic's involvement in international research and career promotion (Pokrovsky 2006). In the best case, Russia follows the so-called »scenario of islands of excellence« (Global SSH 2008:3); however, for Russia with its huge potential in SSH it might be more fruitful to select the general »scenario of excellence« (similar to the United States) and try to involve more centers and universities in such activities. Recently, according to the study by Kuhn and Okamoto (2009), international research accounts for approximately 25 percent of all scholarly research activities. Among scholars surveyed almost 30 percent mentioned that international research helps to liberate the academic from the world of their ivory towers. We can assume that these towers still matters for all scholars, but only a few can get rid of them. Internationalization of SSH will definitely contribute in this direction more than many other means (Titarenko 2009). As clearly expressed by the vice rector of the State University-Higher School of Economics, Vadim Radayev, Russian science has to be transparent and open for the international community of scholars. This is a primarily condition for realization of Russian ambitions to play a more important role in the world scholarly community (Radayev 2008a).

Probably, only a new generation of Russian scholars—with good knowledge of foreign languages and degrees from European universities—will participate more actively in international projects (Yurevich 2004). However, this generation would also need the appropriate funds and motivation both from the national and international communities.

International Relationships

The process of globalization has become a subject of SSH study in Russia since the early 1980s. Major attention has been paid to different aspects of globalization—socioeconomic, sociopolitical, socio-ecological, and socio-cultural. Currently, according to official documents of the Russian Academy of Sciences and the Russian Foundation for Humanity, Russian scholars conduct several research projects related to globalization:
1. A program of fundamental research by the Russian Academy of Sciences, Adaptation of Nations and Cultures to Environmental Changes, Social and Technogenic Transformations.[1]
2. A UN project Russia in a Global World;
3. An analytical special program of the Russian Foundation for Humanity, Russia in a Multipolar World: The Image of the Country.[2]

---

1 http://www.ras.ru
2 http://www.rfh.ru

In 2003–2005 and 2007 these projects accounted for 9 percent and 5 percent, respectively, of the completed projects in the field of global economics financed by RFBR. Social science and philosophy accounted for 9 percent of projects in 2003–2004, almost 7 percent of projects in 2005, and 16 percent of projects in 2007.

The international relationships between Russian social scientists and their foreign colleagues are permanently consolidating. This is especially relevant in regard to the scholars from the Commonwealth of Independent States. An example of collaborative scientific projects, financed in 2008 by the RFBR, is a collaborative project with Moldova (financed by the Academy of Sciences of Moldova as well): Social-ecological Dominants of Integrated and Adaptation Processes in the European Part of the Post-Soviet Area in the Context of Globalization (CIS) « (the chairs are, for Russia, I.. Sosunova, and for Moldova, K. M. Manolache). The results of this project were delivered to both countries (scientific communities and the public) and presented at several international congresses.[3]

The RFH and RFBR conduct special project competitions for joint projects of Russian scientists in collaboration with scholars outside Russia. The major players in the field of international SSH projects are the following organizations: the Russian Academy of Sciences (Institute of Sociology, Institute of Psychology, Institute of Social and Political Researches); Russian Foundation for Basic Research; Russian Foundation for Humanities; the UN; the International Research and Exchanges Board (IREX); The Fluorite Foundation; Open Society Institute (the So-

---

[3] The approbation of the project's results successfully conducted at 14 scientific arrangements (with publications), including III All-Russian Sociological Congress (Moscow, October 2008); XXIX International Congress of Psychology (Berlin, Germany, July 2008); XXXVIII World Congress of International Institute of Sociology (Budapest, Hungary, June 2008); International scientific-practical conference, Actual Problems of Ecological Security and National Health (Moscow, April 2008); Inter-institutional scientific-practical conference, Ways of Education's Perfection of Social Sphere Specialists in Conditions of High School (Moscow, May 2008); Forum, Sustainable Development of Central Federal District in Ecology and Economy (Tula, May 2008); International scientific-practical conference, Actual Problems of Foundation and Development of United State of Russia and Belarus (Gelendzhik, June 2008); International scientific-practical conference, The Development of Collaboration and Integration– the Objective Necessity of Sovereignty Consolidation and Welfare Growth of Commonwealth of Independent States Population (Moscow, October 2008); International scientific-practical conference, Social Ecology in Changing Russia and Contiguous States: Theory and Practice (Belgorod, December 2008); 39 World Congress of International Institute of Sociology (Yerevan, Armenia 2009).

ros Foundation); the Carnegie Moscow Center; MacArthur Foundation; State University—Higher School of Economics; the Vladimir Potanin Foundation; Foundation of Science Support; and the others.

The main countries and participants of collaboration in the sphere of social science and humanity, and the number of grants and projects, are presented below.

Grants and Projects

Russian Academy of Science: In 2008 the specialists of the Department of Social Sciences of the Russian Academy of Sciences participated in different programs of scientific research, particularly in fundamental research:
1. Economy and Sociology of Knowledge (19 million rubles, or 558,824 euros from the Russian Academy of Science funds or 1.2 percent of 1.5 billion rubles that were laid out for 2008 to the fundamental programs by the academy);[4]
2. Adaptation of Nations and Cultures to Environmental Changes, Social and Technogenic Transformations (56 million rubles, or nearly 1,650,000 euros—3.7 percent from the Russian Academy of Sciences funds).[5] For comparison, there were 242 million rubles (16 percent of all funds) laid out to the program, Molecular and Cellular Biology, which was the best-funded program. In comparison to the EU Framework Program the amount of funding is insignificant. The overall budget for the Collaborative project FP7-SSH-2009-B (small- or medium-scale focused research projects) for specific cooperation actions dedicated in international cooperation in SSH is 6 million euros.
3. The scientists work in collaboration with the UN on the project, Russia in a Global World (there is no data about funding and a number of projects).

In 2008 the Institute of Sociology has accepted sixteen transnational projects. The partners are Poland, United Kingdom, France, Mongolia, Finland, Germany, Hungary, China, Kazakhstan, and Bulgaria.

In 2007–2008 the Institute of Social and Political Research accepted four transnational projects and conducted thirteen conferences and semi-

---

4 Data is given for 2008, 1 euro is approximately 34 rubles.
5 http://www.ras.ru/presidium/documents/directions.aspx?ID=52e0dff7-0358-48bd-9b70-defb4a3758b9

nars. Partner countries are Belarus, Italy, Japan, Norway, and the United Kingdom.

In 2008 the Institute of Psychology accepted fifteen transnational projects. Their partners are the United States, United Kingdom, France, Spain, Sweden, Finland, Germany, and Vietnam.

Russian Foundation for Basic Research: In 2008 in the sphere of SSH there were seven projects (2.4 percent of total accepted 289 transnational projects) between Russia and France, Mongolia, Taiwan, Moldova, and the United States. Eleven international seminars in social science and humanities of a total of thirty-three were conducted, which is 33 percent of all conferences and seminars, the main countries that participated are Austria, Italy, Switzerland, and the Baltic countries. In 2007 there were no conferences or seminars conducted in the social science and humanities sphere and the number of realized project is only 0.7 percent (2 projects with Moldova and Taiwan of 259 accepted projects).

Russian Foundation for Humanities: In 2008 sixty-six projects were accepted, and nine scientific conferences were conducted. Belarus, Ukraine, Germany, France, Mongolia, Taiwan, and Vietnam are the participants of the collaboration. Compared to 2008, in 2007, seventy-seven projects were accepted. Belarus, Ukraine, Germany, France, Mongolia, and Finland cooperated with the foundation that year.

As one can see, the major Russian partners in SSH are scholars from the CIS (Belarus, Ukraine, Moldova, Kazakhstan), the EU, United States, and some Asian countries (Vietnam, Taiwan, Japan, China, Mongolia). There are no collaborative projects in SSH between Russia and Australia and any nations in South America and Africa.

United Nations Organization: At present the UN mostly cooperates with Russian institutions in spheres of education, demography, health, environment, and economy. Together with Ukraine and Belarus, the UN helps Russia discover and analyze issues devoted to the Chernobyl catastrophe.

State University—Higher School of Economics: The Scientific Foundation of the State University collaborates with institutions and universities from more than twenty countries as well as with international organizations such as the Organization for Economic Cooperation and Development (OECD), the International Labor Organization (ILO), and others. This is the main participant of the transnational projects in the field of economics in Russia.

The new form of research establishment is the realization of transnational projects of Russian scientists under the supervision and in collaboration with foreign scientific foundations, for example, the following:

Open Society Institute (The Soros Foundation): In 2008, the program of 2007 devoted to Cultural Development between Russia and European Union was continuing with the total budget of near 2 million euros, the minimum sum of a grant is 75,000 euro, the maximum 300,000 euro. The number of grants is not known and the information about Russian partners is not publicly available.

The Fulbright Foundation: There are a few different programs that provide Russian researchers travel grants or grants for MA degrees, seminars, and others (totally, near forty grants per year). The Fulbright-Kennan program is devoted only to the SSH field (five to six grants per year). The number of the participants from Russia is also closed, but Russian scholars participate annually.

Stimulation of the processes of internationalization of research activities and collaboration becomes the integral part of Russian state policy in the sphere of sciences and education. The appropriate international agreements between Russia and the EU, Russia and particular countries have been signed and include the Agreement between the Government of the Russian Federation and the European Commission on the Cooperation in the Field of Science and Technologies (valid till 2009) and the Roadmap for the EU–Russia Common Space of Research and Education including Cultural Aspects (approved at EU–Russia summit in 2005).

The EU runs some special programs, especially for the scientists from Central and Eastern Europe, as well as for countries of the former Soviet Union. The recently closed International Association for the Promotion of Cooperation with Scientists from the Independent States of the Former Soviet Union (INTAS) program was a great plus for internationalization of SSH in this region, including Russia. Framework programs could not substitute for the closing of INTAS, as they are not especially aimed to the countries of this region.

According to N. Didenko, coordinator of the Euroscience Regional Section in Russia, the following steps are necessary to promote Russian participation in transnational programs:
1. An Internet-based working space with all information in Russian.
2. The cooperation within the scientific Network worldwide.

3. Special Guides for writing applications prepared in Russian for Russian scientists.
4. Simplification of the formal procedure in the FP projects.

However, these steps are in the hands of Western scholars and scientific organizations. So, without mutual steps taken together and a common desire to cooperate fruitfully, the situation can hardly been improved. The changes that have so far occurred in Russian science are not sufficient for the international arena in SSH.

Conclusion

Russian SSH are constantly involved in the process of internationalization; however, the level of involvement is lower than Russia's potential in this sphere.

Fortunately, the stimulation of the process of internationalization of research activities and collaboration projects has become an integral part of Russian state policy. Russia has twelve main players in the sphere of SSH. The main countries of collaboration are from the CIS (Belarus, Ukraine, Moldova, Kazakhstan), the EU, United States. and some from Asia. However, in reality only a part of the Russian SSH community is oriented to the internationalization of research and wish to be included into the international scientific community on a regular basis (for regular comparative research). The Russian sphere of international SSH more resembles rare islands of excellence in the ocean of national and regional research in this field. In many cases, these »islands of excellence« receive the state funds as well as the international grants because they belong to a small group of the elite centers and universities (such as MSU, State University-Higher School of Economics, Institute of Economics RAN). Their status and brand names help them to represent the Russian scientific community, and they follow the policy »winner takes all«.

Many other institutions and centers have small funds to develop infrastructure and run scientific projects. Unfortunately, the scientific careers of scholars do not depend on the level of involvement in the international projects or publications abroad. That is why young scholars are not well motivated to stay at the Russian universities and research centers; they prefer private firms running election surveys and public opinion polls. Therefore, it is insufficient to measure Russian SSH by the level of international publications; it is necessary to use other methods and first recognize the importance of cooperation between the European Union and Russia on a larger scale

Research and technology development is not a primary interest of the state in the current economic crisis. This fact contributes into the low participation of Russian scholars in prestigious international projects, such as FP7: they can't meet the Western transnational cooperation requirements, their knowledge of foreign languages is low, and their traditional principles of research differ from the Western principles. For a perfect Russian scholar it is easier to find employment abroad rather than improve the research situation at home. That is why some scholars emigrated and serve now as a bridge between Russian and Western scholars (Boris Doctorov, Vladimir Shlapentoch, Georgy Derlugian). They help bring Russian social sciences to the international social community of scholars. Sustainable development of Russia would help to establish a sustainable financial support for its SSH.

In order to improve the level of internationalization of Russian SSH, some new forms of cooperation with the EU partners can be implemented: long-term exchanges for advanced scholars from both sides, more space for personal cooperation, summer schools, and seminars, longer terms of international projects, and a greater time period for implementation of results. The EU countries can develop cooperation with Russia (as well as with other CIS) in different ways by focusing at the personal rather than the institutional level. Taking into account the high level of bureaucracy and corruption in Russia, it is not realistic to expect that institutional cooperation will work properly and help the talented Russian scholars to participate. Formal institutions are often weak in Russia, and they are hierarchical bodies in most cases, so that primarily the administration (directors and their circles) is involved in TNC as the official representatives while the advanced scholars are kept out of this cooperation or play a very limited and subordinated role. Thus freelance relations can allow Russian scholars to be more open in their research and conclusions. Even from the issue of transparency, freelance relations are better as it is easier to control persons rather than institutions.

Finally we should comment about Russian participation in the process of internationalization of SSH. Although Russia is a »third country« from the EU classification, it does not really mean that Russia is a »Third world country« with the level of science on a developing stage. That is why it is hardly possible to agree with the conclusion of an international research team on Global SSH that Russia is in a »lock-in situation of low resources and poor policy support« (Global SSH 2008: 5). There are some excellent SSH projects and highly skilled social scholars in Russia. Russia is too big, and its regions are too divergent, to be so simply classified. Its potential is still much higher than the current results; however, before making a »final conclusion« it is necessary to conduct complex research from different perspectives—Russian, CIS, EU, U.S., etc. The results of such research would depend on the starting point of observers as well as their desire to discover accurate data.

However, more funds, political effort, goodwill, and time are needed to make Russian SSH an integrated part of the global SSH.

References

Barabanov, O.N./ Lebedeva, M.M. (2002): Globalization and education in the modern world. In A.V. Torkunov (ed.), Globalization: human dimension: MGIMO (U) Ministry of Foreign Affairs of Russian Federation, M.: ROSSPAN. [In Russian]

Global SSH (2008): Newsletter No. 3, September. www.globalsocialscience.org. [Date of access: 15.12.2008]

Guseva, I.I. (2006): Social science and humanities: lessons of crisis. Power, 4. [In Russian].

Ivanov, A.E. (1991): Higher school in Russia in the end of XIX – beginning of XX century. Moscow. [In Russian].

Kinelev, V.G. (1995): Higher education in Russia: historical essays to 1917. Moscow: Moscow State University. [In Russian]

Kuhn, M./Okamoto, K. (2009). Through International Collaborations towards a new multipolar SSH world order. Global SSH Work Package Three Report.

Kulyabko, E.S. (1962): Lomonosov and Educational Activity of Saint-Petersburg Academy of Sciences. Moscow: Academy of Science, USSR, Institute of Natural Science and Technics. [In Russian].

Pipiya, L. (2007): Measuring SSH Potential. Global SSH databook. Moscow: Centre for Science Development Studies, Russian Academy of Science.

Osipov, G.V. (ed.) (1999): Social and Social-Political Situation in Russia in 1998. Moscow: Nauka. [In Russian]

Panfilova, T.V./Ashin, G.K. (2006): Perspectives of Higher Education in Russia: Reformation or Liquidation? In Social Science and Modernity, 6. [In Russian].

Pipiya, L. (2008): Internationalization of Social Sciences and Humanities: Challenges for Russia? Presentation at the Global SSH workshop »The Internationalisation of Social Sciences in a Global Comparative perspective—Challenges for SSH Research and Research policies«, Paris, April 23-24, 2008.

Pokrovsky, N.E. (2006): What is Happening With Humanitarian Education? Sociological Studies, 12, 95-97. [In Russian].

Radayev, V.V. (2008a): About the Program for Russian Sociology. Sociological Studies, 7, 24–33. [In Russian]

Radayev, V.V. (2008b): Embarrassing Questions of Russian Sociology. http://www.hse.ru [Date of access 10.10.2008]. [In Russian]

Rozov, N.S. (2006): Cycling of Russian Political History as a Sickness: is it Possible to Recover? Political Studies, 2. [In Russian]

Sokolov, M.M. (2009): Russian Sociologists on the International and National Market of Ideas. Sociological Studies, 1, 144-152. [In Russian]

Titarenko, L.G. (2009): Contemporary Theoretical Sociology: Thoughts After the Congress. Sociological Studies, 1. [In Russian]

Yudin, B.G. (1993): History of Soviet Science as a Process of Secondary Institutionalization. Philosophical Research: Science and Totalitarity, 3. [In Russian].

Yurevich, A.V. (2004): Social Science and Humanities in Modern Russia: Adaptation to the Social Context. Working Paper WP6/2004/02. Moscow: State University–Higher School of Economics. [In Russian].

Social Sciences and Humanities in Ukrainian Society: The Difficulty of Integration into International Structures

IGOR YEGOROV

Introduction: Economic Crisis and the Soviet Heritage

The national scientific system of Ukraine was formed for the most part during the Soviet Union times. The scientific and engineering complex, encompassing numerous research institutes, higher educational establishments, design bureaus, scientific and engineering departments of enterprises, association of inventors and innovators, and so forth, can be considered as its core. On the whole, according to different estimations, Ukraine encompassed nearly 13–15 percent of the scientific and engineering potential of the Soviet Union, but the research centers of the republic had one important feature: Ukraine accommodated nearly 20 percent of the experimental research equipment of the USSR (GKNT 1990). In some areas of science and technology Ukrainian research organizations had strong positions. This is particularly true for electric welding, new materials, transport aviation, and the development of the specialized software. It was not by accident that the Academy of Sciences of Ukrainian SSR was selected as the leading organization responsible for the new materials development within the scope of the Complex Program of Scientific and Engineering Development of Mutual Economic Assistance Council member states in 1985. The majority of Ukrainian research organizations, including the leading institutes of the

Academy of Sciences, had a clear technological orientation (Paton 1993).

However, it should be noted that Ukrainian specialists, like the majority of specialists in the Soviet Union, did not maintain active cooperation with colleagues from the most-developed foreign countries for the well-known reasons, which resulted in overestimations of results of their own research and development. For instance, the poll of leading scientists about the R&D performance in Ukraine in the early1990s revealed that the majority of them (more than 90 percent) believed that those exceeded or matched the world level. Regrettably, evidence does not support those estimations. Leaving aside the average technology level of developments, only one of ten Ukrainian scientists (with the degree of doctor or candidate of science) had works published abroad in early 1990s, and only three Ukrainian scientists were in the list of one hundred Soviet authors most frequently quoted in the internationally acknowledged journals (Klochko 1994). Such seclusion of the Ukrainian scientific community can be explained by noting that the majority of research and development was aimed at creating powerful military potential. According to the data of a former member of the Ukrainian Parliament, academician P. Kysly, even in the Academy of Sciences more than half of all R&D was ordered by the Soviet military in late 1980s (Kysly 1991). Concerning the so-called branch institutes, where the lion's share of funding and material resources was allocated to, the picture was yet more remarkable: developments in whole industries were classified in different ways and subject of restricted access.

During the Soviet period, Ukrainian social sciences and humanities have been developed under the supervision and sometimes complete control of the Communist Party and the Soviet security services.

Independence proclaimed in August 1991 sparked hope in the Ukrainian elite that liberated from supervision of the Soviet control bodies, the rates of socioeconomic and scientific development of a new independent state would soar (Romaniv 1991). Alas, it never happened. Actual results of the first years of independence turned out very different from these optimistic predictions.

The years of establishing a market economy in Ukraine demonstrated that the state does not have sufficient material resources to preserve science in the condition that it was in the Soviet times. Substantial reduction of funding for scientific research occurred during the period of market transformations. There are no long-term orders placed for R&D projects; science was deprived of prestige and the status of scientists eroded. These changes resulted in a gradual reduction in the number of research establishments and the collapse of many state research institutes. Big research institutions breaking into small ones became a reality. Mobile research teams fully or partly funded by overseas sponsors mushroomed; many scientists became primarily oriented to the interests of foreign customers, and their activities began to focus on the search for grants and other opportunities to participate in Western research projects. A lot of young and middle-aged scientists left their academic establishments and industry institutes and swapped their activities for more profitable ones, many of them emigrating. This caused a deepening of the age gap between different groups of scientists, which was accompanied by a considerable shortage of thirty- to forty-year-old specialists—the most active part in terms of creative capacities (Egorov 2000).

Ukraine is gradually acquiring a system that is basically oriented to the import of scientific and engineering results instead of making its own science and technology products. Negative changes that happened in the country's economy in 1990s facilitate the process.

The Ukrainian economy suffered from a deep crisis in the 1990s. According to World Bank data, Ukraine was the only post-Soviet state whose economy did not grow in a single of year of the decade 1990–1999. Official GDP for this period decreased by more than 60 percent, and whole industries (for example, electronics) virtually vanished. The level of domestic demand for numerous products, and especially high-tech ones, fell dramatically (Dyker/Radosevic 1999). Only Yugoslavia, which suffered a severe international blockade, had worse economic indexes.

In the 2000s the situation has started to change. In 2007 the Ukrainian economy attained a level of almost 90 percent of the 1989 GDP, but that process was not accompanied by fundamental structural changes. The mining sector and industries related to the primary processing of mineral products remained dominant in the economy. The domestic market did not create corresponding demand for science-oriented industries, especially in machine-building, the pharmaceutical industry, and high-tech services. The modular assembly of cars or computers of imported components to satisfy domestic needs created a lot of economic activity, but according to the current international classifications, such activities are classified as low-tech production that do not require high qualifications and R&D.

For Ukrainian science, the changes in the overall economic situation brought about a stabilization in terms of both the number of researchers, and the amounts of R&D funding. Yet this new condition does nothing to overcome the crisis in Ukrainian science. The scientific system remains unreformed, whilst dynamics of its numerous characteristic features are negative.

Considerable erosion of the personnel potential of Ukrainian science occurred in the 1990s. Overall employment in R&D was reduced almost by 60 percent, and taking into account the intense growth of fictitious employment and the impossibility of conducting R&D because of the lack of money for equipment and materials, and the actual cutback of labor spending in science, the figure was even greater. The expectations of a vigorous expansion of graduate studies programs to recruit youth into scientific study were not justified. The increase of the number of defended dissertations is clearly inconsistent with the growth of the number of graduate students. In addition, many young scientists after defending dissertations choose careers outside research or emigrated.

There is a trend in Ukraine toward reduction of general employment in science and technology. In particular, it is worth mentioning that although a major reduction of employment decline was observed in the first half of 1990s, trends in R&D employment remain negative even in conditions of the economic growth from 1999 to 2005.

This intricate socioeconomic situation of specialists in Ukrainian science largely determines a fashion whereby main (or academic) responsibilities are combined with other activities that, unfortunately, are not research or training activities. The number of scientists working two or more jobs doubled in 2005 in comparison with 1991 and amounted to more than sixty-five thousand specialists. In our view, this reflects the crisis in existing institutional structures of Ukrainian science, in which there is a failure to organize a comprehensive research process and provide decent remuneration for research fellows.

Concerning distribution of scientists by different specialities, the representatives of engineering sciences maintain their dominance in Ukraine, accounting for 55 percent of overall R&D employment. The situation still remains almost the same compared with 1991 when such specialists made up nearly 74 percent of this category. However, the percentage of representatives of engineering sciences will keep falling.

In 1999 for the first time during the years of independence the number of doctors of science who carried out R&D in science and technology organizations in Ukraine fell by almost 8.4 percent compared to 1998. This is related primarily to the new standards for retirement plans adopted for research fellows, which allows a research fellow to retire on a pension of approximately 80 percent of his or her wages. Nonetheless, the average age of doctors of science employed in R&D grew in the 1990s by a year every 2–3 years, and in 2006, this average was 62 years, that is higher, than »retirement ceiling« (which is 60 years in Ukraine). Obviously, subsequent aging of doctors of science will inevitable bring about a decline of creative potential of this category of scientists. A similar trend can be also seen in the age dynamics of candidates of sciences.[1]

Apparently, the situations with the personnel in science are alike in different regions of Ukraine. Comprehensive data analysis of the age structure of academic institutes of the Kharkiv region reveals that it is very close to the average national structure: 61 percent of research fellows are over fifty. Specialists under forty makes up less than 23 percent. Science specialists under the age of forty amount to a meager 8.1 percent of the total number of specialists, and doctors of science under age fifty account for 2.14 percent of the overall number of the doctors.

This age crisis in science will be hanging over Ukraine in the years to come. It is practically impossible to solve it, yet the consequences can be somewhat alleviated. The point is that there is a »big gap« in the group of specialists (candidates of science) in the thirty to fifty year age group. These are the people who were the most active over the last decade in »defecting« from science. Added to senior generations leaving active involvement with science, the shortage of skilled specialists in science will be even more striking. The processes of the age imbalance of personnel can be halted by implementing urgent measures; however, the problem is complicated by the fact that it is very difficult to resume research activities after several years of break because of the very specificity of this activity. In particular, measures undertaken by the government to increase the pay of research fellows cannot radically change the situation, although it may make the crisis less acute.

The Ukraine possesses a network of research organizations that in the aggregate are intended to ensure an adequate level of science and technology in the whole state and every single industry of the national economy. The following are the basic types of organizations:
- Design bureaus.
- Development organizations.
- Higher educational establishments.
- Research and development and design departments at industrial enterprises.

Since 1991, the organizational component of scientific potential of Ukraine has undergone changes that became especially noticeable in recent years.

---

[1] There is a two-degrees system in Ukraine, inherited from the Soviet times. The first degree is *candidate of sciences*, the second is the *doctor of science*. The first degree is approximately equal to a PhD degree. At least, as a rule, a holder of a candidate degree is elegible for participation in EU or American scientific programs, opened for PhDs holders.

First, the overall number of research organizations in Ukraine gradually increased until 1999, and then decreased to some extent, and in 2005 there were 1,456 of them (see Databook data on Ukraine), compared to 1991, when there were 1,344 research organizations and institutes). This can be accounted for by the fact that performance of some »old« research establishments in the new environment was not good enough, therefore new and more market-oriented spin-offs were organized.

Second, structural changes in specialization of Ukrainian research organizations are underway. For example, there were only 43 design and development organizations in 2005, whilst in 1991 there were 88 of those. The share of social sciences and history (SSH) institutions is going up (see Databook, data on Ukraine). The total number of SSH institutions reached 243 in 2006 according to the official statistical data.

In terms of the number of organizations, the academic sector of Ukrainian science (more than 300 organizations) takes second place, just after the sector of branch (mainly technologically oriented) institutes. In 2005 this sector's share grew by more than 1 percent. This trend is robust enough—in 2000–5, the share of the academic sector grew by 1–2 percent per year. The academic sector encompasses the National Academy of Sciences of Ukraine (NAS) and academies of sciences of a particular sectors or disciplines—the Academy of Agrarian Sciences of Ukraine (AAS), Academy of Medical Sciences of Ukraine (AMS), the Academy of Pedagogical Sciences of Ukraine, (APN), the Academy of Legal Sciences of Ukraine, the Academy of Arts of Ukraine (AMOu), all of which are public research organizations.

In the early 1990s a number of important think-tanks were created, such as the Institute for Strategic Studies, the Center for Economic Reforms Studies, the Institute for Conversion Studies, and so on. Some of them are undertaking projects for different ministries and state agencies, others are operating as independent or semi-independent organizations, but usually with strong financial support from various entrepreneurial organizations, powerful financial and economic groups, or foreign foundations. They have become an essential supplement to the system of state-sponsored social science institutions, which sometimes helped to make important corrections to the decision-making process at the highest level of the state governance.

Key Changes in the Social Sciences and Humanities (SSH)

The key result for SSH in Ukraine was the shift from mainly theoretical work to the practical services in the interests of different political forces. This process has been accompanied by creation of numerous new organizations that are working in the areas of consultancy and of conducting surveys on orders of political parties and companies. It is interesting that the State Committee of Statistics does not consider the bulk of such organizations as »scientific« ones. In many cases they have no scientific projects »registered« by the State Institute of Scientific Information, which is an important precondition for being considered as a scientific institution.

It is well-known that social sciences in national republics of the Soviet Union were under »double pressure«—from local authorities and from Moscow ideologues, who tried to weed out all signs of so-called »bourgeoisie nationalism«.

This resulted in the practice of selecting as directors of almost all research institutes persons who had worked in the science or ideological departments of the republican Communist Party. All promotions to the members of the Academy of Sciences in social sciences were under strict control from Communist ideologues. The best specialists (for instance, the famous economists Nikolai Fedorenko and Sergei Glaziev) have preferred to work not in Ukraine but in Moscow, where ideological pressure was not so strong, and the room for creativity in research was wider. As Gabovych and Kuznetsov note, this has created a situation in which the highest elite of Ukrainian social sciences has been composed mainly of *apparatchiks and their »clones«,* not the best scholars (Gabovych/Kuznetsov 2007).

This does not mean that the country had no good professionals in the social sciences but their number was limited. The tendency was the following: the closer the discipline to the ideology, the smaller number of really creative and honest specialists it had. That is why Ukrainian scholars in such disciplines as archeology, linguistics, and philology have received higher recognition from their foreign colleagues than philosophers or economists.

As some researchers pointed out (Kneen 1994), important changes occurred in the Soviet scientific community during the last decades of the Soviet Union. Such qualities as commitment to the Communist ideology, obedience, and skill in »spinning« presentation of results became very important, and sometimes dominant, factors in promotion to the officially recognized scientific elite of the USSR.

Forced cultivation of the Communist ideology and the fear of losing their positions after national independence was won, have led many scholars to shift their political positions. They have started to serve the new regime. The problem was that in the 1990s, the new Ukrainian authorities had no particular ideology, but, as with President Yeltsin in Russia, were mainly interested in keeping their power and in justifying their actions as the new legitimate rulers. In the case of social sciences, new authorities could find firm support from the former »servants of ideological front« of the Communist Party.

On the other hand, thanks to the political changes, the ideological function of science has been seriously weakened. In a rapidly changing sociopolitical environment the role of really well-grounded social research is growing.

However, at the same time, almost all the social sciences institutes, which were responsible for the development of »Marxist-Leninist« theory and supplied Communist leaders with »scientific arguments« for their political activities in the Soviet period, have been preserved. It is evident that most of them need a lot of basic transformations to meet the needs of today's Ukraine. Surprisingly, the most ideologically oriented institutes have demonstrated the best adaptive qualities, because they needed to act more decisively to persuade people to forget their notorious practice in the past. Thus the Higher Communist Party School, which was created specially to train the Soviet Ukrainian nomenklatura, has been renamed to the National Academy of Management and become a pioneer in introduction of the paid system of education among the universities and other learning institutes. Former Communist party apparatchiks have even started to name their university the »Ukrainian Harvard«.

A similar situation exists on the »micro« level, with departments of Communist Party History and of Marxist-Leninist Political Economy being renamed Political History and Economics and Management Departments respectively.

On the other hand, some representatives of nationalistic political forces have shared the power with old scientific »nomenklatura«. Not surprisingly, both groups have focused their activities to justify the newly emerged regime. This was especially important under the conditions of shrinking financial support for research in Ukraine. So, in the early 1990s both groups started to argue that Ukraine had the best position among all post-Soviet states for economic development, bearing in mind the volume of per head production of steel, wheat, and some other products. The second argument was that Ukraine has scored the highest points from the group of German experts from Deutsche Bank among the all post-Soviet states. If fact, this conclusion of German experts was based on poorly grounded estimates made by politically oriented Ukrainian scholars (Pavlovsky 1992). Information about Ukrainian advantages has been disseminated by the media, and it has not been challenged by the opponents within the country. At the same time, serious Western analysts argued that these judgments had almost nothing with reality. In August 1991, the *Economist* (13–21) published an article that highlighted the unwillingness of the Ukrainian political establishment to quickly liberalize its economy and break up with Russia.

Articles about the »colonial policy« of Russia with completely falsified data have emerged in Ukrainian scholarly journals. Usually, these articles had had strong support from nationalist politicians and even top managers within the scientific system. Examples are numerous but we could focus on some vivid ones. In 1995–96 the leading journal of the Academy of Sciences of Ukraine, *Visnyk of the National Academy of Sciences of Ukraine,* published a book by Yuri Kanygin *Put Ariev* (*The Way of Aryans*) (Kanygin 1995).[2] For many years Professor Kanygin wrote books and articles (including articles in the Communist Party leading newspaper *Pravda*) on the need to strengthen the »party leadership« in cybernetics and on the problems of informatics in the period of the building up of Communist society. This book was completely different. The author has tried to prove a number of »newly opened« facts, such as the emergence of the Ukrainian nation from legendary Indian Aryans, close relations between Ukrainian and Jewish peoples in ancient times, their joint struggle against Egyptian monarchs and even about family relations between the founder of the Ukrainian capital Kiev (legendary prince Kiy) and Attila, who fought the Rome Empire in the fifth century. On the other hand, Mr. Kanygin emphasizes that the Russian and Ukrainian peoples have completely different origins, as the Russians are the products of some Caucasian tribes and the Egyptian army, which disappeared more than two thousand years ago in its raid to the north. Russians were depicted as a second-class but cruel people, who treated Ukrainians as slaves. All these things could be considered as a bad joke but the problem is that this »scholar« after the publication in the academic journal has published his book three (!) times in the Ukraine with financial support from some politicians and academic circles. He also had the opportunity to share his views on the first channel of the Ukrainian national radio. Only after a critical article about Kanygin's book by the famous Ukrainian archaeologist Petro Tolochko did the Academy of

---

2 The book was published in the Russian language—the only exception I can remember is this journal in the 1990s.

Sciences cease its support of Kanygin. But Kanygin is not the only scholar who has tried to disseminate absolutely ungrounded and falsified information and thus to stimulate ethnic mistrust in the multinational Ukrainian society. In the first half of the 1990s, some Ukrainian scholarly journals were full of articles that blamed Russians (not the Bolshevik—or Communist—regime!) for supposed ethnic cleansing and hatred toward Ukrainian people. They pay no attention to the fact that at least 3 million Ukrainian citizens were members of the Communist Party in 1985, and native Ukrainians were represented well among the highest Soviet leadership. Objective consideration of the real events has been replaced by politically oriented studies that served specific political forces. Some scholars have modified their views slightly to reach the »objectivity«. But this also had nothing to do with the real studies of the complex phenomena of modern history and contemporary life in Ukraine. So in a number of publications inhabitants of eastern Ukraine are described as carriers of predominantly negative features, while representatives of the western part of the country—by contrast—are depicted as highly moral defenders of democracy and European values (Kolodiy 2006; Ryabchuk 2001).

Another important factor that has a decisive impact on Ukrainian social sciences and humanities is the influence of the Ukrainian Diaspora and American-based research centers (influence of European research centers is visible but it is not so strong). American programs for scholar exchange (Fulbright, IREX, JFDP and others) have had a growing impact on the Ukrainian scientific community, as more and more specialists receive the opportunity to visit leading research centers in the United States.

It is also worth mentioning that some sociologists try to study processes in the Ukrainian society from realistic positions. A number of centers, which used proven methods of Western sociology and economic analysis, have emerged in Ukraine. But these centers are serving political processes by doing predominantly applied research.

In accordance with the law on scientific activities, science and technology expenditures are secured expenditure items of the state budget of Ukraine. Scientific studies are funded from the budget pursuant to the basic and program-oriented procedures. Basic funding is made available to carry out:

- fundamental scientific research
- research most essential for the state, including national security and defence R&D
- development of S&T infrastructure

- preservation of scientific objects of national property
- research personnel training

The list of scientific institutions and higher educational establishments to which the budgetary funding is made available to carry out S&T activities is approved by the cabinet of ministers of Ukraine.

Budgetary expenditures on R&D has dropped lately because of a gap between the rates of growth of such expenditures and nominal rates of GDP growth and due to the deficient implementation of government obligations in science and technology. Thus in 1998 the government defrayed R&D and innovation expenditures only to the amount of 50 percent from the planned level and 77 percent in 1999. Even in some years of rapid economic growth (2000–2003) the level of R&D allocations from the state budget was less than 100 percent. Only in 2004 did state expenditures on R&D reach their planned level.

As to the structure of R&D funding in Ukraine, the parts of public and private sectors remained virtually unchanged from 1996 to 2006: private sector's share grew by 2 percent. The R&D area remains mainly in the state's responsibility. Ukrainian entrepreneurs are not yet interested in forming their own scientific base (that is scientific departments of private firms), preferring to use services of state organizations (mainly former branch sectors), thus a share of such orders is growing steadily.

The academic sector and higher education sector are funded from the state budget mainly, whereas R&D organizations that associated with industry sectors (the bulk of them are formally subordinated to the different ministries and state agencies) are funded subject to the agreements with the customers.

In 2005 the general amount of research funding for *humanities* made up less than 1 percent of all funds allocated to R&D in Ukraine. The funds on research in *social sciences* in 2006 reached 4.4 percent of the general amount of funding of R&D nationwide.

On the whole, it should be noted that relatively high wages in social sciences of Ukraine indicate the need of society for these studies whereas the limited number of organizations and researchers employed this sphere proves that public demand is not yet satisfied.

As we mentioned above, in recent years nongovernmental research organizations in social sciences have emerged in Ukraine. They have alternative views on how to solve many important social problems. At the same time, it is worth mentioning that the sphere of their interests is rather limited. They examine some problems of current political life, sociology, and economics (to some extent). A number of other disciplines, such as linguistics or archeology, have not attracted the interest of non-governmental organizations.

Foreign funds usually provide money for politically sound projects or to specific programs. For instance, at the moment, the biggest, the Vidrodzhennya Foundation, supports women studies, Roma and Tartar studies, and research in journalism only. Along with a shrinking support of social sciences and humanities from the EU and theUnited States, this creates a situation of highly politicized social sciences, with very few really well-grounded, long-term projects.

Cooperation with Foreign Partners in SSH

After the gaining independence, Ukraine has started to cooperate actively in the S&T area with numerous countries such as the United States, Canada, Germany, and Russia, and international organizations such as the UN, UNIDO, NATO, and the like. However, the key priority is given to cooperation with EU member-states with an eye toward the Euro integration processes.

Multiple and vigorous partnership relations that Ukraine established with a number of EU member-states differ in their scope and nature. This partnership covers the whole range of S&T and innovation cooperation, such as information exchange, information support for international S&T activities, rendering conferences, workshops, fairs, training sessions, exchange programs, S&T consultancies, and joint projects in the fields of fundamental and applied research. Partially, these activities are supported by the private foundations, such as Volkswagen Foundation (Co-operation with Central and East European countries program), Ebert Foundation, and some others.

An essential role in the international integrative process for Ukraine is assigned to cooperation with the United States via relevant international foundations and programs. The Civilian Research and Development Foundation (CRDF) and Science and Technology Center in Ukraine (STCU) play a special part in developing S&T relations between Ukraine and United States. Programs of the funds are intended for Ukrainian researchers who participate in joint R&D projects, including those aimed at implementing research findings, exchange programs, and holding scientific conventions. Although these foundations support natural and technical sciences, they do sociological surveys of the important segments of Ukrainian research system and initiate discussions on the perspectives of the Ukrainian science as a whole.

Traditionally, an important place in international cooperation is assigned to Russia and other Newly Independent States (NIS)but, first of all, in technical and natural sciences.

Recent positive trends in the development of international cooperation as a whole combined with the rising confidence inside the Ukrainian scientific community encourage investments in national science development, innovation growth, S&T potential preservation and growth, provided Ukraine complies with the agreements and contracts concluded and observes its liabilities, such as paying fees (or its share) in the international organizations.

Nowadays Ukraine trails only Russia in terms of cooperation within the scope of NATO research programs. At home, Ukrainian researchers participated in sixteen S&T projects of the research program, eight of those within the framework of the sub-program Science for Peace, intended to support the implementation of scientific developments consistent with economic growth and market environment. The projects encompassed the areas of nanotechnologies, purification of waste waters, environmental monitoring, and new materials and energy sources, that is, the areas most topical in the worldwide context. Also the Ukrainian government established a think-tank to coordinate joint Ukraine-NATO efforts within the Science Program of NATO and NATO's Committee on the Challenges of Modern Society (CCMS) concerned with the issues of science and environment protection.

To enhance the effectiveness of project proposals and encourage participation in the Science Program of NATO, Ukraine supports particular research projects at the preparatory stage. These include projects on computer networks and environment protection. Today, Ukraine is more concerned with the quality rather than the quantity of grants. In order to make effective use of NATO grants and enhance co-operation in the future, the Ukrainian government currently processes national applications rapidly. The country also aspires to become better represented in international advisory boards.

The framework programs that unite EU member-states are one of the forms of cooperation that became available to Ukraine starting in 1994, during the Third framework program (FP3), when Ukraine signed the Agreement about Partnership and Collaboration with the EU.

Participation of the Ukrainian researchers in projects, initiatives, and numerous measures related to implementation of the noted programs is regulated by the intergovernmental agreements. Partnership and Collaboration Agreement between European Union and its member-states on one side, and Ukraine on the other, entered into force on March 1, 1998, after ratification by the Supreme Council of Ukraine. This document legally substantiates expansion of co-operation in trade, industrial, scientific, and administrative areas. In particular, article 58 of the agreement provides for collaboration in the area of science and technology. It is stated thereby that such collaboration includes »exchange of science and technology information, joint activity at industry of scientific research and technical developments (RTD), activity from professional training and programs for researcher's mobility, researchers and engineering personnel involved in RTD on either side«.

During 2002–6, the Sixth Framework program (FP6)—basic mechanism of coordination and funding of the European S&T programs was in operation in Europe—included the Special program FP6-2002-INCO-Russia+NIS/SSA-4, which was used to facilitate collaboration between the Europe, Ukraine, and Russia and other countries of the former Soviet Union in different areas of science and exchange of knowledge. These programs indicate that Europe understands the necessity of deeper cooperation with the scientists of the NIS in order to be competitive with Japan and the United States in S&T.

The Ministry of Education and Science of Ukraine and its educational establishments cooperate with the European Union within the framework of the Program of Trans-European Cooperation in Higher Education—Tempus (TACIS)—and with the EU member-states on a bilateral basis (specifically with their central state authorities responsible for education and educational establishments) in particular, according to the Agreement on Partnership and Cooperation between Ukraine and European Union signed on June, 16, 1994.

Major priorities implemented within these projects are directly related to SSH: introduction of state-of-the-art educational technologies, development and application of the advanced methods of educational process management, and restitution of close cooperation between universities and industries, which provides for stimulation of innovation activity.

It is important to note that almost all priorities the Ministry of Education and Science of Ukraine proposes are within the scope of the 2004–6 TEMPUS program for Ukraine:

1. Management of Ukrainian higher educational establishments. Academic priorities:
- International relations and European studies
- Law, including the European Law
- Economics and Banking
- Introduction of resource- and power-effective technologies in higher educational establishments of Ukraine

2. Human Resource Management (culture, art, education, training, business and public administration, agriculture, tourism, environmental studies, journalism). Joint European projects on curriculum development will facilitate the development of:
- distance learning and new technologies of teaching
- modern European languages
- professional development of administrators and teachers
- information technologies in education and library management

3. Institutional changes:
- professional development of pedagogical and scientific-pedagogical personnel as well as representatives of professional associations at local, regional and national levels
- Civil education
- social work with the purpose of prevention of drug addiction, HIV/AIDS among student youth

4. Establishment of networks. Establishment of the National Education Centers:
- inclusion into world network of national informative centers on recognition of education credentials ENIC/NARIC
- economics of education
- organization of employment of graduate students of higher educational establishments
- generalization, analysis, systematization, and development of databank of curricula, educational courses, practical exercises, innovative training methods developed in Ukrainian higher educational establishments during implementation of the project will promote reforms in Ukrainian higher education

On the other hand, Ukrainian social sciences are not »visible« in the international publications. As Web of Science shows, Ukrainian authors had no more than 125 publications in internationally recognized journals per year from 1996 to 2005. This is too small for such a large country.

Ukrainian statistics collects data on visits abroad of the Ukrainian scholars. For SSH, the number of visits grew in 2000–5 by 31 percent from 1,700 to 2,250 per year. This amount does not include data for independent think-tanks and, in many cases, long-term visits on foreign research stipends. More than two thirds of the visits are related to the participation in the conferences and seminars, 20–25 percent—with training, and only tiny share—with joint research.

According to official statistics, only 121 foreign grants were received by the representatives of Ukrainian SSH in 2006. These data, again, reflect only information from the research institutes, which are registered by the State Committee of Statistics. Individual grants or grants for independent think tanks are not included properly into this amount.

In many cases, international activities of the researchers are not considered as assets of the organizations, especially in provincial universities, where authoritarian traditions are strong. The author, as a member of Ukrainian-American Commission in 1999–2002 for Fulbright scholar's selection, has faced a number of cases in which authorities from provincial universities have tried to prevent scholars from the these universities from taking part in the preselection interviews. This has resulted in switching the commission's work to Saturdays and Sundays to give chance for contenders to take part in the competition.

National Approaches to the Evaluation of Scientific Capabilities

There are standard procedures for the evaluation of scientific activities established in Ukraine. The general estimates of the institutes are made on the base of the reviews, made by panels (commissions), consisting of the leading scholars in corresponding areas. These evaluations are regular (once in five years), and usually are positive for their participants. At least, there is no information about the closure of research institutes that received low marks.

Individual researchers are assessed by the special commissions of key specialists within their research institutes on the base of individual reports once every four to five years. This procedure is rather bureaucratic, and it is rarely based on quantitative indicators. High-quality research results are not critically important for the evaluation of university professors, as they are heavily involved in the learning process. This reduces the need to publish in prestigious journals and receive an international recognition. The number of publications in prestigious international journals is not used widely as the main indicator of scientific productivity.

According to data from the State Committee of Statistics, total number of scientific publications in Ukraine in 1991–2006 had an unwavering upward trend, even though there was a substantial cutback of R&D employment. An increase in printed output was remarkable in 2000–2006 (an average growth rate of more than 10 percent).

In addition to aggregate amounts of scientific publications growing rapidly, the increase of the number of publications on specific subjects in some regions of Ukraine are noteworthy. So, in 2000–2002 the number of papers on philology and linguistics jumped by a factor of three (!) times, despite a decrease in the number of researchers in these areas. There is no reasonable explanation for this phenomenon. Thus it is essential to note that this process occurred against the background of continuous underfunding of research activity: in recent years, a lion's share of R&D allocations was spent not on scientific experiments but on personnel remuneration and payment of public utility costs.

One of the most important reasons for such a state of affairs is the virtually nonexistent control over the results of statistical information processing regarding publication activity. The total number of publications of Ukrainian authors, mentioned in the international databases, remains stable during the recent years. This is an alarming signal, as the total number of world publications has grown robustly. The publications of Ukrainian authors on social sciences and humanities in referenced journals have become »statistically visible« in recent years only.

To some extent, a stable number of publications in leading journals could be explained by the exclusion of several Ukrainian journals from the corresponding databases in the 1990sand 2000s. This happened not because of somebody's wicked intention, but owing to transparent criteria applied, among which the level of quotation of scientific journals is the top priority. It should be noted here that the interest of foreign colleagues in Ukrainian journals was waning against the background of increasing number of scientific publications. For instance, seventy-one scientific journals were published in Ukraine in 1991, whereas the number of journals and other periodic publications (periodic collections of works or conference proceedings) exceeded a thousand in the first half of 2000s. This number actually includes various collections of scientific papers and abstracts that were turned into periodicals. Most of these journals and collections of articles are published by the local universities and they have very limited distribution among scholars.

Another problem is the relatively small number of specialized journals in different sciences. As it was shown in Kavunenko/Khorevin/Luzan (2006), the bulk of all SSH Ukrainian journals have no one specialization but publish articles on different disciplines.

Gabovych and Kuznetsov have analyzed publications on SSH by the specialists from the national Academy of Sciences in 1998–2003. They revealed that (Gabovych 2007: 66), among 2,536 publications of the institutes of philosophy, history, and law only 4 were in English. It is difficult to say how reliable and exact these data are but the general level of publications in internationally recognized journals remains low among Ukrainian scholars. The same authors have analyzed international publications of the 48 leading Ukrainian academics, who possess different administrative positions (directors and deputy-directors of the institutes, rectors, deans and so on). The results were disappointing: total number of their publications during 1991–2006 in the rated academic journals was only thirty-nine!

Cronyism and conformism dominate the Ukrainian scientific landscape. A number of key positions in the social sciences and humanities institutes are still in the hands of the old-style nomenklatura, which uses privileges, obtained in the Soviet times, to redistribute budget money in their interests. Very few such »scholars« have publications in the West, and knowledge of foreign languages remains poor in these institutes, especially among the older generations of researchers. Directors of the state institutes and research centers do not even consider the possibility of including foreign specialists into commissions, which evaluate the quality of scientific work of their organizations.

At the same time, the growing number of »alternative« (nonstate) research centers reflects the tendency to create new, mobile, and Western-oriented organizations, which are becoming more and more influential in the Ukrainian scientific community. Very often these research institutes and think-tanks receive the support of different Western funds.

Challenges and Prospects for the Support of SSH in Ukraine

The problems confronting the social sciences and humanities in Ukraine can be divided into four broad categories: structural, intellectual, personal, political and ethical. These categories are linked but are also distinct. Each must be considered in turn before we proceed to consider existing models and a new strategy.

Structural Problems

Here the problem is quite clear: the Soviet-type institutional infrastructure for supporting scholarship in the social sciences and the humanities has largely collapsed. While it is true that there are islands of success and of excellence, the overall picture is mixed.

The general economic degradation in Ukraine in the 1990s has eroded salaries, often impoverishing scholars and institutions. Some individuals and institutions have found nonacademic means of support.

The »internal brain drain« in many of these research institutes away from academic life, broadly defined, is often greater than the »external brain drain« abroad.

Economic decline has also eroded investment in the institutional infrastructure for social science and humanities research. Libraries and archival repositories find themselves in especially difficult positions. Libraries in Ukraine are faced with small budgets, lack of funds for international journal subscriptions, and delays in interlibrary loan exchanges, which force them into ad hoc short-term arrangements. Libraries continue to be confronted with the almost insoluble problem of book preservation at the same time that they struggle to catch up with advances made in digital equipment and databases.

As one research shows, a substantial number of representatives of the old generation do not use the Internet or even computers: among those who are more than sixty, this share is 45 percent, while in social sciences it exceeds 65 percent (Isakova/Levchenko 2007).

Of equal importance, the ability of scholars and institutions to remain in contact with one another has declined as transportation and communication costs have risen. The Internet has provided something of a counterweight to the overall decline in scholarly infrastructure, although access to the broadband internet is far from universal, especially in the small regional universities and colleges.

Administrative and bureaucratic expectations often remain strikingly inflexible, old-style managers prove increasingly ineffectual in a changing environment, and corruption degrades many otherwise worthy endeavors. In addition, the process of accreditation for new programs, institutions, and degrees is difficult. Barriers to the interconnection of research and teaching remain high in many institutions. As we mentioned above, scholars also complain of restrictive teaching loads, which relegate professional development to a very low priority.

Some of the worst inefficiencies of the old system remain while new phenomena are slow to evolve. The simultaneous devaluation of old fields and assertion of new disciplines, research strategies, and methods creates a number of additional vexing organizational and institutional issues that simply cannot be managed effectively. While these shifts in disciplinary focus and methodology are inevitable, they only exacerbate the structural constraints hindering the development of the social sciences and humanities in the country.

Intellectual Challenges

There are, of course, many positive aspects of the changes in Ukraine over the past decade and a half for the social sciences and humanities. For example, the country's academic community is more open to the broader international community than ever before. Moreover, a number of private research and training institutions have broken the monopoly of the Soviet state over the social science and humanities.

Most important, and far too rarely mentioned, the research agenda generated by the post-Soviet transition offers an opportunity to study some of the most fundamental issues of many social science and humanities disciplines. Some subfields, such as opinion polling, have taken on new life.

Another aspect of the problem is related to conformism in consideration of important political problems of the country. Thus in 2003–4, leading Ukrainian legal scholars from the Law Academy supported »constitutional reform«, proposed by the former President Leonid Kuchma, but changed their mind two days later, when President Kuchma has decided to postpone this reform.

Recently another vivid example of conformism has taken place. President Victor Yuschenko is a great supporter of the idea of rehabilitation of the nationalistic guerrillas (UPA) that has been formed by the Organization of Ukrainian Nationalists in the Second World War. This problem is not new for the country. In the early 1990s, a special commission, consisting of parliamentary members and scholars worked on this problem but its members could not find evidence that nationalists regularly fought Nazis in Ukraine in 1941–45, despite some episodes of clashes between small groups of UPA and the Nazi troops that took place in some (remote) parts of the country. These episodes are related to initiatives of the local commanders, and were not the results of coordinated efforts of UPA leaders. On the contrary, UPA took an active part in fighting the Soviet Army and Polish guerrillas during and after the Second World War, which is why they were considered Nazi collaborators. UPA was also responsible for the ethnic cleansing of the Polish population in western Ukraine and eastern Poland, when hundreds of thousands of civilians were killed or forced to move from their homes in the 1940s.

After the change of political power in late 2004, a commission of »respected scholars«, most of whom received their academic degrees by serving Communist ideology, decided to support the proposition of President Yuschenko to consider UPA as a legitimate, nationalist fighting force, and it has recommended the preparation of a special law on this issue. Parliament has not supported this recommendation, but after another change of »political weather«, it is possible that these changes in evaluation of UPA's role in Ukrainian history will occur.

Recent gains in archival access, international contacts, and free publication are themselves threatened by the structural impediments mentioned above. As a result, an often private struggle for survival inhibits the capacity of scholars and institutions to respond to the intellectual agenda of a lifetime.

Personal Barriers

This is the point at which personal considerations similarly impinge on intellectual endeavor.

On a more personal level, previous belief systems have collapsed, early optimism has turned to disillusionment, and wrenching social change undermines life and career strategies.

In many cases professional interaction has diminished, contributing to a lack of a sense of belonging to a group. In some areas, younger scholars are left without mentors as senior scholars have left academia or are virtually unavailable as they pursue other endeavors. This leaves even dedicated younger intellectuals without critical support and guidance. As the status of intellectuals declines, there is a corresponding diminished sense of mission for those engaged in intellectual pursuits. But the level of freedom in research is definitely higher than it was decade and a half ago. But, as we mentioned above, views, expressed by the representatives of cronyism are not challenged at all in many traditional research institutions.

We even can see how academics formerly punished by the Communist regime are trying to rewrite history and express ideas that have nothing with scientific approach. So, the famous Ukrainian philologist Ivan Dzuba writes books on Russian and Soviet history. The quality of the studies is relatively poor but his position as a former dissident, hero of Ukraine, and »guru« in social sciences gives him the power to publish these works without any criticism from his peers. For instance, in one recent book, Mr. Dzuba, on the basis of memories of a very few of Stalin's opponents, tries to prove that Georgian-born Stalin (Dzhugashvili) was Russian in his mentality (Dzuba 2003). We do not know any Ukrainian historian who challenged this opinion, despite the poor proof offered to support Dzuba's conclusions, which are based on two to three arbitrary selected sources only. In addition, all these sources are memories of Stalin's opponents. Now, to blame »others« for all Ukrainian problems, Dzuba could prove that all other prominent Bolsheviks were of Russian origin or they had »Russian mentality«. He could expect no negative reaction on his writings from his Ukrainian fellow scholars.

It is not even worth mentioning that politically oriented Ukrainian scholars (Korzh 2006) consider the ideologically motivated struggle of Bolsheviks against the peasantry in the period of so-called »collectivization« »ethnocide« perpetrated by Russians and Jews.

Some »scholars« go even further. In early 2005, MAUP, one of the biggest private Ukrainian universities, organized an »Anti-Zionist Congress« in Kiev, which was attended by some members of the Ukrainian Parliament.

Such negative attitude to other nations, expressed by representatives of the official science, leads to dissemination of hostility among ordinary people within the country and to neighbouring states. Specialists from the nongovernmental organization (NGO) Razumkov Center admit that the attitude to other nations in Ukrainian society is worsening in recent years, despite even some improvements in economic performance in early 2000s.

Political and Ethical Obstacles

Domestic and international politics in the Ukraine also pose obstacles that can threaten intellectual pursuits and Western efforts to provide support. The challenges posed by international relations are significant. Growing anti-Western sentiment in some eastern and southern regions of the country creates new obstacles for Western agencies attempting to implement their research programs. On another front, the continued functioning of a Soviet-style academic bureaucracy in many places makes engagement with institutional structures problematic at best.

Lack of political culture and the habit of noncritically accepting propositions of political sponsors have led to numerous quasi-scientific mystifications and even to the sharpening of latent conflicts in Ukrainian society. Recent events are related to the evaluation of the »golodomor« (artificial famine) of early 1930s and Pereyaslavskaya Rada of 1654, when scholars tried to »satisfy« both former president Kuchma and the opposition forces at the same time. Emotional estimates, not research results, dominated discussions in scientific literature.

Dependence on the state authorities and intention to preserve their positions force social scientists to violate moral principles in selection of members of the scientific elite. Almost 80 percent of the members of the Ukrainian Parliament have received scientific degrees in social sciences, most of them after elections to the Parliament. Skepticism is widespread among »ordinary« scientists about this placement of high-ranking politicians in the National Academy of Sciences without any real scientific achievement. »Selling« of scientific degrees and professorships in exchange of political dividends and extra money from the state budget has become an ordinary practice in the Ukrainian scientific community.

Ukrainian society is losing trust in the objectivity of research results in social sciences and humanities and cannot find answers (or advice) to the key questions of its development. Such a situation does not contribute to the solving of internal conflicts of the country.

Conclusions

It is really difficult to select disciplines that received the best results in recent years. Sociology has accepted new research instruments, and this has opened the way for more adequate studies of social changes in the country. At the same time, it should be noted that the sociologists, with their »applied« projects, were closely involved in the political fighting that took place in the country in recent years.

It is possible to conclude that the political sciences have been a failure despite significant investments and efforts. There is no peer-reviewed political science journal in Ukraine, and few political science articles are published internationally by Ukrainians.

Gender studies has emerged as an academic field with some interesting studies and results but this discipline has not received an official recognition from the side of Ministry of Education and Science and other state agencies.

It is evident that the situation has to be changed. First of all, competitive principles of domestic funds distribution have to be implemented. Currently this procedure is very limited and it is under the control of the directors of the research institutes and university rectors. Foreign specialists have to be participants of the grant committees.

The second step could be connected with development of special partnership programs between Ukrainian and Western scholars, as it was with numerous programs between Central and Eastern European (CEE) and EU countries in 1990s.

Third, Western scholars have to stop their support of researchers on a purely political basis. There are a number of cases in which directors of Ukrainian institutes with poor scientific reputation have been invited to international scientific forums or even elected to scientific academies abroad. Ukrainian Diasporas in the EU countries and especially in the United States have to have less influence on selection of research projects and programs of support. (For instance, as a result of pressure from the Diaspora, the American state agencies discriminatively support book publications in social science in the Ukrainian language only, despite the fact that the Russian language is also a common language of scientific communication in the country, especially in the eastern and southern regions of Ukraine).

But, of course, the key for healthy development of social sciences and their growing contribution to the resolution of social problems will be the understanding of the need for more independence from the state authorities and improving ethical standards within the Ukrainian scientific community. Falsifications of data and plagiarism have to receive a corresponding reaction from the scientific community, despite possible »repressions« by the authorities or »psychological pressure« from the influential opposition leaders.

Unfortunately, in Ukraine, it seems that both existing authorities and the opposition consider social scientists as their servants, not partners in solving the complex social and economic problems the country faces.

References

Dyker, D./Radosevic, S. (1999): Innovation and Structural Change in Post-Socialist Countries: A Quantitative Approach. Dordrecht, Boston, London: Kluwer Academic Publishers.

Dzuba, I.P. (2003): Tridtsiat Rokiv zi Stalynim. Piyatdesiat Rokiv bez Stalina (The Trap: Thirty Years with Stalin. Fifty Years without Stalin).Kyiv: Krinitsa. [In Ukrainian]

Egorov, I. (2000): Dinamika kadrovogo potentsiala ukrainskoi nauki v 1990-e gody (Dynamics of Cadre Potential of the Ukrainian Science in 1990s). Problemy nauki, 11, 31-38. [In Russian]

Gabovych, O./Kuznetsov, V. (2007): Secrets of the Native Non-Natural Science. Ukrainian Journal Economist, 1, 65-71.

GKNT (1990): Skhema Razvytya y Razmeschenya Otrasly Narodnogo Khozyastva SSSR Nauka y Nauchnoe Obsluzhyvanye (The Long-Term Plan of Development and Deployment of the Branch of the National Economy »Science and Science Service« of the USSR), two volumes. Moscow. [In Russian].

Isakova, N./Levchenko, O. (2007): New Information and Communication Technologies in Scientific Activities. Problems of Science, 12, 2-8. [In Russian]

Kanygin, Y. (1995): Put Ariev (The Way of Aryans: Ukraine in the Spiritual History of the Mankind) Kiev. [In Russian]

Kavunenko, L./Khorevin, V./Luzan, K. (2006): Comparative Analysis of Journals on Social Sciences and Humanities in Ukraine and the World. Scientometrics, 66 (1), 123-132.

Klochko, Y. (1994): Utechka Specialistov yz Naychnyh Organizaciy Ukrayny (Brain Drain from Ukrainian Research Organizations). Science and Science of Science, 1/2, 173-180. [In Russian]

Kneen, P. (1994): The Soviet Scientific Legacy: Some Differences in Interpretation. Science and Public Policy, 20 (4), 251-260.

Kolodiy, A. (2006): Ukrainskyi Regionalizm yak Stan Kulturno- Politichnoi Poliarizovanosty (Ukrainian Regionalism as a Situation of Cultural and Political Polarization). Agora, 4, 69-91. [In Ukrainian]

Korzh, V. (2006): Ukrainske Suspilstvo: Realii, Mifi chi Gipnotichni Sny? (Ukrainian Society: Realities, Myths and Hypnosis Dreams). Viche, 21/22, 34-39. [In Ukrainian]

Kysly, P. (1991): Problemy Finansuvannya Nauki v Ukraini (Problems of Financing of the Ukrainian Science). Visnyk Academiy Nauk Ukrayny, 10, 8-10. [In Ukrainian]

Paton, B. (1993): Pro Osnovny Zavdannya Akademyi Nauk na Suchasnomu Etapy (On the Main Tasks of the Academy of Sciences at the Current Stage). Visnyk Akademiy Nauk Ukrayny, 8, 3-11. [in Ukrainian]

Pavlovsky, M. (1992): Pro Economichnu Polityku Ukraynskoi Derzhavy (On Economic Policy of the Ukrainian State). Rozbudova Derzhavy, 4, 10-13. [In Ukrainian]

Romaniv, O.M. (1991): Nayka y Problemy Ukraynskoy Derzhavnosty (The Science and the Problems of the Ukrainian Statehood). Visnyk Akademyi Nauk Ukrayny, 10, 10-16. [In Ukrainian]

Ryabchuk, M. (2001): Dvi Ukrayni (Two Ukraines). http://Krytyka.kiev.ua/articles/s4-10-2001.html [In Ukrainian]

Challenges for Research and Research Policies in Belarus as a Post-Communist »Developing Country«

LARISSA TITARENKO

Recent Historical Development of the Social Science and Humanities Embedded in the History of Belarus

The Republic of Belarus appeared on the global map as a truly independent nation-state only in 1991, after the breakdown of the Soviet Union. Before this date, for a long time Belarus has been a part of many other states. In medieval times, the Great Duchy of Lithuania was a common motherland for current Belarusians, Lithuanians, and some Russians. Then, in the sixteenth century, this state merged with Poland, and the territory of Belarus became a part of Polish state for more than two centuries. Needless to say that social sciences were not developed in Belarus during these periods, and education and science were not available to the population. At the end of the eighteenth century, after three partitions of Poland, Belarusian lands were incorporated into the Russian Empire. Since then Belarus has become a part of the Russian Empire; this situation has lasted until 1917. During those years, there was no institution of higher education or scientific institutions in Belarus.

In 1917 two revolutions (February and October) totally destroyed the czarist government and radically changed the political regime. The socialistic Republic of Belarus was founded and soon became a part of the Soviet Union. During the period of the Soviet power, the first and only national university in Belarus was opened in Minsk in 1921 An Academy of Sciences, with several social science institutions, such as the Institute of History, Institute of Language and Literature, and the Institute of Philosophy and Law, was also founded during this period. One can say that Belarusian social sciences were born during these years and the humanities were reborn. The first national schools of sciences (such as School of Economics, School of Sociology, School of History) were also created in Belarus during this period.

However, during the Soviet period, the coordination of the scientific development in Belarus was conducted from Moscow—the only Soviet center of power (including the power in the sphere of social sciences). Usually only those international projects that were approved by Moscow were permitted in Belarus (Laptenok 1967; Babosov/Sokolova/Rusetskaya 1987; Babosov/Sokolova/Ageyeva 1988; Babosov/Ageyeva/Krukoysky 1989). The majority of scientific contacts were arranged between the scholars from different Soviet socialist republics, while, owing to the Marxist-Leninist concepts of ideological struggle and class differences, the contacts on the global (and international) level were limited.

A new page in the history of social sciences and humanities in Belarus has been launched in 1991: the new perspectives for international contacts immediately were constructed. During the first years of independence, 1991–94, Western influence was significant. For example, several projects in social science and humanities (SSH) were funded by the Soros Foundation in Belarus, IREX, as well as by some other foreign private foundations, primarily American (Manaev 1999). Foreign states also financed some research programs for scholar exchanges (Fulbright Program, DAAD). Belarusian scholars began to visit foreign countries rather often. However, their involvement in the Western (or international) projects was usually limited to collecting empirical data for their Western colleagues. Theories and applied research methods were selected by Western scholars. Therefore, the Western scholars explained and interpreted to the public what was going on in Belarus and the whole post-Soviet region.

Since late 1994, when the population of the Republic of Belarus elected the first president of the country, Alexander Lukashenko, the focus of social-political transformation as well as the scientific politicsslowly changed (Vardomatsky 1995). The Western influence was minimized everywhere, the foreign foundations were closed (or significantly limited in their activities) in Belarus, and the exchange of scholars was slowed down. Because of political isolation of Belarus on the international arena (after the political scandals related to the United States ambassador in Belarus who left the country in 1996, followed by the ambassadors from all the EU countries), the country was discriminated against in scientific contacts within the so-called Framework Programs in Europe and almost all exchange programs sponsored by the United States. Only those scholars who openly stayed in the opposition to the political regime, could enjoy the benefits of fellowships and scholarship abroad (Manaev 2005; Silitski 2005).

In parallel to this process, initiated by the West at the end of the twentieth century, Belarus elaborated its own »reversal policy« to the West. Thus, by the beginning of the twenty-first century Western social scholars were not allowed to do independent research in Belarus, and those Belarusian scholars who supported the political opposition were dismissed or kicked out of the country (or their research institutions were closed). For example, the Independent Institute of Socioeconomic and Political Studies (IISEPS), founded in 1992 and run by Dr. Oleg Manaev, which functioned for several years on the U.S. Center for International Pivate Enterprise (CIPE) grants, was closed in Minsk in 2005 (a bit later it was reopened in Lithuania; however, its founder and first director moved to the United States). Some private educational establishments with close connections to the West were also closed. One of the best private universities, the European Humanitarian University (EHU), mainly funded by the EU countries since its foundation in 1993, was closed in Minsk and therefore moved to Vilnius in 2004, together with several famous Belarusian social scholars. Currently EHU works in Lithuania, where Belarusian scholars (Anatoli Mikhailov, Grigory Minenkov, Almira Ousmanova, Vladimir Furs) do research in the fields of political science, history, anthropology, european studies, and humanities.

As a result, the Belarusian SSH is not really positive in regard to the research processes of globalization and internationalization (Yevelkin/Artyukhin/Korshunov 2004). Overall, approximately 10 percent of all Belarusian scholars are employed in the sphere of SSH (a bit less than three thousand people, among them approximately two thirds have a degree of doctor or candidate (*nauk*—an equivalent of Ph.D.). Out of them 70 percent are in social sciences, and 30 percent in humanities, according to the official statistics. More than two thirds of all social scholars live and work in Minsk where the social and scientific conditions are much better than in the provinces. This creates an enormous disproportion in size and quality of research. The Belarusian state does not have enough resources to finance the social research itself, while the sources from outside are strongly limited (for aforementioned political reasons). The average salary in science is lower than in industrial sector, while the average salary in SSH is even lower than in the natural sciences. Therefore research funds are always in demand. Western countries either do not want to invest in SSH for joint transnational research (TNR) in Belarus for topics that are officially approved, or these countries can more easily cooperate with Russia, Ukraine, or other Newly Independent States (NIS) where the state control is not so strong as in Belarus.

Currently, all the research institutions involved in social sciences in Belarus can be divided into three different groups (Zaiko 2001; Titarenko 2002):

- The institutions (both educational and research) that officially belong to the presidential administration and therefore have several preferences over the other institutions (the Academy of Management, the Institute of Social and Political Research, and some small, official research firms such as Ecoom).
- The institutions and groups that directly or indirectly support the political opposition and have therefore experienced difficulties in their activities (the IISEPS played a major role among them; after IISEPS was closed in Minsk, the rest of these institutions could function only through the Internet—mainly, their websites. There are several Internet journals such as *Fragments, Frontiers, Belarus Partisan, New Europe*, etc.
- The state institutions that try to be objective in their research. This group includes all the institutions of the social sciences and humanities within the National Academy of Sciences, Belarusian State University, as well as other state universities. They try to meet the traditional academic research goals in their scientific activities and investigate different spheres of interest, including national, regional, international, and global.

However, it is not easy to perform this mission of social sciences, as the lion's share of finance comes from the state. Therefore the state institutions and their administrators have to take into account their financial dependence on the state and therefore do not touch some political topics if their research (especially the results) can be interpreted as somehow negative for the image of the state. This indirect state control is a serious obstacle for the construction of social knowledge. However, if the research topics are far from the topics that are primarily under the state control (such as freedom of media, elections, and democracy), the research results can be quite critical, or at least precise.

In regard to these three groups of the institutions, three different mechanisms of production of social knowledge in contemporary Belarus can be selected.

The first mechanism refers to the privileged institutions: they have unlimited access to the sources of information, they have much more financial resources than other institutions, and they are usually allowed to research the touchiest topics related to the social-political sphere of Belarusian society. Their mission is to reflect the reality as it is and then help policymakers by providing them with the relevant information. As a rule, the empirical data produced by these institutions and their analytical reports are closed for the public. The scholars employed by these institutions need special permission to publish any results or participate in the conferences. Their major consumer is the presidential administration. Therefore, their research activities and their scientific results are mostly invisible and beyond public control. They do not have international networks: at best they can have contacts with the similar institutions within Belarus or the Commonwealth of Independent States (CIS). They produce more or less objective knowledge; however, they keep this knowledge as a »secret« while providing the public with some mythological ideas and limited data.

Usually, the scholars from these institutions are oriented to the national topics, so, they are not involved in international or transnational research and are not interested in the real transnational cooperation, even if their rhetoric may display some interest in international topics (Danilov 1998; Kotlyarov/Zemlyakov 2004).

Their major paradigms are nationally oriented: they are derived from the Belarusian national state ideology (this is not Marxism any more; however, some Marxist ideas are incorporated in it). These scholars officially reject the possibility of applying Western paradigms to Belarus.

The second mechanism of production of social knowledge has been constructed and used by the groups of the aforementioned oppositional social scholars. They also have some »secret« sources of information and therefore may produce the social information that is not known for the public (some of their data is used for political purposes only). The major difference is that the goal of the oppositional mechanism of knowledge production is to distribute their critical results to the public. The more they distribute the better for them. Quite often this distribution is a condition for receiving a grant (Manaev 2007). Being politically biased (as well as the first group) these scholars may not disclose some data if it is not »politically correct« from the view of the opposition. Thus, before the national parliamentary elections in Belarus at the turn of the millennium, the Internet resources of this sort usually published a lot of materials against the official powers and »scientifically predicted« the political changes after the elections. Until now such changes, however, did not happen as a result of the opposition activities, while the explanations of such mistakes were usually simple and trivial: »the results have been manipulated«.

Nevertheless, these counterregime sources of information are extremely useful for social scholars who are interested in comparative analysis of different views on the same situation (either in Belarus or in other countries). The existence of such sources of information can't be stopped in the current society because the Internet is not under the total control of any power. These courses are usually sponsored from abroad, because there is no financial support for them within the country. However, they are not very famous beyond the boundaries of the political opposition, the national intellectuals, and some foreign politicians. Therefore, this mechanism of social knowledge can't be considered as the »model« or »basis« for development of the international or cross-national research.

Their major topics of research are also national: in other cases, these scholars may loose their audience and can't play the role of the counterpart of the official social science.

These paradigms for an explanation of the current situation in Belarus and the world are foreign made (usually, the American social theories are the major sources of their theoretical framework).

The third mechanism of production of the social knowledge exists within the academia. Social scholars run the surveys with a limited budget; however, they try to be objective as long as they can and cover as many topics of research as possible (Rubanov 2000; Novikova 2001; Sokolova 2006). The state usually formally controls the topics and the money spending; therefore these organizations can't arrange broader research within or beyond the nation-state. At the same time it is possible for academics to participate in the international and transnational research with any foreign countries (Artyukhin/Zaytcev 2005).

The topics of theoretical research that do not dramatically depend on the financial situation may be focused on everything: from the local to national and global phenomena. That is why the scholars from the academia are the »core« persons for international or transnational surveys in Belarus. If they are lucky to be a part of the international (transnational) team, they usually have to accept the Western paradigms of research because the team leaders and coordinators are Westerners (Grishchenko 2004). The primarily role that Belarusian social scholars play in the international research is to collect the empirical data, to do the field study. As the labor cost is relatively low in Belarus, this division of labor is very convenient for the Western scholars: they use the data collected by their Belarusian's colleagues to prove the Western paradigms and Western theories.

Belarusian scholars usually work very hard if they are accepted for the international research (Rotman/Haepfer/Tumanov 2003; Titarenko 2004). Their names can be found in some reports (Euro Barometer, INTAS, European Value Studies, etc.) on the Internet. They always expect to analyze their own data in TNR. However, as a rule, they act as simple low-paid employees within these TNR projects. In some cases they agree to compromise their responsibility within the international team if the Westerners expect this behavior. In contrast to Belarusian natural sciences scholars who would prefer to do research abroad, Belarusian social science scholars prefer to collect the empirical data at home being employed in the joint (Western) project, because many of them do not know foreign languages, do not publish abroad, and clearly understand their dependency on the Western colleagues.

The only way to change this situation is to strive for equal cooperation between the scholars from Eastern Europe (in this case, Belarus) and the Western scholars, and therefore assure the Westerners that their eastern European partners employ high levels of professionalism and possess appropriate knowledgeof their research topics.

Current Situation and the Main Research Topics in SSH

According to the methods of the state financial support for research, only those topics that meet the officially approved »Priorities for the Fundamental and Applied Research for the Republic of Belarus for 2006–2010« for all spheres of science have been selected for funding as »the major and the most important« in the country. This means that all the scientific research has to follow these government-selected and clearly indicated directions. Briefly, the dominant topics are the following:
- Economics: sustainable development of the socially oriented market economy in Belarus, innovation, competitiveness, defense of economic interests of Belarus on the international level, new methods of economic and technical prognoses of Belarusian social and economic development.
- Philosophy: theoretical support of the Belarusian model of social development, preservation and strengthening of national identity of Belarusian population under the current conditions of globalization

and internationalization, methodology of research in the spheres of natural and social sciences in Belarus.
- History: history of Belarus from the medieval times till now, historical reinterpretation of several »touchy« events in the past in accordance with the current views, international context of Belarusian historical development.
- Political Science: theoretical support of the current political system in Belarus, including its structural components; peace policy non-alignment policy; Belarus and its international affairs, national interests of Belarus.
- Literature: modern Belarusian, linguistics and history, language and national identity, Slavic languages, role of the national literature in the spiritual development of a personality, moral education of the younger generation.
- Law: improvement of criminal legislation, struggle against the human trafficking, corruption, legal basis for the current Belarusian international affairs, legal justification of the Union of Belarus and Russia, improvement of the system of social security.
- Sociology: social stability, consolidation of interests of all social groups in Belarus, labor motivation under the current socioeconomic conditions, strengthening the family, motivation of fertility, gender equality, quality of life in Belarus, and education.

Other problems can be funded from the national private foundations or foreign sources (in the SSH the latter is not visible because Belarus has a status of an ICPC country). That is why Belarusian scholars are open to any international cooperation, especially within the European region.

However, as has been mentioned, currently Belarusian scholars participate only in three projects within the EU Seventh Framework Program (FP7), while there were no SSH projects at all with Belarusians within the EU Sixth Framework Programs (FP6). As for some private Western foundations and universities, Belarusian participation is more significant, especially in the sphere of the humanities. The cooperation is more visible on a bilateral level (between two particular universities or two particular scholars).

To illustrate the current situation of Belarusian involvement in the global SSH projects, some data from an empirical survey conducted in spring 2007 are provided. The questionnaire was distributed within the specialized electronic network of scholars, so that everybody could take part in this survey. All the questions were translated into Russian, and it took approximately twenty minutes to answer the questionnaire online. thirty-two questionnaires have been completed and then analyzed. Our hypothesis is that these respondents represent only the most advanced part of social scholars in Belarus, while those who did not participate had nothing to report on their TNR activities. It is impossible to prove this hypothesis statistically as the sample was so small. However, logically, the hypothesis seems to be correct, even if we take into account that not all the scholars who are really involved in TNR, participated in this survey.

Among those who took part in our survey, more than a half worked in big state universities (overall, 70 percent were employed in the institutions of higher education), almost a quarter—in research institutions; only 6 percent were employed in the private sector. Two thirds of the respondents were between forty and sixty. It is important to mention that the gender proportion was almost 50:50 while in general two thirds of those who are employed in SSH are women. It means that the proportion of male scholars in TNR is much higher than the proportion of female in comparison with the gender representation in the national science. Three quarters of respondents had scientific degrees of candidate (40 percent) or doctor (34 percent) of sciences. Probably, this is a reflection of the unequal practice under which younger scholars without a degree are welcome to TNR. Also, it can be a proof that the younger researchers are not interested in TNR. However, this fact may simply reflect the current situation: the average age of scholars in SSH in Belarus is close to fifty.

Among those who participate in TNR almost 90 percent knows one or more foreign languages, especially English (60 percent) or Polish (35 percent). Poland is the border country for Belarus, which is the reason why a big part of research cooperation is represented by Polish-Belarusian ties and projects.

Half of the respondents indicated Russian as their mother tongue, while all of them were citizens of Belarus. That is why cooperation with Russian scholars is the simplest, and some respondents are primarily involved in Russian-Belarusian research programs or publish in Russia. Also, Belarusian scholars more easily adjust themselves to the level of scholarly demands to the publications in Russia (because of the common Soviet background) than in the West.

Overall, almost 30 percent of the respondents said they had publications in foreign languages. We can assume that they did not mean the most prestigious journals, as there are no names of Belarusian social scholars among the first hundred of scholars mentioned in the Science Citation Index on the international websites. However, as the financial support of Belarusian scholars does not depend on the number of papers published abroad, a good part of Belarusian scholars are not interested in such publications at all; only those scholars who are personally devoted to TNR still participate in the preparation of foreign publications (either as a part of research team or as single authors). Overall, 15 percent said they published in English. Polish and Ukrainian were mentioned as foreign languages of publications second (5 percent each), and German was the fourth foreign language (2–3 percent). Only 2 percent of scholars said they have publications only in their native language.

Regarding the current practical involvement in TNR, almost half of the sample mentioned that they participate in exchange programs and TNR projects, while more attend international conferences from time to time. In most cases, such activities constitute 25 percent of all the scholars' activities.

The most popular region where Belarusian scholars did TNR is Europe (more than 60 percent). Unlike Russians, only a few mentioned the TNR with Americans (on the same level as with Africans). Currently, the research interests are focused only on Europe, that is, the most industrially developed countries, and not the »Third World« countries. Probably, to some extent, this focus reflects the self-evaluation of Belarusian scholars: they view themselves as equal among the other Europeans. In the future they also want to run TNR in this region.

How Belarus and Belarusian SSH Are Affected by and Respond to Globalization

Global Challenges

Primarily, the global challenges for research mean the changes in the world economy that influence—directly and indirectly—the financial situation for research in a particular »third« country. For example, the inflation of U.S. dollars (USD) actually decreased the financial possibilities of investing in research (on the global level and national level) in the CIS. If, for example, the Belarusian budget was equal to $10 billion USD in 2007, the purchasing power of this budget was much higher than the same 10 bln USD in 2008 in Belarus, because the purchasing power of USD fell. The new high prices for oil and gas brought enormous difficulties for the poor developing countries, and Belarus is not an exception. Even Russian scholars worry about their relatively low salaries and low level of research finance, while the situation in Russia is much better than in Belarus or other post-Soviet republics (with the exception of the Baltic states). The current global financial crisis put Belarusian science on the edge of survival: approximately one third of the staff was fired, while the payment level of the rest was reduced.

All the SSH fields have got much less finance in Belarus in 2007 and 2008 because of inflation and a limited budget for social sciences (although it was not openly declared by the government). This is an additional reason for Belarusians to pursue international cooperation. Any global financial support for science is extremely welcome by scholars in Belarus, because these scholars would like to be involved in the international projects and receive financial support for their research.

Regardless of all negative aspects of the financial and political situation, the fact is that Belarus has qualified personnel for potentially successful research in SSH. There are many scholars with international experience and some younger scholars without experience but with good knowledge of foreign languages and a strong desire for international collaboration. This is a small group, but their motivation for TNR is strong.

Thus, in the aforementioned survey, more than a half of the respondents said that TNR occupied 25 percent of their research activities, 35 percent that TNR activities constitute 50 percent of their activities, and 5 percent mentioned 75 percent. On a national level, of course, all three groups taken together will constitute maybe 10-15 percent scholars, or even less.

The same research data show that Belarusian scholars are strongly motivated to be involved in international/transnational research. The major reasons to participate in such research, according to our data, are the following:
- »to learn from others« (25 percent),
- »to match with the internationalization« (22 percent),
- »intellectual curiosity« (6 percent),
- »to promote my academic career« (6 percent).

Belarus, like many other countries, is interested in the new »global« topics such as global illegal migration, human trafficking, unemployment, ecological and moral issues, and the like. Belarusian scholars can contribute their knowledge to transnational projects on these global issues. The problem is that Belarusian scholars do not have enough resources for SSH. Also, they are often ignored by their potential Western partners. Meanwhile, it is common in many countries that the national governments do not finance comparative studies on a large: such projects are financed by non-governmental funds. However, the major goal of Belarusian scholars now is to increase cooperation and the number of joint projects. This is the official policy of almost all post-Soviet countries, including Belarus.

Another aspect of the global challenges is quite different: currently, most topics of research can't be discussed and analyzed on the national or regional level without an analysis on the global level. So, at least several countries of a region have to be involved in solving practical issues. Research is becoming global, international, and transnational.

Currently, the national government does not know much about such research. If the government and financial bodies do not understand these new global features, and therefore do not actively support and stimulate the Belarusian involvement in transnational research, it will mean the failure to meet this challenge. So, it is necessary to establish some special European Union or Council of Europe policies to explain the situation to the third countries and not leave these countries to struggle alone with their problems.

Western support can be viewed in the international promotion of Belarusian involvement in TNR projects. Currently, the government does not even know the results of many TNR projects. According to the aforementioned survey, the major recipient of the TNR results is the academia (35 percent); civil society comes next (22 percent), and politicians have been mentioned as major recipients of TNR data by 5 percent of the scholars.

Significant Western assistance can be provided by the establishment of close relationship between the EU research organizations and Belarusian research institutions with a goal to stimulate the joint participation in the EU projects (unfortunately, only a small number of the EU Seventh Framework Programs [FP7] calls are directed toward the »third countries«, and a limited number of SSH projects can be financed, so that many EU teams prefer not to cooperate with the unknown partners from the developing countries like Belarus). Therefore, scholars from Belarus can hardly present their results at the international conferences and compare the data with other countries, because the methodologies and samples, among other things, are different.

Regional Challenges

These challenges are also very important. For example, the Eastern-European post-Soviet region of which Belarus is a part differs from the Central-European region where such countries as Poland or the Czech Republic play a role. Scholars from each region can easily understand each other and cooperate in Russian because of their common background (if they are not politically biased). It makes sense to work together in the Central-Eastern European region regardless of the fact that some countries belong to the EU and some other countries do not. Actually, in the post-Soviet region all the scholars are experiencing the financial shortage in their research. That is why it is necessary for Belarusian scholars to cooperate with the EU or other Western partners in any topic that is available at this moment.

Sometimes the regional needs promote some particular theories and methods as the most important and appropriate for the moment, while on the global level these methods and theories cannot be used. This does not mean that regional methods or theories are bad: the problem is to learn more about all the possibilities and have good explanation why a particular theory was taken. Regional methodology can be often be used for special topics of research. In any case, all the methods and paradigms have to be known to all participants, be well explained, and compared with other possible methods and approaches.

Therefore, Belarusian scholars have to read the current journals and follow the first-line results. Regional national methodology is not enough if it does not allow comparison of the research results from different countries.

National Challenges

Of course, Belarus as a post-Soviet developing country aims to improve the standard of living, and Belarusian scholars are oriented to perform better research, contribute to their country benefits, and improve the system of education, among other goals. In order to achieve the national goals, Belarusian scholars need international support.

For many years after 1991, the International Association for Cooperation with the Scholars from the Newly Independent States (INTAS) programs perfectly functioned and helped former Soviet scholars to survive: scholars stayed in their native countries, did research with Western colleagues, and had some extra money to survive. INTAS established the workable model of the Eastern-Western communication, transfer of knowledge and research technology, and exchange of views. Thousands of post-Soviet scholars participated in INTAS projects.

The decision to close INTAS programs was a big loss for Belarus and the CIS in general. This program provided Belarusian scholars with some international experience, increased their self-esteem, created a possibility to communicate with colleagues from other countries, and kept these scholars more independent from their national government. The end of INTAS financial support for informational research and cooperation resulted in the strong decrease of Belarusian involvement in the new international projects in SSH.

Unfortunately, the FP7 could not substitute INTAS. There are several reasons why these programs are not equal.
- Belarusian sources of information on FP7 are limited (because of bureaucracy, complexity of websites, lack of skills how to find information on these websites, etc).
- Belarusian involvement in FP7 is low, although officially there are no obstacles. Belarusian scholars do not have enough knowledge and skills to apply for a grant and therefore to participate.
- Since 2007, the amount of information on FP7 increased, and new rules for participation were introduced for new calls. Although there was a lot of useful information in CORDIS website, scholars were not familiar with how to find it. They needed some help, guidance, translation, and explanation of the possibilities.
- Unlike the EU countries, Belarus does not have state support for participation in the majority of international projects. Only one informational office exists for FP6/7, which can't assist all the scholars in need.

From 2006 to 2007 the total number of projects where Belarusian scholars were included has increased from nineteen to forty-eight. However, only six out of forty-eight projects received funding from the EU. The level of success was higher than in Russia; however, it was much lower than for the EU countries. Unfortunately, the number of SSH projects was small: there were 5 applications in SSH, and only three successful projects. In the 2008 SSH call, no Belarusian partners were selected for competition, while the government insists on their participation regardless of the complicated reality.

NCP-SSH Construction as a Response to the Challenges

Before 2007, there was only one, national National Contact Point (NCP) in Belarus. In the winter of 2007 the Belarusian government decided to construct eight thematic National Contact Points according to the priorities of FP7. Following the recommendation of the EU, Belarus created the new NCPs with a hope that NCPs will become effective informative and communicative tools for scholars.

Construction of NCP-SSH is the first step in the process of strengthening international European cooperation. Among third countries, only Russia, Belarus, and Egypt constructed the national network of NCPs to meet the EU recommendations and improve their international cooperation.

Belarus is a country that belongs to the EU International Cooperation Partner Countries (ICPC). The priorities for NCP in SSH include the creation of international network and assistance for particular organization in their contacts within FP7. With the help of NCP-SSH scholars can receive fresh information from FP7 website (seminars, calls, trainings, reports, etc.), so that they can participate in many activities (in case they have resources). The scholars and their organizations also receive regular messages from the national office regarding everything (there are many electronic addresses for distribution of such information). The scholars actively participate in the trainings organized by the FP6/7 projects. However, the most difficult issues are still the partner search and selection of the appropriate project to apply.

More information about Belarusian involvement in the international network can be found on the special website www.net4society.eu.

How Does Higher Education in Belarus Respond to and Prepare for Internationalization?

The process of internationalization of the system of higher education (HE) in Belarus can be divided in three unequal periods.

The first period covers the years after the breakdown of the Soviet Union when the old Soviet system of HE continued its functioning without any serious changes.

Since the mid 1990's, a period of reforms began. Its goal was to make the HE in Belarus similar to the Western countries and to meet the Bologna criteria. Thus the new degrees (BA and MA) were introduced in the most universities, the programs of education were reformed, new subjects (similar to Western subjects) appeared, and the like. This period covered almost a decade with some clear positive and negative results.

However, a few years ago, a reverse tendency became stronger: to turn back to the pre-Bologna system. The BA degree disappeared from the university, and the previous, major degree, »specialist«, returned. Also, a one-year MA degree became a norm. The whole Bologna system was rejected (if not totally, then substantially).

Currently, the Western innovations in HE are not welcome; on the contrary, the national initiatives prevail in social sciences. For example, many university professors teach the students of social sciences in the same way they did a decade ago or more. They do not use Western paradigms (with some exceptions, of course), as the national paradigms seem to be »more appropriate«. There are still many scholars from the Soviet period employed at the universities (see Professors 2001).

As a result, Belarus is behind the West in the educational issues: many Belarusian students are not familiar with famous Western theories and special methods of research. Only the students in some universities in Minsk, the capital city, where the majority of professors are employed, can learn more and be familiar with the modern social sciences. Additionally, some students are involved in Erasmus-Mundus exchange program and therefore familiar with the Western staff.

There are several approaches to HE in Belarus, and the most powerful is the so-called »national«: its aim is to keep a unique Belarusian system of HE, not adapted to Western standards. For this reason, students from Belarus can hardly be accepted for traditional BA and MA programs within the EU countries, and the system of credits is also not used in Belarus, so that the students' exchange is not useful for Belarusian students. The exception is the EHU: situated in Vilnius it is still oriented to Belarusian youth.

Because of the different approach to HE in Belarus it is difficult to compare it to any EU country.

However, this feature does not create any obstacle for involvement of young scholars from Belarus in the transnational research: first, Belarus is a part of Erasmus-Mundus program; second, the students study foreign languages and their knowledge is much better than it was ten to twenty years ago; third, if the students work in the international team they earn experience from the practice. All in all, Belarus is a »normal« potential partner for any transnational research.

Research Policies in SSH

According to the aforementioned survey, only 20 percent of scholars believe that research policy to support the TNR in SSH exists in Belarus while 25 percent believe the opposite. In reality, the research policies are elaborated by the branches of the government (State Committee on Science and Technology, Ministry of Education); to some extend, they determine the real participation of Belarusian scholars in TNR (however, not totally).

First of all, officially there are no obstacles for international research in Belarus. The government does not help much in establishing contacts (no preferences for them) but does not prevent such projects and contacts. At the same time, the State Committee on Science and Technology wants scholars to increase contacts, conduct joint transnational research, and bring new knowledge to our country.

However, secondly, Belarusian fiscal legislation differs from the legislations in the EU countries. Therefore the scholars experience additional problems with managing expenses and paying taxes when they have joint projects with the EU countries. I am sure that the more projects we have, the easier will be the mechanism to solve these issues (mainly, to change the national rules). The EU can contribute to this process of inclusion of third countries into the EU–funded projects (FP7) if more calls are especially prepared and oriented to the participation of the third countries (with such topics as migration, democracy, new markets, morality, and ecology).

Research policy in Belarus is based on the governmental approach. As in the past, the government has the right to decide which topics and spheres of science to finance. So, the state-owned universities and academic institutions have the lion's share of research money from the state that dictates the priorities. Usually, SSH are financed on the basis of »getting the rest« (always not enough). For this reason, individual scholars and the institutions search for extra research money from different foundations to whom can apply for funds. As for the national level of foundations, there are not many, and therefore, the funds are insufficient. As for foreign foundations (including those funded by the EU) they function in a way that is still not understood by our scholars. The major problem is that the initiative belongs to the EU countries: the third countries are always in a position to be selected by someone else. Thus, if our scholars have some preliminary contacts with foreign scholars, there is a chance to be included in some FP7 projects. Unfortunately, the brain drain of the 1990s took some promising young scholars out of the country. If they still have connections with our scholars, it is limited to their previous fellows. (This is a typical model for research in natural sciences.) Belarusian SSH scholars can't write proposals for TNR projects, and they do not have extra money to hire managers for their international projects. It means that the scholars have to be managers for their own projects, that is, to combine scholarly and managerial competencies. Therefore, Belarusian scholars evaluated such competencies as very important for the involvement in TNR projects, as well as social competencies. In such situations when Belarusian scholars combine the skills for science and management, they often loose as managers. Only additional trainings can help to improve their skills, while extra money for TNR projects would pay for managers and breach this »vicious circle«.

What are the problems?
- A relatively low level of knowledge of foreign languages. According to our survey, half of the respondents know one foreign language—preferably English. This, however, it is not enough for TNR. Only a few scholars mentioned that they »often publish abroad« while more than 60 percent publish papers only in the native language. For this reason, instead of scholars with Ph.D.s in social scientists, some scholars with Ph.D. or MA degrees in languages can profit: they are more likely to do TNR because they can easy communicate with foreigners. Needless to say, these newcomers to SSH can't substitute for the trained social scholars, and the final performance of the former group differs greatly from the results of the latter group (of »normal« social scholars). A second inappropriate possibility is to invite social scholars who do not know a foreign language and then communicate with the help of professional translators. However, any research work can be successful only in case of the mutual personal understanding between the partners, therefore the kind of communication by »translation« is negative. Additionally, this approach does not stimulate the scholars to learn languages for their professional needs.
- A lack of a middle-age generation of researchers in Belarus who are ready for TNR (Artyukhin/Zaytcev 2005). Social sciences represent mostly the so-called »old« generation of scholars (50+), socialized and educated in Soviet times. Additionally, there is a smaller group of the younger postgraduates (25–30), educated in the independent Belarus after the breakdown of the Soviet Union, who still need more experience and knowledge to be equal partners in TNR. Probably, it is necessary to combine the old and the young scholars in any serious TNR projects; sometimes it is not possible because of the aforementioned problem with foreign languages.
- Belarusian »technical« level in SSH (equipment, IT) differs from the Western level of equipment and skills. In the natural sciences our scholars can buy and use the up-to-date technology. In SSH, in the most cases the scholars use the old (out-of-date) computers and old software and statistical programs, and the communication possibilities are limited. That is why it can be difficult for Western colleagues to communicate with scholars from Belarus (and other similar countries) on a practical level or collect data in the same way as they do at home country.
- Belarusian legislation on science is not well defined, especially in the issues related to the international projects. It is necessary to know everything, or »have ties with authorities«, or take the risk of

mistakes when being involved in the TNR. Therefore many scholars try to do research that avoids international contacts. Also, there are some national restrictions in spending money from the international projects (the taxes are higher than in the EU, the overhead rates differ, etc.).
- The »diplomatic« issues (such as visas). Citizens of Belarus—even the well-known scholars who travel a lot—need visas to visit the majority of foreign countries. The process of getting a visa is long and complicated: lots of documents have to be collected, lots of time spent for queues to the embassies, and lots of money from the projects or out of the researchers' own pockets spent for travel. Probably, for this reason the EU scholars would rather prefer not to include the scholars from the third countries whose participation always creates extra problems of this kind or provides some of them a workplace in the West (White/McAlliste/Light/Löwenhardt 2002; Korosteleva/Marsh/Lawson 2002). The communication can be much easier if the European Commission makes some special procedures for those scholars who are involved in the international projects. This is an open question that is important for the majority of international projects in Belarus.

References

Artyukhin, M./Zaytcev, A. (2005): Republic of Belarus Within the New Tendencies of International Intellectual Migration. Sotsiologiya, 3, 50-56. [In Russian]

Babosov, Y./Sokolova, G./Rusetskaya, V. (1987): Social Structure of Soviet Society: The Working Class. Minsk: Nauka i tekhnika. [In Russian]

Babosov, Y./Sokolova, G./Ageyeva, L. (1988): Social Structure of Soviet Society: The Intelligentsia. Minsk: Nauka i tekhnika. [In Russian]

Babosov, Y./Ageyeva, L./Krukoysky A. (1989): Social Structure of Soviet Society: The Peasantry. Minsk: Nauka i tekhnika. [In Russian]

Danilov, A. (1998): Power and Society: in Search of a New Harmony Minsk: Universitetskoye. [In Russian]

Grishchenko, J. (2004): The Elite of Belarus in Search for the Global-Local Invariant of Socio-Cultural Development, Democracy and Local Governance (Ten Empirical Studies). Pultusk, VSH, 52-71.

Korosteleva, E./Marsh, R./Lawson, C. (Eds.) (2002): Contemporary Belarus: Between Democracy and Dictatorship. London: Routledge.

Kotlyarov, I./Zemlyakov, L. (2004): Religious Denominations in the Republic of Belarus. Minsk: Institute of Management. [In Russian]

Laptenok, S. (1967): Moral and Family. Minsk: BSU. [In Russian]

Manaev, O. (Ed.) (1999): Youth and Civil Society: the Belarusian Model. Minsk: V.M. Skakyn Publisher. [In Russian]

Manaev, O. (Ed.) (2007): Belarus and »Wide Europe«: In Search of Geopolitic Self-Determination. Novosibirsk: Vodoley.

Manaev, O. (2005): Emerging of Civil Society in Independent Belarus. Sociological Experience: 1991-2000. Riga: Layma.

Net4Society (2009). Trans-national Co-operation Among National Contact Points for Socio-economic Sciences and the Humanities. www.net4society.eu: [Date of last access 25.08.2009]

Novikova, L. (2001): Religiosity in Belarus at the Turn of the Millennium: Trends and Peculiarities of Manifestation (Sociological Aspect). Minsk: BTN-inform. [In Russian]

Rotman, D./Haepfer, C./Tumanov, S. (Eds.) (2003): Lifestyle and Health of the Population in the Newly Independent States. Minsk: self-published. [In Russian]

Rubanov, A. (2000): Mechanisms of Mass Behavior. Minsk: Pravo i ekonomika. [In Russian]

Silitski, V. (2005): Preempting Democracy: The Case of Belarus. Journal of Democracy, 16 (4), 83-97.

Sokolova, G. (2006): Belarusian Labor Market: Tendencies of Development and Social Mechanisms for Regulation. Minsk: Pravo i ekonomika. [In Russian]

Titarenko, L. (2002): Belarus: Fears, Hopes, and Paradoxes of the Transformation. In: E. Shiraev/V. Shlapentokh (Eds.), Fears in Post Communist Society: A Comparative Perspective (pp. 81-96). New York: Palgrave Press.

Titarenko, L. (2004): The World of Values of Contemporary Belarusians: Gender Aspect. Minsk: BSU. [In Russian]

Vardomatsky, A. (1995): Election of the First President of Belarus. Sociological Studies, 9, 45-49. [In Russian]

White, S./McAllister, I./Light, M./Löwenhardt, J. (2002): A European or a Slavic Choice? Foreign Policy and Public Attitudes in Post-Soviet Europe. Europe-Asia Studies, 54 (2), 181-202.

Yevelkin, G./Artyukhin, M./Korshunov, G. (2004): Atlas of Science of the Republic of Belarus. Minsk: Tekhnoprint. [In Russian]

Zaiko, L. (Ed.) (2001): Belarus: On the Way to the Third Millenium. Collection of Articles. Minsk: Belarusian Think Tanks, FilServplus. [In Russian]

Challenges of International Collaboration in the Social Sciences

DORIS WEIDEMANN

Introduction

Internationalization is neither an abstract process nor an institutional inevitability. It is based on the willingness and ability of individuals to strike up international contacts, to initiate and maintain international collaborations. As the contributions to this volume illustrate, social scientists operate in local contexts that differ widely with respect to the availability of resources, international orientation, and research priorities. Researchers who are collaborating internationally are faced with the challenge of achieving understanding across these—and other—differences while having to organize the joint research endeavor and to uphold commitment to a common cause.

Research collaboration is not a well-defined term. I shall use it here as referring to the interaction of individuals with a (partially) shared goal that is directed at the exchange and/or production of scientific knowledge, provided that the interaction is sustained for a significant period of time and accompanied by a certain sense of cohesion of the collaborating group. Collaboration may vary with respect to the intensity of contact between partners, media of communication (face-to-face or virtual), the nature and organization of tasks (sharing or dividing tasks), the role of collaborators, etc. A central element of research collaboration is that »divergent perspectives are brought together to address a shared question or object in the interest of producing knowledge« (Cornish/Zittoun/Gillespie 2007: paragraph 18). Divergences might be across countries, across theoretical or methodological orientations, or across disciplines (ibid.).

While »intercultural collaboration« may be the more adequate term when it comes to conceptualizing the diversity of perspectives—in fact all of the above dimensions may productively be framed as »cultures«— I shall prefer the term »international collaboration«, which accentuates the »cross-country«-perspective. This choice of terminology not only replicates the logic of country studies that underlies this entire volume but also highlights those factors and constraints that result from the national orientation and organization that pertains to much of the social sciences. While cross-cultural diversity is an important factor in science collaboration, the dimension of the »national« possesses important explanatory power with respect to understanding the challenges of global research collaboration. Therefore, I shall use »international collaboration« as the general term for collaboration across national borders, while explicitly acknowledging the importance of different disciplinary, gender, generational, religious, or linguistic cultures.

When concentrating on the interpersonal dimension of international research collaboration, some neglect of political and institutional aspects is inevitable. It is, however, important to consider the general context that international social research projects are operating in. I will therefore briefly review some findings on the modus operandi of social science research that influence international collaborative enterprises. Among these, general rules of »academic culture«, disciplinary orientation, the importance of national frameworks and international power structures seem to me the most important ones.

Academic Culture: How Social Sciences Work

Disciplinary Cultures

With a humorous and slightly provocative undertone Becher and Trowler (2001) refer to scientific communities and research disciplines as »academic tribes« and their »territories«. The ethnographic vocabulary emphasizes the observation that academic activity is guided by collective rules, norms, and aims that set the specific (cultural) context of producing scientific knowledge. Some of the fundamental ideals and values of academic culture have been formulated by Robert Merton (1942/1985 in Felt et al. 1995: 60–63) who listed *universalism* (scientific contributions should be judged on the basis of scientific criteria only), *disinterestedness* (researchers pursue the aim of advancing scientific knowledge, not the aim of personal gains), *organized skepticism* (scientific findings should be validated by a critical scientific community) and *communism* (scientific findings should be made public) as constitutive characteristics of science. While Merton's views have been criticized as overly idealistic and inappropriate for today's market-oriented scientific research (e.g. Etzkowitz 1993), these ideals have survived in collective (self-) representations of scientists and may thus well be regarded as an essential part of academic culture. It is not the whole picture, however. Competition and the importance of intellectual ownership as basis for reputation are other elements of academic life. An inherent paradox consists in the fact that scientific reputation is based on the recognition by the same peer group that also acts as competitor in the scientific game. Merton's picture of a scientific community that harmoniously strives to advance scientific knowledge therefore needs to be supplemented by a strong element of competition and fight for prestige. As a result, one of the highly visible characteristics of academic culture is the great importance that is placed on rankings of institutions, journals, and individuals. As Becher and Trowler muse: »One of the striking features of academic life is that nearly everything is graded in more or less subtle ways« (Becher/Trowler 2001: 81).

Many of the rules and values of academic life aim to back the claim to validity of academic findings and are therefore centrally linked to the scientific goal to produce »true statements«. Still, mechanisms of scholarly life are not strictly rational nor are they directed at the mentioned aim exclusively. Sociology of science, gender research, and laboratory studies (e.g. Knorr-Cetina 2002) have pointed out the importance of epistemic cultures, that preconfigure topics and modes of research as well as interpretation of findings, and have shown that despite a strong claim to the validity of meritocratic criteria, status and recognition in the academic community strongly rest on (male) sex, (Western) origin, and affiliation with (famous) institutions.

Becher's and Trowler's finding show that despite strong calls for interdisciplinarity, disciplines continue to be the general basic framework of »normal« academic work in the social sciences (see also Wallerstein et al. 1996 who take this a major point of criticism). Their analysis of »academic territories« discerns noteworthy differences between academic disciplines that concern epistemological and methodical approaches, choice of research topics, communication practices, modes of collaboration, etc. However, the increasingly acknowledged fact that empirical findings of natural and social sciences alike are, as a rule, »underdetermined« has partly attenuated formerly strong differentiations between »hard« and »soft« sciences. The insight that empirical data always need a certain amount of interpretation and agreement in order to »make sense«, has somewhat lessened the gap between conceptions of the natural sciences and humanities. Pointing out specific characteristics of the social sciences therefore accentuates trends rather than categorical differences—especially because of the broad range of social science orientations that reach from close affiliation with neuroscience on the one hand to hermeneutic approaches on the other.

With respect to our topic, the following findings appear as especially relevant: Becher and Trowler observe that in comparison to the natural sciences, social research is operating in a less competitive climate with less pressure to immediate publication. Because research questions often address complex issues that are not easily divisible in single tasks, research collaborations are less common than in the natural sciences and exist more frequently in applied than in theoretical social research. Consequently, the image of the individual scholar who singularly works at his or her desk is very much in place. Merton's image of science as a disinterested search for universal truth is, again, mitigated by the observation that intellectual controversies are rarely carried out in a direct manner. Instead, researchers of different schools of thought tend to meet among themselves and publish in separate media. While most of these characteristics do not prepare promising ground for international collaboration, it is the inherent national orientation that is probably the most prominent feature of the social sciences.

National Orientation

Because social science disciplines evolved alongside the creation and in reaction to the needs of modern nation states in nineteenth-century Europe, national frameworks are deeply engrained in research fields and concerns. Wallerstein et al. (1996: 80) remark: »The social sciences have been very state-centric in the sense that states formed the supposedly self-evident frameworks within which the processes analyzed by the social scientists took place«. As a result, social science thinking firmly rests on nation-oriented categories and most research is restricted to the institutions, society, or citizens of a specific state. Even if research is carried out internationally, states remain the unit of comparison, thereby rather confirming than transcending the category of the national.

A national orientation is also visible in career patterns: different from careers in the natural sciences, »most careers in the humanities and social sciences have proven rather resistant to internationalization« (Jarausch 2005: 32). Longer research periods abroad make researchers miss important chances of network building in their home country and do not contribute to scientific prestige. Generally speaking, international experience is only regarded a plus in area studies where scientific expertise rests on intimate knowledge of the culture and language of the target area (ibid.).

Maybe as a result of both of the above observations, publications are also in large parts targeting national audiences. Different from the natural sciences that have adopted English as major language of publication in the social sciences and humanities there is a noticeable preference to read and write in local languages (Yitzhaki 1998).

Taken together, all these trends firmly root social scientists in their national environments and constitute an unfavorable precondition for international collaborations.

Academic Power Structures

It is characteristic of the current international power structures that some academic communities do not match the above description. In many countries of the lesser-developed world, social sciences were a transplant from the United States and Europe and have therefore featured a strong international orientation from their beginning (cf. the contributions in this volume). In these countries, research topics, publications, and careers are closely modeled at and dependent on Western countries—usually the United States. While this does not mean that the general national framework that Wallerstein et al. (1996) observe is absent from the social science in these countries, it does, however, imply an intimate knowledge of other world regions (the developed ones)—sometimes to the detriment of an understanding of the home environment.

Reflections on the global science system usually employ a center-periphery model that identifies »the West« as academic center and most parts of the »non-West« as periphery. Alatas (2003) introduces the notion of »social science powers« to designate countries that occupy center position (United States, Great Britain, and France) and explains:

»These are defined as countries which (1) generate large outputs of social science research in the form of scientific papers; (2) have a global reach of the ideas and information contained in these works; (3) have the ability to influence the social sciences of countries due to the consumption of the works originating in the powers; and (4) command a great deal of recognition, respect and prestige both at home and abroad.« (Alatas 2003: 602)

Alatas extends the common bipolar model of »the West« and »non-West« (or center and periphery) to include a third category that he calls »semi-peripheral social science powers«—a category that he reserves for countries, such as Australia, Japan, Germany, or the Netherlands (ibid., 606) that occupy an important but not dominating position. As examples from European research contexts will show, this differentiation provides an essential conceptual basis for understanding many of the European debates on internationalization and the difficulties of international collaboration that involve researchers from these »semi-peripheral« research communities.

International power structures affect cooperation in many and significant ways (Alatas 2003; 2006): (1) Because affluent countries sponsor most international research activities they dominate international research agendas. (2) International inequality is reflected in academic division of labor that usually reserves theorizing for Western partners and has partners from peripheral countries concentrate on collecting empirical data. (3) Following the dominant position especially of the United States and Great Britain, English has become the international language of science, which plays to a further advantage of English native speakers and the spread of Anglo-Saxon theories (also see below). Finally (4), power structures also influence project partnerships: Often, peripheral research communities entertain collaborations predominantly with Western counterparts. As a result, interregional ties are often weaker than ties to faraway United States or Europe.

Even though following decolonization the nature and degree of »Western« academic dominance have been widely discussed, its effects have neither been mitigated nor are researchers from the social science powers usually aware of them.

The above mentioned issues constitute the general context and preconditions of international research collaboration. They shape actual collaborative research, its development and outcomes. In the following, I will move from context to the level of actual research collaborations and discuss some of its prominent challenges: the need for translation, epistemological and methodological questions, the organization of joint research and intercultural communication problems (also see Weidemann 2007).

Challenges of International Cooperation

Translation

Since the Second World War English has become the international language of science (Ammon 2001). Even countries with strong national science traditions have in recent decades been switching publication languages of science journals from national languages (such as German or French) to English. Though differences can be noted between regions and disciplines—in Europe, English was adopted as language of science earlier in the northern countries than in the south and more widely in the natural sciences than in the social sciences and humanities (Carli/Calaresu 2006: 31)—the general trend is uncontested. It would seem that the ubiquitous use of English makes translation obsolete. Yet, as social sciences are deeply embedded in idiosyncratic social realities and depend on natural languages to a far greater extent than the natural sciences, linguistic diversity and the need for translation constitute one of the most acute and unsolved problems of international collaboration.

Following the *interpretive turn* with a new focus on the concept of »text«, translation has in recent decades been reconceptualized in order to include a broader variety of interpretative acts. Reflections on translation have transcended the original confines of literary and translation studies and have become a central referential term in much of (culturally oriented) social science—so much so that some scholars observe a veritable »translational turn« (Bachmann-Medick 2007).

Applying this extended understanding of translation to scientific activities reveals translation as a ubiquitous as well as delicate issue that affects empirical research as much as theory building. It uncovers the often overlooked fact that »data collection« is not the neutral and objective activity the term suggests but involves manifold translation procedures that transform life experiences of actual persons into »scientific« terminology, reports, and figures. Translation is not a technical process but entails interpretation, selection, reformulation in new (cultural) context. If carried out with a low level of self-reflexivity it is prone to produce subjugations under its own priorities, needs, and value judgment—even in the case of matter-of-fact scientific reports. In his analysis of early geographical journeys—Lapérouse's trip in 1787 to the then little-known area of Sakhalin—Dutton (2005) shows that the implicit understanding of what counts as »scientific knowledge« serves as a forceful filter in collecting and translating »indigenous« knowledge of local people. With the aim of gathering reliable data on the area's geography that could be used to enhance topographical maps of the time, Lapérouse recorded accounts of local inhabitants and translated their stories into »scientific reports«. As a result, »lengthy tales of gods and legends were ›translated‹ into more ›rational‹ instrumental accounts« (Dutton 2005: 92). Dutton identifies »extraction, mobility, and an ability to combine with and reconfigure other elements of an existing story« as central elements of »science in action« and points out the close relation of this endeavor with imperialistic aims and methods (ibid. 94). Translation turns out to be both at the core of the scientific activity and the imperialist enterprise.

Discussions of Eurocentrism and orientalism in »Western« translations of non-Western histories, knowledge, and practices are well established and shall not be repeated here. Yet the degree to which translation problems are engrained in analysis of social »realities« still often remains hidden. As Chakrabarty demonstrates, processes of »extraction, mobility and reconfiguration« even apply to studies that are based on allegedly »innocent« sociological concepts, such as the notion of »labor«:

»›Work‹ or ›labor‹ are words deeply implicated in the production of universal sociologies. Labor is one of the key categories in the imagination of capitalism itself. In the same way that we think of capitalism as coming into being in all sorts of contexts, we also imagine the modern category ›work‹ or ›labor‹ as emerging in all kinds of histories. . . . Yet the fact is that the modern word ›labor,‹ as every historian of labor in India would know, translates into a general category a whole host of words and practices with divergent and different associations. What complicates the story further is the fact that in a society such as the Indian, human activity (including what one would, sociologically speaking, regard as labor) is often associated with the presence and agency of gods and spirits in the very process of labor. *Hathiyar puja* or the ›worship of tools‹, for example, is a common and familiar festival in many north Indian factories. How do we—and I mean narrators of the pasts of the subaltern classes in India—handle this problem of the presence of the divine or the supernatural in the history of labor as we render this enchanted world into our disenchanted prose—a rendering required, let us say, in the interest of social justice? And how do we, in doing this, retain the subaltern (in whose activity gods or spirits present themselves) as the subjects of history?« (Chakrabarty 2008: 76–77)

While the concept of »labor« arguably finds closer equivalents in European languages than in the Indian context, there's widespread uneasiness about the prevalence of foreign (English!) terminology even on the European continent. Homogenization of »the West« often obscures the fact that translation problems also occur in inter-European or U.S.-European collaboration. Even if there are corresponding terms for things such as

»childcare«, »aging« or »unemployment«, due to specific cultural and historical developments, social phenomena usually take different forms in different countries. As a result, terminology not only refers to different social practices but also reflects different value systems and features different connotations. Translation that ignores this leads to inappropriate assimilation of foreign practice and meaning to own ideas.

European researchers' misgivings, however, do not concern the ill-translated voices of the »subaltern« (or »sample groups« in general) but the fear that in the face of ever-growing pervasiveness of English local science traditions become marginalized. In their view the need to communicate and publish in English leads to a dominance of Anglo-Saxon concepts that blurs distinctions and impoverishes scientific thinking and theory:

»The use of English as a language of scientific production, or the requirement that anything produced in another language must be translatable into English, may put a brake on scientific creation, and may eventually impoverish it. The example of German sociology is a case in point. Abandoning this tradition, or maintaining it solely as far as its concepts can be translated into English, obliges a rethinking of the tradition itself from a different interpretative perspective.« (Siguan 2001: 68)

Switching to English may distort meaning even if expressions can claim joint ancestry and exist in similar variants in different European languages. Local usage and connotations may still create significant nuances that influence scientific thinking and theory building. The terms *equivalence, equivalence*, and the German *Aequivalenz* may serve as an example. As Snell-Hornby (1986) points out, these terms are, paradoxically, not equivalent. Whereas the English and French words have entered common language use roughly five hundred years ago, the German word has only been in use since the late nineteenth century and has found its way into the humanities and social sciences via the natural sciences and formal logic. It therefore implies a stricter understanding of conditions of »similarity« than its English and French counterparts. It is easy to see how such differences will affect the formulation of scientific theories that build on the concept of equivalence (ibid.).

In international projects the difficulty and sometimes outright impossibility of »accurate« and »true« translation is persistently perceived as an annoying problem. Where literal translation appears unsatisfying, translation strategies involve additional explanations that contextualize the foreign concept or leaving central concepts altogether untranslated (as in the case of Tönnies' notions *Gesellschaft* and *Gemeinschaft*).

Yet, the concentration on »smooth« solutions misses the potential that translation problems also provide for social research. In fact, the frictions and disparities that emerge during attempts of translation serve as empirical data in their own right: it is here that (otherwise) implicit meaning systems become visible and exchange of world views is initiated. Translational space irritates and creates new options and, as Bauman observes, it also transforms the participants: »No act of translation leaves either of the partners intact. Both emerge from the encounter changed, different at the end of act from what they were at its beginning« (Baumann 1999: xiviii in Bachmann-Medick 2007: 253–54). It is this transformative potential that can make the need for translation a productive resource for international research. With respect to the example of the tool-worshiping workers and drawing on Vincente Raphael and Gayatri Spivak, Chakrabarty extends a similar thought:

»An ambiguity must mark the translation of the tool-worshiping jute worker's labor in to the universal category ›labor‹: it must be enough like the secular category ›labor‹ to make sense, yet the presence and plurality of gods and spirits in it must also make it ›enough unlike to shock‹. There remains something of a ›scandal‹—of the shocking—in every translation, and it is only through a relationship of intimacy to both languages that we are aware of the degree of this scandal.« (Chakrabarty 2008: 89)

Translation, we may conclude, requires sensitive, dedicated and creative solutions that are »disturbing« enough to prevent unilateral assimilation and to create curiosity about the »other« meaning system. Translation is thus an important resource of international collaboration, but it does not absolve researchers from the responsibility of acquiring foreign language skills. In the end, »it is only through a relationship of intimacy« with foreign languages that different meaning systems become accessible.

Extending Epistemological Ground

Taking reflections on the difficulties and chances of translation a little further, one is confronted with a considerable host of epistemological questions. As Western approaches and theories are increasingly felt to be inadequate to demands and social reality in non-Western environments, claims for alternative, »indigenous« concepts of knowledge and science have made their appearance in different parts of the world. The reflections on the adequate representation of north Indian workers' may be regarded an example of this search for alternative epistemologies.

While part of this exploration is directed toward rediscovering knowledge traditions that were discontinued on the adaption of Western style social sciences, indigenization movements are generally rooted within the framework of modern science. As Chakrabarty remarks, European thought has become a firmly established part of the intellectual inventory (and manifest politics) in places all over the earth, and has thus become »everybody's heritage«—though one that should be treated with reflective care. He argues, »European thought is at once both indispensible and inadequate in helping us to think through the experiences of political modernity in non-Western nations, and provincializing Europe becomes the task of exploring how this thought—which is now everybody's heritage and which affect us all—may be renewed from the margins« (Chakrabarty 2008: 16).

The task that Chakrabarty calls »provincializing Europe« involves exposing the European roots and biases of theories and concepts. Following Chen (1998) it may also be called »decolonizing« the social sciences and further implies making social sciences (politically) relevant to local social concerns, rediscovering and integrating discontinued knowledge traditions, and linking researchers of non-Western science communities in this endeavor.

This vision is not at all unfamiliar to Europeans who themselves feel an overly strong influence of American social theories. The history of German psychology is an enlightening example of how former research traditions have been abandoned in favor of American approaches in a European country (Danziger 2009). And Danziger's observation demonstrates well the dominant position of the United States vis à vis this and other Western (as compared to East Bloc) countries during the decades following the end of World War II, »In the West, if psychologists outside the United States remained poorly informed about the developments there, they were at risk of suffering some loss of professional status, whereas American psychologists habitually ignored work done elsewhere with complete impunity« (Danziger 2009: 212).

This dominance of U.S. research raised concerns among European scholars even some decades back, and in the case of psychology, attempts to »Europeanize« social psychology go back to the 1970s. It is a case in point, though, that the employed terminology is utterly different: European research communities may deplore the scientific hegemony of the United States, yet they do not consider themselves peripheral, and Europeanization would hardly be considered »indigenization«. European researchers may lament the general disinterest of U.S. researchers in their contributions; this does not, however, prevent them from exhibiting the same pattern of ignorance of research done outside the Western world. An impetus that could be directed at a joint search for better scientific models and theories is thus largely lost to local struggles for academic influence.

Despite some benevolent interest in »indigenous« social sciences the pattern of ignorance is still more or less in place today. As a result, and in stark contrast to scholars from emerging communities who are usually well informed about »Western« science traditions and culture, representatives of the established social science communities often still lack basic knowledge of other communities' potential, goals, and scientific contributions. International collaborations therefore often leave the task of translating foreign realities to Westerners exclusively to partners from emerging communities. Only if they are able to speak English (or less frequently French or German) and only when they are well read in Western literature, and well connected to Western research communities is collaboration at all possible. Rare are the Western social scientists who speak and read Hindi, Arabic, or Chinese. It is obvious that this situation is utterly unsatisfying, and with growing capacity of the emerging communities, less and less tolerated. The fact that the search for alternative epistemologies is mainly initiated by non-Western scholars should not affirm Western indifference. After all, it is mainly the self-contained Western research communities who are called to extend epistemological ground.

Comparative Methodology

The Eurocentrism that is apparent in models and theories is mirrored by the nonchalant assumption that methods and categories that are valid in one's own context will be equally applicable to foreign settings as well. Probably even more telling is the fact that methodological aspects of cross-national research do not receive much attention in mainstream academic education at all. Nor are results of cross-cultural research usually much questioned.

Without entering the complex field of related methodological considerations, I will only hint at the most vexing problems of cross-cultural research: methods and data collection procedures must ensure the equivalence of constructs, operationalization, and measurement. This does not only lead to obvious problems (how to standardize items when the collection of some personal data is illegal in some countries, how to carry out questionnaire research with illiterate subjects) but also to fundamental methodological considerations: if social science concepts are imbued with cultural context the definition of the actual object of comparison and of the referential dimension, the *tertium comparationis* against which comparisons takes place, is not at all trivial (Matthes 1992).

The history of social research is replete with failed cross-cultural research. Early intelligence testing (that consistently »proved« low mental capacities of non-whites and women) is a particularly embarrassing example but more recent cases of ethnocentric comparative research could be quoted as well. The interpretation of cross-culturally gained data against the matrix of one's own values and understandings can probably be considered the norm rather than the exception. As our own research into the practice of transnational social research in Europe shows, researchers are not usually well trained for carrying out international comparative research and often unprepared for methodological problems they encounter (Kuhn/Weidemann 2005; also: Somekh/Pear-son 2002).

Intercultural Communication

The huge variety of backgrounds, skills, and knowledge of researchers is one of the assets of international research collaboration. Yet, while diversity initiates creativity, it also reduces group coherence and requires special efforts of establishing a shared frame of interaction. »Building common ground« is unanimously quoted as the initial and fundamental challenge of establishing international research projects in the EU context (Kuhn/Weidemann 2005; Cornish et al. 2007). Only if partners arrive at a shared understanding of goals, tasks, roles, procedures, and scientific approach can work be carried out jointly. As any researcher who has ever been part of such an endeavor can testify, negotiating a shared understanding is a time consuming process that accompanies projects during their entire life span. Sometimes a shared understanding is never achieved.[1]

Still, the benefits (and sometimes the sheer necessity) of international research collaboration are beyond doubt. In their analysis of experiences gained in five EU-funded research projects, Cornish et al. consider two benefits of collaboration as especially noteworthy: first, the observation that collaboration *enhances reflexivity*. The presence of different perspectives helps to identify cultural perspectives that otherwise

---

1　Somekh and Pearson (2002: 500) provide the example of an EU project that used the method of »action research«. Only after their collaboration started did partners discover their utterly different understanding of this method: While the English partners held the use of control groups to be in conflict with the method, to the Spanish and Greek partners is was a central part of it. As the authors report, this difference was never really overcome during the entire life span of the project.

would go unnoticed. They note that »the presence of a collaborator's different perspective can facilitate reflection on one's own perspective, as it is problematised by the contrast with that of the other« (Cornish et al. 2007: 24); A second potential benefit of collaborative research is *novelty*: collaboration may bring up new questions and new solutions to problems. Creativity and synergy are often quoted as prominent benefits of multicultural groups, but often positive effects of joint work are countered (and often outweighed) by coordination losses.

Whereas the special dynamics of cross-cultural communication have received a lot of attention in management studies, there is a remarkable scarcity of research in the field of international science collaboration. It is easy to anticipate that cultural variation in human cognition and behavior, such as differences of handling conflict and politeness, of direct and indirect communication styles, of handling hierarchies and social relations, also apply to the social dynamics of international research teams. Yet the normative claim that »rational« scientific thinking and acting be unaffected by cultural diversity has prevented a systematic analysis of those values, norms, and principles that underlie scientific action and thinking in different science communities. As a result, international science collaboration handles cultural diversity intuitively at best.

Cultural differences do not necessarily result in miscommunication. Yet research shows that because we are usually not sufficiently aware of our own and foreign cultural standards, misunderstandings and conflicts are frequent in intercultural interactions. These are further aggravated by a tendency for self-serving attributions, stereotypes, and emotional discomfort that may result from the confrontation with behavior that deviates from own cultural standards and expectations.

These general observations also apply to cross-cultural collaboration in the sciences. In international research projects, for example, partners may have different—partly culturally shaped—understandings of

- form, function, and frequency of meetings;
- the importance of schedules and deadlines;
- project members hierarchies;
- monitoring and steering projects (how much control or flexibility is required);
- how to present information (e.g., how often and how much of it, in »entertaining« or »matter of fact«-style)
- how to write scientific reports and articles, and even of
- fundamental understandings of the overall purpose and nature of science.

As the following examples demonstrate, shared membership in the »academic community« ensures common knowledge of the general format of scientific communication (such as conferences, meetings, presentations, reports, journal articles) but does not override local cultural interaction scripts. Misunderstandings frequently occur with respect to the following aspects:

Time management, schedules, and deadlines: Because time is a background variable to scientific activity that does not usually catch our focused attention, differences in handling it are—when they occur—both unexpected and irritating. Cultural differences concern the importance that is attached to punctuality and schedules, of »proper« ways to organize work, or of »appropriate« lengths of meetings, breaks, and presentations. Hall's (2003) differentiation of »monochromic« and »polychromic« ways of handling time explains the discomfort that results from interactions between people of both orientations: because polychronic people are used to flexible schedules and master the art of handling different things simultaneously, monochronic people (who value appointments, punctuality, and aspire to finish one task before taking on the next) appear to them as inflexible, unimaginative, and overly task-oriented. The reverse view perceives polychronic people as badly organized, unreliable, and unprofessional. In pan-European collaborations national stereotypes usually contain some of these aspects, and experienced research teams will agree beforehand whether to have a German partner set up and enforce the agenda (the stereotype being that of the monochronic northern European who will act with utmost precision and rigidity), and whether to have a French-style lunch (i.e. have a two-hour break at a restaurant) or a quicker alternative. Even if partners are aware of cultural differences, styles of handling time are usually deeply engrained and not easily adjusted to different standards. Frictions—especially regarding deadlines—are therefore common.

Hierarchies: In EU-research projects another frequently observed problem results from different views of the project coordinator's role. The project coordinator is the person who is legally responsible for the project vis-à-vis the European Commission, but apart from this function, which sets him/her apart from other project partners, his/her role is not precisely defined. Interpretations of the coordinator's duties and responsibilities therefore vary widely and show distinctly cultural patterns: a »flat hierarchy« approach implies the view that this position has not many implications beyond the legal part. Therefore all project partners are expected to share equal responsibility for carrying out their part of the scientific work. According to this approach, having the coordinator guide, supervise, or control other partners' research would imply doubt about their professional skills. In contrast to this is what may be called a »strong hierarchy« approach: the understanding that the leading role of the coordinator naturally gives him/her the authority and the duty to oversee partners' work progress. The resulting problem is easily anticipated: if the project coordination has a »low hierarchy« orientation, she will confirm the overall project schedule on the first meeting and, after that, rely on everyone's professional self-monitoring. Partners who hold a »strong hierarchy« orientation confirm the schedule and will wait for further instructions—that never come. When the deadline arrives, the »low hierarchy« coordinator will be surprised that not much work has been carried out, or to discover that partners are not happy with the project. When unaware of different hierarchy orientations, the coordinator is likely to regard partners as lazy or unprofessional—while project partners will blame him or her for amateurish project management. After all, the coordinator never gave precise instructions and apparently did not care much about their contributions and well-being. Analogous differences exist with respect to the coordinator's role in conflict-resolution and decision making.

Misunderstandings concerning hierarchies also occur when status cannot be discerned cross-culturally: even in environments that do not stress hierarchies very much, academic life intrinsically builds on them and the initiated will be able to judge personal status from individual behavior, speech, or clothing. Because hierarchy markers differ across cultures, signs will be much more difficult to read for outsiders, and mistakes may occur in settings when first-name use, casual clothing, young age, or poor English language skills are mistaken for low academic status.

Communication styles: Language communities form specific communicative styles that regulate the format of speech acts (such as requests, apologies, complaints, etc.) and form general patterns that have been described by categories, such as »direct« versus »indirect«, »fact-oriented« versus »person-oriented«, or »self-assertive« versus »self-effacing«. Communicative styles also affect argumentative patterns (»deductive« versus »inductive«) and expressions of politeness.[2] Because communica-

---

2   There are a large number of comparative studies on different communication styles. Instructive examples can be found in contributions by Wierzbicka (1996) who compares English and Chinese »cultural scripts« from a linguistic perspective and Hall (1989) who introduced the popular distinction of »high context« and »low context« communication.

tion styles are acquired during the course of enculturation and thus have become part of our »normal« behavior, they are still partly effective when speaking a foreign language. They are certainly visible in international science collaboration even when a lingua franca is used. Examples include different presentation formats (that follow an inductive argumentative pattern according to Chinese and a deductive pattern according to European tradition), or expressions of disagreement (that are voiced in a »direct« manner in some northern European countries, and »indirectly« or altogether nonverbally in many other places). Negative stereotypes of »rude«, »superficial«, or »inscrutable« project partners often rest on different communication styles that are misinterpreted cross-culturally.

Academic writing styles: Different communication styles also pertain to written texts. In fact, differences of academic writing styles are among the best-documented aspects of cross-cultural collaboration (see e.g. the contributions in Duszak 1997). As they are arguably the standard format for reporting academic findings, journal articles have received much attention. Journal articles constitute a particular text genre with a standardized structure and specific features (such as brevity, objectivity, simplicity), which was established alongside the introduction of empirical method in the natural sciences and that—due to the historical context of its inception—closely follows Anglo-Saxon communication standards (Moessner 2006). As linguistic research shows, national variants of academic texts deviate considerably from this standard format. Differences concern the overall structure of the text (linear or allowing for excursus), argumentative styles, contents of abstracts and summaries, or vocabulary (close to everyday language or using abstract terminology).

In contrast to spoken language that exhibits much tolerance for deviating formats, publishers enforce text formats of journal articles quite rigidly. An Anglicization of academic writing styles has therefore been set in many countries despite local discomfort with the »new« style that is sometimes perceived as not precise enough (Duszak 2005).

Different writing styles are also visible in project reports and may create confusion in research as much as in publication projects. The following account of Mack and Loeffler (2003) illustrates well the stereotypes of different European working and writing styles, »The reports of the Greek and Italian partners were handed in late. The British partner provided a very short contribution that resembled a mere list. The French report turned out long and wordy. Only the Swedish partner gave a report that matched the expectations of the German project coordinator« (cf. Mack/Loeffler, p. 406, own translation). This anecdotal description is at the same time a telling example of ethnocentric comparison that evaluates writing styles against German standards.

Toward a New Kind of International Collaboration

What can be learned from the above observations? Drawing on knowledge about the functioning of »normal« social research and the particular challenges of international collaboration, we have listed a host of obstacles that are in the way of productive international discourse. It would seem that the social sciences that are intimately embedded in cultural contexts provide an especially difficult field for solving some the abovementioned methodological and theoretical problems. On the other hand, there also exist supportive factors that in part result from the specific properties of academic culture and partly from social scientists' academic competences in particular. Among these are the following observations:

- Norms of collaboration, free communication, and sharing of information are part of overall academic culture, as is a general »climate of trust« that does not need to be established initial to the cooperation (as in economically motivated partnerships) but can be taken for granted (Laudel 1999).
- Shared disciplinary cultures provide partially shared conceptual frameworks that may serve as joint starting points. Yet international collaborations are at the same particularly effective in overcoming overly narrow disciplinary confines (Kuhn/Weidemann 2005).
- Social scientists are equipped with methodical and theoretical knowledge to analyze social realities, communication, and interaction that may be applied to the scientific cooperation itself. However, experiences show that this does not yet guarantee success as »difficulties arising in the research team may stem from exactly the same kind of conflicts the team is trying to examine« (Thomas 1999: 529). In any case, norms of self-reflexivity serve as a resource in the interaction process.
- Finally, the fact that partners from (semi-) peripheral research communities often possess good English language skills and knowledge of research carried out in the Western world constitutes an extremely important prerequisite of collaboration.

Building on all of the above, we may now draw some conclusions for future paths toward more fruitful forms of international collaboration.

Questions we need to ask include the following: What should be the aims and characteristics of such collaboration? What would be the personal prerequisites and competencies that enable scholars to enter such collaborations? And what implications does all that have for higher education and future research? In the following, I will try to sketch some preliminary outlooks.

Polycentric Research

»If the researcher cannot be ›neutral‹ and if time and space are internal variables in the analysis, then if follows that the task of restructuring the social sciences must be one that results from the interaction of scholars coming from every clime and perspective (and taking into account gender, race, class, and linguistic culture), and that this worldwide interaction be a real one and not a mere formal courtesy masking the imposition of the views of segment of world scientists. It will not be at all easy to organize such worldwide interaction in a meaningful way. It is thus a further obstacle in our path. However, overcoming this obstacle may be the key to overcoming all the others.« (Wallerstein et al. 1996: 76-77)

Calls to »restructure« or »unthink« social sciences are not new (Wallerstein et al. 1996; Wallerstein 2001). Neither are summons for »better« forms of international collaboration or »true internationalization« (see the introductory chapter), summons to » decolonize«, »provincialize« or »renew« social research (Burawoy 2005; Chakrabarty 2008; Chen 1998).

It seems to me that despite different individual programs and focuses some shared concerns can be identified. Among them are (1) the recognition of cultural foundations of science, which includes the questioning of U.S./European theories, methods and findings as »universal«, (2) to discontinue the »nineteenth-century evangelist legacy of comparative studies, which offset the practices of one civilization against the philosophical or normative concerns of another« (Nandy 1998), and (3) the establishment of nonhegemonic academic relationships. As all these concerns aim to overcome the current model of theoretical and practical hegemony of an academic »center« over »peripheral« research communities, I would like to borrow Danziger's term and to summarize this approach as a call for *polycentric social science*.[3] Polycentric social

---

3 Danziger criticizes the homogenized historiography of modern psychology and proposes a »polycentric history« that takes the multiple locations where psychological knowledge is produced seriously and that recognizes »the relationship between the results of psychological knowledge production

science may not be exactly compatible with the above vision (by Wallerstein et al. 1996) of a »real worldwide interaction« that—once achieved—will result in overcoming all other obstacles toward better social science. Non-Western intellectuals rightly distrust such visions of a harmonious unity: Ashis Nandy reminds his readers that the West »has an exceedingly poor capacity to live with strangers. It has to try to either overwhelm or proselyte them« (1998: 143) and concludes that »a conversation of cultures subverts itself when its goal becomes a culturally integrated world, not a pluricultural universe where each culture can hope to live in dignity with its own distinctiveness« (ibid.: 148).

Polycentric social science would have to allow for plurality of viewpoints that are not always capable of being integrated in a harmonious universally valid model. Yet, because it rests on dialogue and on identifying interrelations among experiences and interpretations it would also not be likely to succumb to relativism. In a globally linked world, it would »multiply the directions of knowledge flow« and be able to produce social theories of higher validity, much as Chen describes: »A more collaborative, collective, and comparative intellectual practice may well be the possible mode of knowledge production to multiply the directions of knowledge flow, and to better understand the changing shape of the world, if interventions in the global-local dialectic can be more effectively inserted« (Chen 1998: 4).

Polycentric social science needs to take the task of translation seriously as well as issues of intercultural *Verstehen* and comparison. It would have to develop a sensitivity of cultural foundations of human action that also implies awareness of cultural factors in research collaboration itself. And it would have to overcome the above-mentioned tendency to avoid controversy by keeping discourses apart. While it does require a different general approach to science collaboration, polycentric science is not achieved by simple determination to »be polycentric«. In order to create new modes of international collaboration, researchers need skills and knowledge that are not currently developed by our higher education systems.

---

and the local context for that production« (ibid.: 220). A polycentric history thus focuses on context, on the interrelations among locations, the »migration« of concepts, and the emergence of common understanding. It also »entails an enhanced link between historical reflection and current practice that is likely to reduce the high level of ethnocentrism that disfigures so much of what passes for core psychology« (ibid.: 223).

Skills and Competences Required of Individual Researchers

Taking the above reflections seriously leaves us with the uncomfortable insight that solutions require much effort. I shall leave the important conclusions aside that need to be drawn with respect to international research infrastructures and higher education and focus on the individual scholar only.

Engaging in international collaboration would require the following skills, competences and attitudes:
- *Profound knowledge about foreign research contributions and science traditions, societies and cultural reference frames.* Scholars need to transcend the national framework of disciplines and familiarize themselves with research carried out elsewhere, extending the scope of interest to (other) emerging research communities.
- *Horizontal academic relationships in order to overcome the current center-periphery structure of academic exchange.* This and the above aim partly rest on firsthand overseas experience and international research sojourns, and require scholars from established research communities to undertake study visits to emerging communities.
- *Foreign language skills* are an inevitable prerequisite of such an involvement with different research communities and knowledge traditions. While many social scientists outside the Anglophone world are at least bilingual, competence in non-European languages is still limited. Researchers need to develop both active and passive language competencies in multiple languages. Wallerstein et al. (1996) urge:

»A world in which all social scientists had working control of several major scholarly languages would be a world in which better social science was done. Knowledge of languages opens the mind of the scholar to other ways of organizing knowledge. It might go a considerable distance towards creating a working and fruitful understanding of the unending tensions of the antinomy of universalism and particularism. But multilingualism will only thrive if it becomes organizationally as well as intellectually legitimated: through the real use of multiple languages in pedagogy; through the real use of multiple languages in scientific meetings.« (Wallerstein et al. (1996, 89)

The familiarity with different languages also contributes to a sensitive approach to translation problems and lingua franca communication in international research teams.

- *Knowledge of comparative cross-cultural research methodology and methods* and experience with carrying out cross-cultural studies. Methodological skills are vital to international collaboration, regardless of whether they employ a quantitative or a qualitative approach to empirical research.
- *Intercultural sensitivity and ability of perspective taking*. The awareness of cultural difference, knowledge of specific cultural contexts and behavioral skills in cross-cultural situations are also referred to as intercultural competence. Whether it can be acquired by focused training or rests on firsthand experience with different (epistemic, ethnic, national) cultures is a matter of debate. However, the ability of perspective taking and culture-sensitive interpretation can be considered a fundamental skill for any kind of empirical social research.
- *Awareness of academic power structures*: A sensitivity toward culture should not deflect attention from manifest power relations. Researchers need to be aware of mechanisms that express and produce dominance and inequality.
- *Skills to effectively moderate scientific collaborations*. These include cognitive as well as social skills to integrate different perspectives and approaches.

All of these aspects would contribute to overcoming the Eurocentric perspective that is so characteristic of current social sciences.

Conclusion

Taking a look at the above list of skills required of the international scholar, it becomes apparent that very little of it is part of current curricula. Probably, many of these skills and knowledge can only be acquired in culturally diverse and interdisciplinary environments with much firsthand experience of (international) collaboration and with decided interest to overcome the national outlook of social science degree courses. The internationalization of faculty, student bodies, and curricula is therefore equally important as extended international sojourns. Internationally oriented textbooks and integration of foreign-language literature are part of this mission.[4] It is self-evident that intercultural knowledge and experience does not only benefit the few students who later on engage in in-

---

4 There has been a notable recent increase in (English) textbooks and readers that aim to internationalize debates (in the field of psychology, e.g., Stevens/Wedding 2004, Brock 2009; in the field of cultural studies, e.g., Chen/Huat 2007).

ternational research but all students whose professional and private lives confront them with cultural diversity.

While multinational organizations have chosen to offer their staff intercultural trainings and provide special support to their multinational teams, scientists consider these measures unnecessary. This might of course be true, but qualified judgment would have to rest on empirical research of international science cooperation processes and their outcomes. Nobody has yet estimated process losses in international science projects that result from poor cooperation, nor has—to my knowledge—»scientific excellence« yet been considered to include those social skills and competencies that researchers need and develop when running international research projects that follow a polycentric orientation.

References

Alatas, S.F. (2003): Academic Dependency and the Global Division of Labour in the Social Sciences. In: Current Sociology, 51 (6), 599–613.
Alatas, S.F. (2006): Alternative Discourses in Asian Social Science. Responses to Eurocentrism. New Delhi: Sage.
Ammon, U. (Ed.) (2001): The Dominance of English as a Language of Science: Effects on Other Languages and Language Communities. Berlin, New York: de Gruyter.
Bachmann-Medick, D. (2007): Cultural Turns. Neuorientierungen in den Kulturwissenschaften. Hamburg: Rowohlt.
Becher, T./Trowler, P.R. (2001): Academic Tribes and Territories. Intellectual Enquiry and the Culture of Disciplines. Ballmoor: SRHE and Open University Press.
Brock, A.C. (Ed.) (2009): Internationalizing the History of Psychology. New York, London: New York University Press.
Burawoy, M. (2005): Provincializing the Social Sciences. In: G. Steinmetz (Ed.), The Politics of Method in the Human Sciences. Positivism and Its Epistemological Others (pp. 508-525). Durham, London: Duke University Press.
Carli, A./Calaresu, E. (2006): Die Sprachen der Wissenschaft. Die wissenschaftliche Kommunikation im heutigen Trend zur monokulturellen Einsprachigkeit. In: Sociolinguistica, 20, 22–48.
Chakrabarty, D. (2008): Provincializing Europe: Postcolonial Thought and Historical Difference. Reissue, with a new preface by the author. Princeton: Princeton University Press.

Chen, K.-H. (1998): The Decolonization Question. In: K.-H. Chen (Ed.), Trajectories. Inter-Asia Cultural Studies (pp. 1-53). London, New York: Routledge.

Chen, K.-H./Huat, C.B. (Eds.) (2007): The Inter-Asia Cultural Studies Reader. London, New York: Routledge.

Cornish, F./Zittoun, T./Gillespie, A. (2007): A Cultural Psychological Reflection on Collaborative Research. Conference Essay: ESF Exploratory Workshop on Collaborative Case Studies for a European Cultural Psychology. In: Forum Qualitative Sozialforschung/Forum: Qualitative Social Research, 8 (3), 21.

Danziger, K. (2009): Universalism and Indigenization in the History of Modern Psychology. In: A.C. Brock (Ed.): Internationalizing the History of Psychology (pp. 208-225). New York, London: New York University Press.

Duszak, A. (Ed.) (1997): Culture and Styles of Academic Discourse. Berlin, New York: de Gruyter.

Duszak, A. (2005): Between Styles and Values: An Academic Community in Transition. In: G. Cortese/A. Duszak (Eds.), Identity, Community, Discourse. English in Intercultural Settings (pp. 69-94). Bern: Peter Lang.

Dutton, M. (2005): The Trick of Words: Asian Studies, Translation, and the Problems of Knowledge. In: G. Steinmetz (Ed.): The Politics of Method in the Human Sciences. Positivism and Its Epistemological Others (pp. 89-125). Durham, London: Duke University Press.

Etzkowitz, H. (1993): Redesigning 'Solomon's House': the University and the Internationalization of Science and Business. In: E. Crawford (Ed.), Denationalizing Science (pp. 263-288). The Contexts of International Scientific practice. Dordrecht: Kluwer.

Felt, U./Nowotny, H./Taschwer, K. (1995): Wissenschaftsforschung. Eine Einführung. Frankfurt a.M., New York: Campus.

Hall, E.T. (1989): Beyond Culture. New York: Anchor. [First published 1976]

Hall, E.T. (2003): Monochronic and Polychronic Time. In: L.A. Samovar/R.E. Porter (Eds.), Intercultural Communication. A Reader (pp. 262-268). Belmont: Thomson Wadsworth.

Jarausch, K. (2005): Challenges of Internationalization: Careers in Humanities and Social Sciences. In: K. Rampelmann (Ed.), What Factors Impact the Internationalization of Scholarship in the Humanities and Social Sciences (pp. 32-43). Arbeits- und Diskussionspapier 3/2005. Bonn.

Knorr Cetina, K. (2002): Die Fabrikation von Erkenntnis. Zur Anthropologie der Naturwissenschaft. 2. Aufl. Frankfurt a.M.: Suhrkamp.

Kuhn, M./Weidemann, D. (2005): Reinterpreting Transnationality— European Transnational Socio-Economic Research in Practice. In: M. Kuhn/S.O. Remoe (Eds.): Building the European Research Area: Socio-Economic Research in Practice (pp. 53-84). New York: Peter Lang.

Laudel, G. (1999): Interdisziplinäre Forschungskooperation: Erfolgsbedingungen der Institution »Sonderforschungsbereich«. Berlin: Sigma.

Mack, A./Loeffler, J. (2003): Gemeinsam forschen in Europa: Projektmanagement in europäischen Teams in Forschung und Entwicklung. In: N. Bergemann/A.L.J. Sourisseaux (Eds.): Interkulturelles Management (pp. 399-415). Berlin, Heidelberg, New York: Springer.

Matthes, J. (1992): The Operation Called »Vergleichen«. In: J. Matthes (Ed.): Zwischen den Kulturen? (pp. 57-99). Göttingen: Schwartz (Soziale Welt, Sonderband 8).

Moessner, L. (2006): The Birth of the Experimental Essay. In: M. Gotti (Ed.): Linguistic Insights. Studies in Language and Communication (pp. 59-77). Bern: Peter Lang.

Nandy, A. (1998): A New Cosmopolitanism: Toward a Dialogue of Asian Civilizations. In: K.-H. Chen (Ed.): Trajectories. Inter-Asia Cultural Studies (pp. 142-149). London, New York: Routledge.

Siguan, M. (2001): English and the Language of Science: on the Unity of Language and the Plurality of Languages. In: U. Ammon (Ed.): The Dominance of English as a Language of Science: Effects on other Languages and Language Communities (pp. 59-69). Berlin, New York: de Gruyter.

Somekh, B./Pearson, M. (2002): Intercultural Learning Arising from Pan-European Collaboration: a Community of Practice with a »Hole in the Middle«. In: British Educational Research Journal, 28 (4), 485-502.

Snell-Hornby, M. (1986): Einleitung: Übersetzen, Sprache, Kultur. In: M. Snell-Hornby (Ed.): Übersetzungswissenschaft - eine Neuorientierung: Zur Integrierung von Theorie und Praxis (pp. 9-29). Tübingen: Francke.

Stevens, M.J./Wedding, D. (Eds.) (2004): Handbook of International Psychology. New York.

Thomas, A. (1999): Comparison of Managing Cultural Diversity in German-Chinese Research and Business Cooperation. In: W.J. Lonner/D. Dinnel/D.K. Forgays/S.A. Hayes (Eds.): Merging Past, Present, and Future in Cross-Cultural Psychology (pp. 520-531). Lisse: Swets & Zeitlinger.

Wallerstein, I. (2001): Unthinking Social Science. The Limits of Nineteenth-Century Paradigms. Second Edition with a New Preface. Philadelphia: Temple University Press.

Wallerstein, I./Juma, C./Fox Keller/Kocka, J./Lecourt, D./Mudimbe, V. Y./Mushakoji, K./Prigogine, I./Taylor, P./Trouillot, M.-R. (1996): Open the Social Sciences. Report of the Gulbenkian Commission on the Restructuring of the Social Sciences. Stanford: Stanford University Press.

Weidemann, D. (2007): Wissenschaft und Forschung. In: J. Straub/A. Weidemann/D. Weidemann (Eds.): Handbuch Interkulturelle Kommunikation und Kompetenz. Grundbegriffe—Theorien—Anwendungsfelder (pp. 667-678). Stuttgart, Weimar: Metzler.

Wierzbicka, A. (1996): Contrastive Sociolinguistics and the Theory of »Cultural Scripts«: Chinese vs English. In: M. Hellinger/U. Ammon (Eds.): Contrastive Sociolinguistics (pp. 313-344). Berlin, New York: de Gruyter.

Yitzhaki, M. (1998): The »Language Preference« in Sociology: Measures of »Language Self-Citation«, »Relative Own-Language Preference Indicator«, and »Mutual Use of Languages«. In: Scientometrics, 41(1-2), 243-254.

Facing a Scientific Multiversalism — Dynamics of International Social Science Knowledge Accumulations in the Era of Globalization

MICHAEL KUHN

Introduction: Through Scientific Universalism to the Universalization of Western Parochialism

This chapter reflects on the concept of scientific universalism that was promoted in the post war era, the international collaborations of social science communities under the Western reign in the social sciences, and the challenges of an emerging multipolar science world. It should be emphasized beforehand that this chapter does not try to exhaustively analyze all the new phenomena of social science collaborations that have emerged in the era of globalization. It rather makes some attempts to highlight new phenomena requiring much more substantial research in the future.

Any relativization of knowledge conflicts with its nature, may this be diverging, opposing knowledge or just the spatial limitations of its validity. The concept of postwar scientific universalism, created in the new world order under the reign of U.S. sciences, expresses this need to overcome spatially limited knowledge via the ideal of universally valid knowledge, known as Mertonian norms (Merton 1973/1942). Aiming at the spatially unlimited validity, scientific knowledge that was generated in the context of individual nation states, seeks to liberate knowledge from the politically set borders of its validity and to become international, ideally universal. In fact, the concept of a scientific universalism was the reigning postwar concept of international social sciences: »After 1945, scientific universalism became the unquestionably strongest form of European universalism, virtually uncontested« (Wallerstein 2006: 51). Based on a presupposed definition of what science is methodologically, thus conceptually contradicting the very intention of geographically nonrelativized knowledge, the concept of universalism has resulted in the universalization of the parochial Western model of social sciences, not only of the Western definition of what scientific knowledge is, but in the entire post war science world.

The invitation the concept of a scientific universalism addressed to scientists around the world, according to which »claims to truth are evaluated in terms of universal or impersonal criteria, and not on the basis of race, class, gender, religion, or nationality« (Merton 1973/1942: 270), not only put the inviting science community into the position of hosting the international scientific dialogue of scholars. Addressing the invitation to individual scholars also uncovers the idea that international collaborations were conceptualized as a matter of individuals and excluded the various political environments in which knowledge is created by individual scientists because these were seen as irrelevant to the search for truth. In doing so, however, the existing differences between the scientists, such as their academic working environment or diverse concepts of knowledge or science, were also ignored. Ironically, the need for this invitation results from the previous exclusion of those scientists the very inventors of this invitation now addressed. More important, the postcolonial invitation to join a scientific universalism even built on the acknowledgement of the existence of the variety of scientific social particularisms their colonial predecessors had introduced as nationally separated and conceptually parochial science communities into the world of science. The political borders scientific universalism is facing and aims to overcome are the result of a conceptual parochialism of social sciences constructed in the context of the emerging nation states.

Rather than abolishing parochial viewpoints the postwar concept of scientific universalism invites particularity as the major concept of international social sciences. It invites to join an established scientific enterprise from which nobody should be excluded—or could escape—taking for granted that what had been defined *methodologically as* what counts as scientific knowledge and what not, before and beyond any content of any knowledge, is the nature of scientific knowledge. The invitation to scientific mankind can thus also be phrased from the perspective of the invited: whatever the so-far excluded scientist might be, think, and have, the methodological definition of science does not exclude anybody from what has been beforehand methodologically defined as what constitutes social science knowledge—strictly beyond any personal or social criteria and thus applied to *everybody* in the world. In other words, the methodological definition of what science is from this time on considers this definition of science as the world's definition of science. Everybody *can* join, but because according to this understanding there is no science beyond this particular concept of science, any researcher who wants to be taken seriously *must* join. What can be defined as science—and what not—is beforehand decided by those who invite the rest of the world to join their concept of scientific knowledge and thus establish their knowledge concepts as all-encompassing.

Epistemologically phrased, intended or not, any discussion about the acknowledgement of what can be considered as scientific knowledge beyond discussing its contents results in defining ex ante what constitutes scientific knowledge and is inevitably exclusive. It applies—purposefully or not—the state of art of the local Western concept of scientific knowledge to the world, thus creating methodologically the global hegemony of local Western science concepts over those parts of the world of science they until now had excluded. This is why Wallerstein rightly critiques the postwar concept of a scientific universalism and notes that »nothing is so ethnocentric, so particularist, as the claim of universalism« (Wallerstein 2006: 40)

After the knowledge that was excluded from international discourses via the knowledge definitions of scientific universalism could no longer be ignored by social sciences once they detected their own interests in non Western knowledge and started to explore the world of knowledge in the emerging era of globalization, this knowledge, not suiting to the Western definitions of scientism, received the attribute of indigenous knowledge. International social sciences were again confronted with the results of the exclusive definition power of scientific universalism.

»Indigenous knowledge (IK) presents us with four related challenges. [...] Ontologically, the forms of knowledge it produces are *sui generis*, each being incommensurable with other indigenous knowledges and with Western science, whose universalistic knowledge claims stand in direct contradiction to it. This in turn leads to an epistemological problem with indigenous knowledge, of how those outside its originating culture can assess its knowledge claims, or more fundamentally what meaning those claims have outside the immediate context of their production.« (Resist 2009: 13)

Why Western knowledge is called »science« and does not belong to the category of indigenous knowledge and why Western science, unlike indigenous knowledge, does not have the same constraints as indigenous knowledge, is the result of the tautological power of the exclusive definition of knowledge through the ideas of a methodological scientific universalism.

However, the scientific universalism is not only confronted with the exclusive results of the definition of the Western parochial concept of science as universal science, some Western social sciences know why an international scientific discourse is after all epistemologically impossible. Just as if the multiplicity of cultures and contexts were not the major *substance* of reflections and discourses in an international contexts, arguing with a context, path or cultural dependency of social phenomena (see: Wallerstein 2006: 77) results in proving the entire relativization of knowledge and the impossibility of any international scientific discourse.

»The concept of science that was outside ›culture‹, that was in a sense more important than culture, became the last domain of justifying the legitimacy of the distribution of power in the modern world. Scientism has been the most subtle mode of ideological justification of the powerful.« (Wallerstein 2006: 77)

If knowledge itself is mystified and locked in its mysterious culture, any international discourse is scientifically proved to be impossible. The irony that the Western superior »science« epistemologically proves its superiority through the proof of the impossibility of internationally sharing knowledge is the scientific price Western sciences pay for the assumption that their knowledge represents mankind. The Non-Western academics must decide if they consider this creation of two incomparable types of sciences, proving the impossibility of scientific knowledge beyond the Western concept of scientific knowledge and thus of an international scientific discourse, as an invitation to join such a discourse.

In any case, the result of the methodologically invaded world of science, the domination of Western knowledge concepts, and Western world interpretations resulting from the Western set of categories form the basis, starting point, and incentive for a new period of international collaboration in the era of globalization.

International Scientific Discourse Patterns in the Era of Globalization

The increasing internationalization of the social world, that is, the penetration of social phenomena with elements received from other national societies attaching to them international dimensions has caused the social sciences to incorporate these international aspects into social knowledge as a part of ordinary academic work. As a consequence, in the era of globalization international collaborations were transformed from an occasional part of individual academic activities to a standard routine of academic work. Moreover, international collaborations more and more shift from the traditional ex-post exchange of knowledge created in the context of nation-based societies toward an international collaborative knowledge production about the international nature of social science phenomena.

International academic activities are no longer exceptional events of individual academics but have become an imbedded and normal part of academic social science communities. Their disciplinary category system, reinforced by the fact that international social sciences activities, science policies so far did not pay that much attention to, became the subject of national science and technology policies and the missions they attributed via their funding programs to international scholarly collaborations.

Discourses are the means for the production of knowledge in the social sciences more than in the natural sciences. While humans share the same biology, social science phenomena incorporate differing, opposing, or conflicting social interests. This applies even more to knowledge productions in globalizing societies that need to surpass the conceptual limitation of local, no-longer-encapsulated societies. While the social nature of social phenomena is mainly constructed through the shaping power of social entities, mostly nation states, in the era of the globalization social science phenomena incorporate social elements they either share with other social entities or that other social entities attach to them.

It is the nature of globalizing societies that they not only extend the international penetration of social phenomena but in doing so also increase the needs for scientific reflections encompassing the particularity of a single societal context and thus to increase the validity of social knowledge via extending the geographical range of a discoursive knowledge production beyond the context of individual societies.

The growing normality of international collaborations between science communities framed by the political missions of local science policies and their funding programs, and the successful reign of the scientific universalism resulting in the global dominance of Western world interpretations, builds the ground for new modes of international collaborations of social knowledge in an era reigned by the *Zeitgeist* of globalization. The dominating Western world interpretation in the era of globalization is the interpretation of knowledge as international economic good and the practical transformation of international discourses into the global market of science and scientists (see Kuhn 2006b).

Though, undoubtedly, international collaborations are still a matter of collaborations between individual, independent scholars, the discourse modes in the era of globalization differ paradigmatically from the collaboration modes in the era of postwar ideas of a scientific universalism. The economized and politicized concepts of scientific knowledge introduced via the global domination of Western knowledge not only transform the interpretations of the social world into the perspectives of politics and economies (see below) but have also generated new international scientific discourse patterns.

Personalization of Scientific Judgments

By stating that »the claimant's own personal or social attributes (e.g. race, class, political and/or religious views) are irrelevant to the validity of truth claims« (Merton 1973/1942: 270 ff), Merton and other proponents of the ideas of scientific universalism suggest the principle of the competition of equal discourse participants. Unlike this equality between academics and the irrelevance of their »personal or social attributes« in the competition for new knowledge, the concept of »scientific excellence« introduces explicit criteria for judging the quality of sciences beyond judging the content of knowledge. »The focus shifts to quantitative comparisons, away from judgments of research proper. In lack of a substantive definition of excellence, it has to be understood as a relational term« (Weingart/Schwechheimer 2007: 6). The quantified degree of scientific excellence may then serve as credits of institutional or individual rankings. The application of concepts stemming from management theories is an example of economized knowledge concepts invading not only the rhetoric of social science thinking; they also affect social science practices, including the practices of international collaborations. The mainstream thinking that subordinates social thought under the ideas of a global market transforms the substance of international collaborations from sharing the content of knowledge into knowledge as the sheer measurement material of external measurement criteria. It is not that scholars further on fully replace their scholarly discourses about knowledge with an external evaluation of its quantified ratings. However, the audited excellence of knowledge introduces criteria into international academic discourses they have to take into account in their collaboration activities, if only to make sure that their own credibility does not suffer because of the lack of credibility of the international partners. The externalization of the judgment about scientific knowledge thus results in the personalization of judgments about knowledge holders—in fact, the polar opposite of the ideas of scientific universalism. Rather than focusing on the content of knowledge the research subject becomes the criterion for judging about his/her knowledge. The personalisation of the discourse about knowledge re-invites all those personal aspects into the international discourse that the ideas of an impersonal search for knowledge tried—for good reasons—to avoid. As a result, the international sharing of knowledge has been transformed into a competition among academic subjects about their academic status.

Competition of Researchers

As a consequence of judging about knowledge via criteria external to the content of knowledge, discourses about the content of knowledge are transformed into a parallel competition of the research individuals. International collaborations must consider that collaborations with other academics are not only about topics of joint interests but also about academic status. »Rather, the objective is to introduce competition among researchers and institutions, and, by sanctioning certain products, to create a hierarchy among them. Excellence has to be scarce. At the top is only room for a few, the steeper and narrower the hierarchy the better« (Weingart/Schwechheimer 2007: 6-7). The introduction of hierarchies into international collaborations affects the criteria for the choice of international partners as well as the modes of their discourse.

Selecting international collaboration partners is no longer a matter only of assessing a partner's scientific competence but is also a matter of his or her scientific status. Because many researchers cannot access or read publications in foreign journals, they cannot easily judge the scientific quality of an international partner, and therefore external judgments about scientific ranking become the preferred way of gauging content quality. In this way, competition between the international partners is introduced as a factor in the joint search of shared knowledge. Thus the partners discourse about the sharing of knowledge is accompanied by their opposition as competitors for status, and their different scientific status interferes with the validity of their knowledge. Knowledge is no longer only validated by scientific criteria but via the hierarchical status of the researchers. Other academics are considered as competitors in the search for knowledge, which serves as a means to climb the hierarchies of scholarly status.

The Privatization of Knowledge

The definition of knowledge as an economic good has applied the idea of exclusive possession of knowledge to the social sciences by extending the concept of copyright toward the »intellectual property rights« on knowledge.

While the notion of copyright only meant to secure the acknowledgement of published knowledge, the idea of intellectual property rights applies the objective of patented knowledge to the social science. Patented knowledge in the social sciences and in international discourses about social knowledge obviously conceptually conflicts with the »communism« of knowledge (Merton 1973/1942) as a precondition of international discourses.

The idea of an individual researcher or institution exclusively possessing knowledge not only contradicts the social nature of knowledge. After making scholars in international collaborations to opposing competitors about their status, the privatisation of knowledge replaces the idea that »the substantive findings of science are a product of social collaboration and are assigned to the community« (Merton 1973/1942: 273) with a paranoia about whether an international partner will appropriate another's knowledge. As Merton phrased it, »Secrecy is the antithesis« of communism (ibid.: 277-78). International discourses about the findings of social thoughts according to the Mertonian norms of universalism defined knowledge as »a product of social collaboration [...] assigned to the community« (ibid.: 273) and limited »the scientist's claim to ›his‹ intellectual ›property‹ [...] to that of recognition and esteem« (ibid.). In the era of globalization and under the reign of conceptualizing knowledge as a private commodity this former openness has been transformed into a mystery mongering of knowledge that has made international discourses focus on the recognition of ownership of knowledge rather than the discursive subject of an »organized skepticism« (ibid.: 277-78)

Diplomatic Acknowledgments of Knowledge

Proponents of the idea of »organized skepticism« assert the need to interrogate assumptions behind thoughts in contrast to those who judge knowledge through the personality of the knowledge holder, which transforms discourses in an international context into the opposite of a rigorous skepticism and a critical review of knowledge.

In the era of globalization and privatized knowledge concepts, national science communities are transformed in representing regional or local ethnocentrisms and are expected to contribute to an international collection of parochial world interpretations. Viewing academics from other science communities as representing regional ethocentristic views, results in the exchange of undisputable particularisms rather than any critical review of knowledge. The accumulation of knowledge in international collaborations is therefore often reduced to an agglomeration of non-shareable parochialisms, acknowledging the impossibility of a scientific skepticism as the starting point and the outcome of the discourse among the privatized local knowledge holders.

The International Accumulation of Alienated Knowledge

International social science knowledge accumulations in the era of globalization are confronted with manifold ambiguities and contradictions. Not only that the effects of Western mainstream interpretations of knowledge as an economic good and science as an international knowledge market counteract international scientific discourses among social science scholars; the most important contradiction in the globalizing world of social sciences might be the discrepancy between the universalization of the Western parochial world interpretations and the need for multiple interpretations of the world created by the very same process of globalization. While transforming the reproductions of the societies of the globe economically into market economies, the same equalization of the economic living conditions raises the attention of those very societies to their particular identities interpreting globalization through the perspective of the role *they* play on the globe, constructed via the roots of their own individual histories and their distinctive cultural and political traditions. However, the mainstream interpretations of the globalized world *are* mainly interpretations through the universalized categories of the Western social sciences and their disciplinary agendas, which therefore can not explain other than the Western sciences views on globalisation.

Different from Alatas, who in earlier studies clearly analyzed how subjectivation through knowledge is a matter of content and not of the number of produced papers (see Alatas 2006: 607), Western scientists seemingly never really understood that the categories they created about and from the Western societies are not applicable to societies in other countries in the world. This applies to the world of politics, economics, and societal phenomena, and—last but not least—to science. The price for the negativity of their thoughts, explaining the non-Western world beyond the West as deviations from civilization, the price for the negativity of their comparisons they pay today by the erosion of trust science communities and societies in non Western societies have in Western social knowledge.

What could better prove the lack of understanding by Western sciences of other concepts of knowledge than their belief in the superiority of their concept of scientific rationality. While they believe that their enlightened and secularized rationalism proves the universalism of Western thoughts, they do not realize that the empiricism of their concept of scientific rationalism, which confides the reality to finally judge the validity of thoughts, illustrates how much they secretly sympathize with the local social reality they are supposed to analyze and which their empirical rationalism though only echoes.

The following sections illustrate with some examples how Western social sciences in the era of globalization fail to explain other societies and leave behind a most distorted and alienated international knowledge production—for either side. The international accumulation of knowledge, developing the international state of art in international collaborations based on Western categories and agendas unconsciously applied to societies beyond the Western world, creates some quite distorted knowledge about these non-Western societies, which is neither very enlightening for the Western sciences nor for science communities and the societies in the collaborating countries.

International Missions of Science Communities

Following the Western model, which applies its own parochial interpretations to the world, the major approach to international cooperation in the era of globalization is constructed as the collection of the complementary parochial views in other countries or regions. Academics collaborating internationally thus become academic diplomats and their work represents the contributions of their national science communities to international knowledge production. What they contribute is defined as the international mission national social science communities are attributed by the views Western socials sciences have created about other science communities and their societies. How international missions are handed down to local science community as a definition of their contribution to an international knowledge production can be briefly shown in the case of Turkey.

The major international mission given to the Turkish social sciences is to carry out research on Islam (Kuhn/Okamoto 2008). Applying a preconceived image about the Turkish social sciences, the Western international social science community that is convinced Turkish scholars see as their own scientific interest in the Turkish society, they transform their own scientific views about and scientific interests in Turkey into a mission the Turkish academia wishes contribute to international research. Any Turkish scholar working internationally can report about invitations to, or the rejection from, international projects, based on whether they do or do not contribute what they are supposed to, according to the mission they have been assigned. As interviews with scholars from Turkey show (ibid.), scientific contributions from Turkish academics beyond the topic of Islam are not usually recognized as subjects about which Turkish scholars can contribute. Knowledge beyond this mission is simply ignored, since Western scholars seem to know better that the Turks themselves what Turkish scholars do and do not know.

Ironically, most of the Turkish scholars seek to collaborate internationally in order to get support for their secular scientific work as opposed to the unwelcome religious intervention that they are confronted with in their home environment. The result of making them international experts in Islamic issues forces them to give the issue more scientific visibility than it would gain without the pressure coming mainly from Western scholars.

As a consequence, because of the lack of support from local science policies for international activities, Turkish scholars have the choice between rejecting the international mission they are given and encountering major difficulties working at all internationally or to make a the topic of Islam more prominent in Turkey than they would prefer, since they consider this topic as their real major problem, also to participate in international academic activities. As a result, Western sciences receive the confirmation, either through the collaboration of Turkish scholars about studies of Islam or because they refuse to do so, that Turkish social scientists are international specialists in research on Islam. Whatever they decide, their mission does not give any picture about the real, pressing social science topics that are relevant for societies such as Turkey, as for example the rapid development of the middle class, but confirms the Westerns stereotypes about Turkey and contributes to the international Western mainstream agenda that has not much in common with local concerns.

The case of Turkey is by no means an exotic example. The reigning Western scientific elites attribute with a clear conscience and belief in their own scientific impartiality missions to research communities around the world. These missions rest on the images Western scholars have of international research communities' scientific expertise and spell out the expectations of what these communities can contribute to an international research agenda.

Serving International Social Science Fashions

Internationally, the research agenda of social science represents the Western mainstream agenda created from and for the intellectual needs of Western societies. Since the scientific progress of the science communities around the world is measured with respect to their contributions to social science fashions of this Western mainstream agenda, the participation of academics from non-Western societies in international research activities results in the alienation of major parts of national science communities from their own research priorities. The case of a distorted and alienated science community that serves international scientific fashions rather than local research needs can be illustrated with the case of Jordan.

While it is obvious at first glance that societies in countries like Jordan are facing other problems than the vexations that come with a knowledge-based society, a major European fashion created by the World Bank think tanks (see Kuhn 2006b), or the issue of »postmodernity«, scholars from Jordan can hardly refuse from taking part in a knowledge production that creates knowledge they cannot apply to phenomena requiring serious social science reflections in their local societies. Neither does the reality of the Jordan society know the issues the Western social sciences eagerly discuss internationally, nor do the analytical frameworks reflecting these and other issues, which obviously follow the development of the Western sciences, correspond with the local phenomena they pretend to analyze. The international mainstream agenda thus inevitably appears to non-Western social scientists, who are confronted with their own local agendas they cannot serve, as a theater of always-changing social science fashions of the international social science elite.

»Analytical frameworks and approaches employed in these studies, especially those conducted by Western researchers, have also changed. Studies conducted in the 1960s, 1970s, and 1980s employed the Marxist and functionalist approaches whereas recent anthropological studies tend to adopt theories of postmodernism, especially the theoretical frameworks developed by Michel Foucault and Pierre Bourdieu.« (Al Husban/Na'amneh in this volume)

However, having to care about their academic careers, especially in the context of a privatized academic environment in which funding for research is monopolized by Western funding agencies, Jordanian scholars, as most scholars in developing countries, have—as a matter of academic survival—to join and contribute to the Western social science fashions. As a result, the top academics from science communities, who are—being as a whole a product of international interventions—much better prepared for international collaborations than their colleagues in Western science communities, work on a research agenda that is alienated from the problems their societies are facing, with the consequence that many of them, if they can, finally emigrate, thus weakening the status of the social science community in their countries and contributing to the questioning of social sciences as a relevant voice.

International Thinkers and Academic Data Suppliers

Any production of knowledge implies certain phases, from the design of research to the publication of the research findings. In international collaborations the unity of the knowledge production process becomes the subject of an international division of labor. Alatas has extensively discussed this phenomenon with regard to the impact this has on the local social science communities (Alatas 2003; 2006: 57 ff). The case of Russia can serve as an illustration for a discussion not only of the implications for the participating science communities but also for how this affects international collaborations and the international accumulation of knowledge.

More than for any other science community, Russian social scientists who work internationally need as a matter of survival the acknowledgement of their academic work through the Western academe. The international Western academe is very present in Russia with its funding organizations and »help«—themselves, with the help of Russian scholars.

A topic made fashionable by the Western academia in Russia is research on poverty. Since the Fifth European Framework Programme science policies in Europe have replaced the concept of »poverty« as a research topic by the notion of »social exclusion«. In spite of the fact that European Trade Unions have warned the European political elite that a growing segment of the European employed working class is not able to survive without social benefits, Western funding organizations have detected poverty as a major research issues that they have supported with massive funding, not in Europe but in Russia. (Zavarukhin/Pipiya 2007:7ff). The motivation to carry out research about poverty in Russia can be easily guessed: providing scientific knowledge about poverty and gender in Russia provides knowledge that can be politically exploited.

For the sake of this political academic mission the Western funding organizations and scholars have established a division of labor in international research teams, in which the Russian scholars collect the material for social science thinking about poverty and the Western scholars think and create the theories about what has been found. Russian researchers—as non-Western researchers in many other instances—are made data collectors who provide the empirical material for reflections that are carried out and published by the Western scholars:

»In international projects, during last years, Russia often represents itself as a supplier of initial scientific ›raw material‹ (data of surveys, results of expeditions, new archival materials and so forth), while the production of scientific ›product‹ itself is carried out somewhere in America or Western Europe. […] That turns the domestic SSH into a mechanism of translating knowledge acquired by foreign science into the domestic social practice where researchers are mainly engaged in application of foreign concepts to our social problems without taking into account the cultural frames of the place of their development.« (Zavarukhin/Pipiya 2007: 7)

To make sure that they deliver data suitable to the political aims of this research agenda, the funding organizations specify the methods by which the »raw material« must be collected (Zavarukhin/Pipiya 2007: 7ff). As the chapters in this book about Latin America and Argentine show, this procedure fixing the methods of data collection as the division of labor is not at all restricted to the Russian social science community.

The result of the Russian data digging serving reflections about these data and publications of the research findings in the West, provide contributions to a Western discourse about Russia, and it is certainly a practiced irony that Russian scholars do not get any acknowledgement for their academic work by the international academe. At best, the Russian academics working internationally contribute to the image they were given by their Western partners as reliable data providers. Their exclusion from the international academe is thus not only reinforced, but the international division of labor between the Western and the non-Western world (not just in Russia) creates two categories of internationally working academics: elite thinkers and international data suppliers.

The emphasis on the need to contextualize knowledge obviously does not count for the non-Western academic suppliers of empirical data. Apparently, the knowledge about the context only the local academics have, and which would be so important for understanding the concepts and the context of poverty, is not needed. The epistemological paradigm of the context dependency of knowledge, which epistemologically attributes the exclusive accession to and possessions of knowledge to local academics, counts only if it protects the mystique of the knowledge of the international scientific elite against any question about the concepts of knowledge they apply to the local social phenomena, possibly raised by the theoretically unknowledgeable local academic data suppliers.

Just as important in the context of an international production of social knowledge, however, is that the established division of labor into an elite of theoreticians and academic data suppliers results in the production of internationally alienated knowledge serving the images Western scholars have about the societies around the world they are only exclusively allowed to think about. The fact that the outcomes of such reflections are published in English and are thus not available for major parts of the Russian social science community could be compensated for by improving lingua franca abilities. However, subsuming the reflections about poverty under the economic and political categories of what Western societies define economically and politically as poverty is already a most contested issue within the Western social science community. Disconnecting the conceptualization of poverty from category systems of the Russian social sciences not only produces knowledge about Russia that cannot contribute to the scientific discourse about poverty or gender in Russia but provides knowledge about the society in Russia to an international discourse that might serve the development of Western stereotyped thinking about the Russian society, but does certainly not provide any substantial insights into the Russian reality.

To an international accumulation of knowledge it contributes distorted knowledge while simultaneously excluding the academic data providers from an international discourse about major issues of globalization.

The International Appropriation of Parochial Knowledge

International knowledge produced via the international collaboration of local research communities is knowledge accumulated for developing the state of art of knowledge created in the context of Westerns societies accumulated in and for the Western science communities.

This international knowledge production does not result in international knowledge responding to the needs of different science communities around the world. Rather than producing global knowledge, this international knowledge accumulation reproduces preconceived interpretations of the societies in the world resulting from the research agenda and the categories of Western societies applied to the world. The globalization of interpreting the world's societies as »knowledge-based economies« might serve as an example for the universalization of a parochial world interpretation.

The universalization of a regional parochialism not only creates split and widely disconnected science communities around the world, dividing and disconnecting local science communities into branches that focus on international and local research, which do not very much benefit from each other already because of the constraints of knowledge that is published, evaluated, and rewarded according to Western quality standards and procedures. It also creates an international social science knowledge production to which academics from non-Western science communities either contribute only as academic pariahs who are excluded from the theoretical development of social sciences providing the substance for scientific reflections elsewhere. If they can contribute knowledge, they contribute knowledge to the Western interpretation of the societies and the topics local science communities are attributed via an international mission, to be rendered into the images Western academics have created about them or their societies. This internationalized knowledge accumulation thus constitutes the emergence of an international science community that can be described as a new type of a global academic ivory tower, creating knowledge alienated from either social reality.

Considered on a global scale, international collaborations based on this international division of academic labor are a global appropriation of knowledge toward the Western science communities, not in an economic sense, but in terms of accumulating knowledge from and with the help of the academic tributary communities. This appropriation of knowledge from the non-Western science communities complements their expropriation through the transformation of the international academe into an international knowledge pool for the global competition about academics.

This international appropriation of social science knowledge constructed on the scientific images created by scientific world interpretations and the accordingly constructed mainstream agenda not only creates any knowledge that could help scholars understand non-Western societies. It rather results in the reproduction of stereotyped images about the world, providing the issues for an international research agenda and thus the perpetuation of the universalization of a local parochial worldview.

Western scholars certainly are convinced that they are only contributing knowledge to an international knowledge production they consider as the state of art of social sciences. A breakout from this circular, self-reproducing paradox of a universalized parochial world interpretation can only be achieved by academics who confront this paradoxical knowledge and expose the fraudulent claims to universal knowledge about the multiplicity of global societies, created through this system of international knowledge accumulation, as the representations of categories constructed in the context, and for the benefit, of local Western societies.

The Erosion of the Categorical Universalism

Western social sciences can neither deliver the promises of the models of Western society to the world nor can they provide interpretations for the world. Thus the hegemony of their world interpretations and their concepts of scientific knowledge are eroding, the more the science communities around the world are urged to provide world interpretations serving the local societies and their views on globalisation. The erosion of the scientific credibility of Western universalized social science categories, which no longer serve the multiplicity of world interpretations are indicating the emergence of a new social science universe.

As international academic activities become normal knowledge production practices, local science communities encounter each other and question the categorical domination of the Western thoughts in international debates.

Economized Knowledge

The more the latest Western knowledge concepts intrude into international collaborations and are applied to the world ignorant of the multiplicity as of the peculiarities of social realities, the more they contribute to their alienation and thus to the erosion of their global scientific credibility. The categorical subsumption of the world into the theory of the knowledge-based economy is the most recent example of an application of a Western ideology to various world societies (Kuhn 2006a, b). Since in particular the EU has adopted this ideology for the political interpretation of the totality of the social reality both within the EU and in the world, social science thought throughout the world has been widely reformulated and adjusted to the set of thoughts constituting this ideology. Consequently, the categories that are applied to the world of academia have been adjusted to this economizing thinking that transforms knowledge into a commodity and subsumes science under the category of a market economy. This economization of Western social science categories leads to a view of the world's societies *as if* they were devoted to nothing but serving the global economy, reinforces this categorical alienation.

The ideology of a knowledge-based society and the interpretation of any societal subject and activity through the categories of managerial theories creates already within Westerns societies a number of theoretical zombies. The theoretical economization of the social world and naturalization of the economy might appear plausible in society sectors, which have been transformed into service agencies for global competitiveness, and thus adjusted to the ideal of the »knowledge-based society«, that guides not only the policy agenda of the EU. Thus, Western higher education has been reinterpreted as if education was buying and selling knowledge, though no academic ever bought any knowledge, including the economistic interpretation of the world, which, though created by think tanks of the World Bank, also was not bought, but intellectually generated.

However, societies in which even economies are only partially structured as a market economy, and still employ parallel economic sectors defying the rules of economic growth, would struggle with social sciences, which, for example, consider anybody who has no job as »unemployed«, no matter if the society does not even know employment as the only way to earn one's living conditions—not to mention the theory interpreting the reason why people are jobless as lacking »employabilities«. Millions of people around the world certainly work hard and could not understand this as employment, since the social relations incorporated in that notion do not exist in their societies as a ruling principle of their economies, as much as the economized categories wish to interpret this as »developing towards a knowledge-based economy«.

The economization of perceptions of social realities has in fact become the latest Western ideology in the new millennium. It has replaced the postwar interpretation of a world developing toward civilization by an interpretation of a world developing toward knowledge-based economies and societies.

Yet the application of economized thoughts of a naturalized market economy to the totality of a political, social, and cultural reality not only conflicts with societies that do not obey the rules of market economy. Applied knowledge concepts imply a fundamental utilitarian and functionalistic perspective on any social transforming social science thinking into technocratic social science engineering, which provides an accordingly practical functionalism as the governing interpretation of the societal subjects. Societies, in which such functionalistic social subjects in reality are not constructed as a function serving the wholeness of a society, the functionalistic thoughts incorporated in the economized categories inevitable fail to understand a world in which rather all kind of dys-functionalities are the societal rule. Western categories and their incorporated utilitarian approach to the social world and their theoretical functionalism mirroring the regulatedness of any social phenomena geared toward an economically defined functionalism are not applicable to societies without such ruled and functionalized social entities.

Politicized Knowledge

Politicized categories are even more fundamentally incorporated into Western social thoughts than the more recent economic utilitarianism. Politicized knowledge concepts and categories have penetrated social sciences since their inception due to the origination of social sciences in the context of the emergence of Western nation states and the particularity of politically conceptualized and governed societies. Beyond Europe and the society in the United States, only Japan is similarly constructed, in fact due to an adaptation of the Western policy concept.

The politically constructed and defined social in these societies is constitutive for the emergence of social sciences, the shaping of the social sciences disciplines and their distinct set of categories. Every single category in any social science discipline reveals its origination from this politically constructed social in European societies, mirroring the conceptualizing of a fundamental legalism of the social.

With regard to international collaborations in the social sciences, the universalization of these politicized social science categories have contributed to the parochialization of European social science knowledge, creating the major categorical and methodological challenges for an international collaborative knowledge production.

Dealing with the epistemological challenges of constructing a joint conceptual basis is the normal starting point for any international academic activities. Yet the fact that science policies have imposed much more radical concepts of politicization into social science, interpreting the world of any social from the perspective of politics, undoubtedly creates new hurdles for international collaborations. Directing research toward a scientific service for a multiplicity of national policy agendas and managing the implied necessity of orienting research toward a multiplicity of locally different national policy needs and the political categories on which they are constructed conflicts already on the first glance with the need to construct a shared categorical basis in international collaborations.

Much more subtle and thus scientifically much more influential and, if not discovered, much more challenging in international discourses is the fact that Western social science categories imply fundamental concepts originating from the politicized interpretation of a legalized social, not only of the sphere of politics, but of the social as a whole. Western social sciences do not know any category that is not constructed via the idealization of a legalized social, in which societal diversifications appears a being extinguished by regulated and harmonized categories. Already the very categorical fundament of *the* social science, the sociology, constituting sociology as a distinctive discipline, the social, represents a generalization about a diversity of subjects constructed as the entity of equalized subjects, just as the notion of social *subjects* implies their nondiscriminability as a their social identity, both categories providing the existence of a reality of such abstractions. The subjects of sociology have thus created all kinds of abstractions constituting the subjects of sociological thoughts: the citizen, the taxpayer, the student, the worker, children, unemployed, adults, employers, politicians, and pupils, all categories of subjects constructed and only understandable through the legalism of their politicized definitions. Answering the question of what a child is beyond its legally constructed political definition is as impossible as understanding all other categories. All these idealized abstractions from the real distinctiveness of individual subjects do exist in the reality of Western societies but are all unthinkable without their real existence.

More than other societies in the world, European societies are societies in which any social phenomena are the product of politically established legalized definitions (see Therborn 2002). There is no category in European social sciences that does not contain the legally defined idealization of any social distinctions, discrepancies, or conflicts. Even the category of a conflict contains in its European connotations its political definition and regulations for the legally ruled harmonization of its execution.

Applying this categorical idealism of a legally idealized social toward any social in societies who do not share the political history and the construction principles of European societies causes epistemological confusions on either side and fundamentally fails to enlighten us about how the realm of the social is constructed in other societies. Most obviously, in societies around the world that are confronted with most fundamental and radical changes in the context of globalization, the European idealism of the social as the harmonized interplay of lawful regulated entities must fail to explain societies in which nothing is what it is in Europe. In the era of globalization, the idealization of the legalism of European social science categories contrasts with disruptive dynamics of societies in an effort to accommodate their societies and their social thoughts to a so far very Western model of globalization.

The paradox in international social science collaborations therefore is: the more European academics proselytize non-European science communities while these societies compare themselves with European society models, the more politicized European categories lose their scientific credibility.

Approaches toward Internationalization in a Multipolar Social Science Universe

The more international collaborations become the normality not only for Western science communities but for all science communities on the world the more it becomes obvious that Western society models do not only not deliver the promises made to the world last but not least via their scientific voices. Like any other knowledge universalized Western thoughts inevitably reveal their categorical roots in Western societies, the more they are applied to the non Western societies around the world. The superior universalized Western knowledge more and more turns out to be a version of the very local knowledge they elsewhere discredit as traditional or indigenous knowledge.

Once no longer only exported to and adopted by the world but seriously challenged to comparatively interpret the different societies on the world, the explanation that the world's societies, politics, and cultures are just variations of the same path toward civilization and the categories from which this explanation is constructed lose their plausibility. Though it is true that the economies of all those societies around the world have been indeed adjusted to the production modes of a globalized capitalism, their social reality has and cannot not be dehistoricized.

It is obvious from the above considerations that Western scientific concepts no longer serve the needs of a multipolar science world. World interpretations exclusively seen through the Western categories, selectively published according to Western knowledge needs, and discussed along Western discourse patterns obviously conflict with the multiplicity of science approaches of the multiplicity of societies in the era of globalization.

The universalized Western categories not only fail in many global societies to interpret the world from their perspective. Applying knowledge paradigms and the Western agenda to social realities they are unable to interpret or explain inevitable urges the creation of alternative science approaches opposing the established categorial universalism based on westerns knowledge paradigms, on Western practices of social science knowledge productions as on the society missions academics are given in the old scientific centers of a globalized world.

More than the conservative Western science communities, who mainly defend their hegemonic position, emerging social science communities increasingly voice the need for collaborative efforts to identify truly universal sciences.

Following the prominent work of Edward Said (1978), the ethnocentrism of Western theories and research has been amply exposed. In addition, the search for the so-called »indigenous« science approaches has highlighted the fact that international cooperation is not only confronted with local variation of universally valid phenomena but has to take local varieties into account that do not exist in »Western« societies and that are thus not represented by »Western« conceptual repertoires. Whereas in the past universality was claimed for theories and findings of »Western« origin, emerging social science communities increasingly voice the need for collaborative efforts to identify true universals that take local phenomena into account.

Taking such an approach seriously implies the willingness to accept the different knowledge cultures not only on the surface-level of different disciplinary and national research practices but also with respect to their epistemological foundations. »Therefore, the recognition and understanding of different knowledge cultures is an increasingly salient norm in the world of research« (Weiler 2006: 10). Not only resistance to the perceived »imposition« of Western approaches, theories, and notion of science is growing. Science communities have already voiced a desire to explore alternative concepts of knowledge and science as they have of alternative scientific discourse and accumulation cultures. No doubt, international social sciences are changing from the unipolar Western universe toward a multipolar science world. The idea of scientific universalism is no longer defined by the exclusive definition power of one social science community but has become itself the subject of multiple interpretations and practices.

Defining what internationalization means for different social science communities is more a matter of political decisions academics make about the way they see their role in international collaborations rather than a scientific discourse about what internationalization is. Whether these definitions will come to play a role depends of the scientific power science communities have or stand to gain in their approach to give internationalization a relevant voice. Which concept of internationality, questioning the reign of the unipolar definition of what science and of what internationality is, will gain a voice in the international sciences, is a question the future will show.

The following section tries to describe some different interpretations of internationalization in a new multipolar science world, more precisely of how local social science communities interpret an international discourse and their role in this while creating a new social science world. The attempt to identify different approaches toward an international discourse is therefore not more than a starting point from which future research could start.

Needless to say, this sketch of trends about the ways science communities position themselves in international collaborations, and how they define what an international discourse is, says nothing about the position individual scientists might have. It does also not say anything about the knowledge contents science communities contribute to an international discourse. The latter issue is after all more important for the progress of knowledge than the way scientific communities interpret international scientific collaborations.

However, the way different social science communities conceptualize international discourse sets the ground for new discourse patterns in a multipolar science world. In the following paragraphs four approaches will be discussed. Given the fact of the still-existing Western predominance, alternative concepts of internationalization are inevitably conceptualized as a response to the reigning Western definition of universalism.

Responding to the Western Parochial Universalism with Alternative Regional Parochialisms

The most powerful new players in the international scientific power relations are mainly the old ones. Within the Western science world the European science community mainly challenges as a whole the leading scientific power of the United States with an alternative Eurocentric worldview. At the same time, the European science community consists of powerful national science communities, which individually don't necessarily play the same role in an international context and compete with each other about the extent to which they can use the scientific power of the EU as a whole for their individual power ambitions. Thus the scientifically influential science communities gathered in the EU act dually in international scientific power affairs: they jointly appear under the umbrella of the EU science and technology policies as a player in the international competition between global regions, mainly challenging the global reign of the U.S. sciences, and they also are concerned about the new scientific world powers in India, China, and Brazil and the future global scientific world hegemony, trying to use the up-and-coming powers in alliances against the United States. Based on the ground provided by the joint EU science policies, the scientifically influential individual science communities such as Germany, France, and the United Kingdom act individually in bilateral contacts with science communities according to their individual national priorities. Within these power games the same individual science communities that collaborate under the umbrella of the EU in competition with the global competitors compete with each other about their individual scientific international influence on other international science communities. In this role of individual global competitors the German and French science communities consider themselves more as joint victims than as promoters of the internationalization of what they consider as an »Anglo-Saxon« science tradition, fending off the dominance of British social sciences within the EU and elsewhere.

Both for the European science community as a whole a well as for the individual European science communities, strengthening both alternative Eurocentric or national parochial worldviews, stressing the European traditions of humanism which »provide the resources that may allow for a meaningful dialogue and mutual interactions on a global scale«. (Global SSH 2009: 5) as well as the particular national traditions of their science communities, is the main driving force and approach positioning the European science communities in international collaborations in the context of a new distribution of scientific world power.

In East Asia, unlike the Korean science community, both the Chinese and the Japanese social science communities seem to be caught by strong local nationalisms, prioritizing the representation of their search and preservation of their national identities as their social science communities' contributions to an international discourse. Unlike their previous approach, which mainly imported knowledge from the West, Japan is now making more efforts politically to support the export of social science knowledge toward an international science community by translating Japanese books into English (see Okamoto in this volume). As the Chinese, the Japanese approach to internationalization therefore seems to be to respond to the domination of the Western parochial world interpretation with an alternative local parochial worldview. This Chinese and the Japanese approach responding to the domination of a Western world interpretation with competing local parochial world views also seems to be the major problem for creating a pan-Asian social science community.

Chinese social sciences seem to be mainly concerned about losing the Chinese identity in a globalized world and act rather defensively trying to add their contrasting worldview to the mainly untouched Western dominating worldview with their own alternative Chinese parochial world interpretation. Similar to Europeans, the prominence attributed to the Chinese cultural traditions in international relations indicates that Chinese social sciences also construct their regional parochial world interpretations on the traditional cultural values rather than on alternative social science theories addressing the different interpretation of the reality of globalization, thus sharing a conceptual conservatism with the European social sciences of competing parochial world interpretations.

Joining the Western Universalism

A second strategy of internationalization consists in a conscious attempt to disconnect from local science traditions in order to join Western mainstream research. The case of Russia is a most instructive example:

After the Soviet empire switched itself off and transformed Russia and the other former Soviet countries into capitalistic economies, the Soviet social science approach seems to have been deleted overnight from academic minds not only in Russia but in major parts of the Western social science world. The extinction was so abrupt and radical that the only way to describe the role the contemporary Russian social science internationally plays today is to compare them with the role they played inside and outside Russia not even a generation ago.

Internationally, historical materialism was the most influential social science approach, certainly even more influential than the Western science models. Extinguished is the fact, that the whole of East Asia, especially China, even though the Chinese had some ideological differences with the Russians, major parts of intellectuals in both parts of Korea, the social sciences in Japan; most intellectuals in Southeast Asia and in India; in Africa; and the whole Latin America; all of whom interpreted the world via the theories of Marxism and Leninism. What is preferably forgotten in Europe is that also major parts of the European social science academe, namely in France, Italy, and Spain, less in Germany and the United Kingdom, not only widely shared the systemic Marxist critique of the Western society model but also adopted Marxism as an alternative mainstream approach in European social sciences.

The fact that Marxism was abolished by a collective act of forgetting and not by serious scientific criticism might have had more influence on the role Western social science internationally play today than Western science has itself. Whatever the scientific substance of the Soviet interpretation of Marxism was, the price Western social sciences pay for the collective act of ignoring an substantive alternative interpretation of the world can be seen in the fact that since then the social sciences have no longer created great epochal social theories any more, although globalization, the global reign of the Western society model inaugurated a new historical era. Instead, confronted with the rise of new social science communities produced by globalization, the reactivation of the scientific traditions of European humanism is the main creative idea Western social sciences recall.

To not forget: discussing the ways the Russian social sciences position themselves toward international collaborations, one should mention that the other price paid for the collective ignorance, the extinction of creative thinking, is the price the social science community in Russia pays more than any other science community. Suspiciously observed as always by the West and massively pushed by Westerns funding agencies which benefit from the lacking support from the Russian policies, the Russian academe tries to prove that social scientists in Russia never were anything other than believers in European humanism. Critiquing creative thinking seems to be the preferable proof for belonging to European humanism. Joining the Western-dominated discourses is therefore the best way to characterize this approach to internationalization.

Opposing a Western Model of Internationality

In many countries in the Islamic world, the social sciences have been strongly driven by the international Western-dominated agenda, which has created an uncomfortable tension between national realities and the adequacy of methodologies to tackle the issues at stake, in particular when dealing with the role of religion and political analysis. Mainly Islam science communities meanwhile entirely refuse to accept the Western knowledge concepts any longer, and create »Islamic« or »Hindu« social sciences based on their indigenous religious-cultural backgrounds, incorporating an explicit opposition to Western knowledge paradigms.

A similar shift seems to have affected research communities in Africa where SSH world order has been confronted by segmented local science communities, which are scattered in different communities and, additionally, divided between those who defend a vision of »catching-up« and propose a deeper inclusion inside the new world order, and those who, at least rhetorically, reject collaborations with Western-dominated communities and defend a refuge in indigenist and nativist alternatives.

Different from these communities, Latin American scholars, who have a longer history of defending »Latin-American« thinking, combine local knowledge with radical rereadings of Western scholarship to challenge Western political and intellectual hegemony. Unlike the Muslim and African opposition against Western scientific hegemony, Latin American scholars are however not so much advocating a radical opposition against a dialogue with the Western science communities.

Opening Internationalization toward a Scientific Multiversalism

The era of globalization also gave birth to yet another internationalization approach that does not complement the universalization of the Western parochialism with a local parochialism or opposes any international collaboration.

Being themselves a product of an approach of internationalisation that had colonized social thoughts any alternative modes counter-colonizing social science thinking with the approach of the colonizers is not the lessons some social sciences communities learned from their history.

Unlike social sciences in Japan, which imported Western social sciences but adjusted it to the mission they were politically given (see Okamoto in this volume), mainly social sciences in former colonies contribute an approach to international scientific collaboration trying to be neither parochial or exclusive, including the science communities that colonized them.

Following the response on the Western parochialism already Said insisted that »the answer on Orientalism is not Occidentalism« (Said 2003: 328) Maybe more than social scientists in Latin America, social scientists in Africa, the Middle East, and countries such as Turkey and Korea, in which intellectuals did not only have to come to an arrangement with the imposed colonial knowledge, but in which the multiplicity of perspectives has become the culture of the sciences, do not consider international collaborations as the competition of parochial world interpretations but as an opportunity to absorb different perspectives toward a discourse about the multiplicity of worldviews.

Not only the policy agendas but the political systems as a whole in these countries are important to the creation of an intellectual creativity, because they are the subject of continuous public controversies, unlike the politically unified noncontroversial Western societal discourses. This political ferment in non-Western countries may better prepare their social sciences for a multiple scientific universalism than the monocultural Western sciences, which are still locked in their cultural heritage of humanism.

»In this sense, the multiplicity of scientific worldviews, if we may use such a term, is part and parcel of every scientific tradition in that the findings of a particular scientist or in a particular field of science are interpreted in a variety of ways that may or may not agree with other interpretations. In fact, this was the case in traditional societies where we have always multiple cosmologies both across and within specific traditions.« (Kalin, no date: 4)

Facing a multiplicity of scientific worldviews seems to be seen not as a problem but an inspiring perspective in the scientific tradition in countries in which social science communities are not obliged to preserve the heritage of a monocultural history and—last but not least—less inclined to and captured by representing any international political mission in international scientific collaborations. Science communities considering the multiplicity of worldviews as their scientific tradition seem to promote much more than others an approach to internationalization that invites multiple concepts of science and a discourse about the multiplicity of world interpretations. Thus, in fact it seems that the former colonized sciences are much better positioned and prepared for a scientific multiversalism than the science communities trying to conserve their parochial worldviews in a globalizing world or those who feel the need to develop an alternative parochialism.

Facing a Scientific Multiversalism

While the reconfiguration of space and power through globalization necessitates multipolar interpretations of the world created through the social sciences, globalization itself challenges the unipolar paradigms of knowledge and science as a shared scientific platform through which the world has been interpreted so far. Not only the views on the world need to be reconfigured, our scientific means to reconfigure our views on the world, the social sciences, and humanities, and especially their international ways to collaboratively create knowledge about globalization are the subject of a fundamental reconfiguration process in the era of globalization.

The reconfiguration of social reality through globalization necessitates the understanding of the peculiar social and cultural prerequisites of social thought; it also requires an international discourse that allows for conflicting interpretations of globalization and of the emerging new science world. If the social sciences and humanities are to be truly universal they must become open to a plurality of cultural realities and schemes of interpretation. In this process of an emerging multipolar science universe social sciences and humanities will become radically reformulated and transformed through multiple dialogues and interactions among the individuals, groups, and institutions that generate and ultimately create a new social science universe. This creation of global social sciences inevitably will have to go through a phase of a multiple scientific universalism, a scientific multiversalism, encountering all paradox incorporated in the epistemological contradiction of a pluralism of universalism.

Whether this results in new globally shared concepts of social science mainly depends on the epistemological abilities allowing all parties to accept their different interpretations of science as a subject of their discourse instead of a precondition to participate in it.

Sharing knowledge about different concepts of science that all claim to be by their nature universally valid can only be the result of a discourse just as in any other social science topic. In an international context the epistemological paradox of a multiple scientific universalism cannot avoid paying tribute to the paradoxical reality of the universalization of one parochial worldview. The multiversalization of social sciences therefore can only accept the multiple interpretations of science as the starting point of a discourse, or the discourse remains in the paradoxical reality of an exclusive definition of knowledge before the discourse, starting the process of sharing with its presupposed results.

Instead, the concept of a scientific multiversalism turns the old postcolonial ideal of a scientific universalism upside down: While the latter mirrors the idealization of a science world methodologically unified by the application of presupposed concept of scientific knowledge imposed on the world, a bottom-up constructed universalism starts from reflections about how different and opposing concepts of social knowledge could proceed toward universal discourses about the globalizing social and its opposing or conflicting interpretations.

Presupposing any ex ante universalism of knowledge inevitably results in parochial particularisms. Whether a multiversalism of social sciences results in universally shared knowledge is the result of the discourse or it is not. In other words: the extent to which such discourses arrive at any universally valid knowledge depends on nothing else but the extent to which such knowledge is shared—as it the case for any knowledge, social or natural, due to the social nature of any knowledge.

The most important contribution Western social sciences could make to globally shared knowledge might be to give up their missionary view on the world and just listen to what others know.

References

Alatas, S.F. (2003): Academic Dependency and the Global Division of Labour in the Social Sciences. Current Sociology, 51 (6), 599-613.

Alatas, S.F. (2006): Alternative Discourses in Asian Social Science. Responses to Eurocentrism. New Delhi: Sage.

Global SSH (2009): Rethinking the Social and Human Sciences, Towards a Framework for Creativity in Global Context, Policy recommendations for funding and support of research in the social sciences and humanities, http://www.globalsocialscience.org. [Date of last access: 14.10.09]

Kalin, I. (no date): Islam and Science: Notes on an Ongoing Debate, www.muslimphilosophy.com. [Date of last access: 14.10.09]

Kuhn, M. (2006a): The »Learning Economy«—The Domestication of Knowledge and Learning for Global Competition. In: M. Kuhn/M. Tomassini/P.R.-J. Simons (Eds.), Towards a Knowledge Based Economy (pp. 19-56). New York: Peter Lang

Kuhn, M. (Ed.) (2006b): New Society Models for a New Millennium: The Learning Society in Europe and Beyond. New York: Peter Lang.

Kuhn, M./Okamoto, K. (2008): Perspectives for Future Research Collaborations between the EU and Turkey in the Social Sciences, http://www.knowwhy.net/publications.html. [Date of last access: 14.10.09]

Kuhn, M./Okamoto, K. (2009): Through International Collaborations Towards a New Multipolar SSH World Order, http://www.knowwhy.net/publications.html. [Date of last access: 14.10.09]

Kuhn, M./Remoe, S. (2005) (Eds.): Building the European Research Area, Socio-Economic Research in Practice. New York: Peter Lang

Merton, R.K. (1973/1942): The Normative Structure of Science. In N.W. Storer (Ed.), The Sociology of Science (pp. 267-278). Chicago: University of Chicago Press.

Resist (2009): Researching Inequality Through Science and Technology, Final Report. http://www.resist-research.net. [Date of last access: 14.10.09]

Said, E.W. (2003/1978): Orientalism. 25th anniversary edition with a new preface by the author. New York: Vintage

Therborn, G., (2002): The World´s Trader, the World's Lawyer: Europe and Global Processes. Paper presented at the EURONET workshop, Naples November 2002.

Visvanathan, S./Kraak, A. (1999): Western Science, Power and the Marginalisation of indigenous modes of knowledge production, Interpretative minutes of the discussion held on »Debates about Knowledge: Developing Country Perspectives« co-hosted by CHET and CSD, Wednesday 7 April 1999. http://chet.org.za. [Date of last access: 14.10.09]

Wallerstein, I. (2001): Unthinking Social Science. The Limits of Nineteenth-Century Paradigms. Second Edition with a New Preface. Philadelphia: Temple University Press.

Wallerstein, I. (2006): European Universalism: The Rhetoric of Power. New York: The New Press.

Wallerstein, I./Juma, C./Fox Keller/Kocka, J./Lecourt, D./Mudimbe, V. Y./Mushakoji, K./Prigogine, I./Taylor, P./Trouillot, M.-R. (1996): Open the Social Sciences. Report of the Gulbenkian Commission on the Restructuring of the Social Sciences. Stanford: Stanford University Press.

Weiler, H.N. (2006) Values, Ethics, and Research: Commerce, Politics, Integrity, and Culture, OECD/IMHE General Conference 2006, http://www.oecd.org. [Date of last access: 14.10.09]

Weingart, P./Schwechheimer, H. (2007): Conceptualizing and Measuring excellence in the social sciences and humanities, http://www.globalsocialscience.org. [Date of last access: 14.10.09]

Zavarukhin, V./Pipiya, L. (2007) Institutional landscape and research policy for SSH in Russia. Unpublished Report of WP3 in the Global SSH project.

Authors

Abdel Hakim K. Al Husban is associate professor in sociocultural anthropology at Yarmouk University, Jordan. His research interests include: political anthropology, tribe and state, urban anthropology, space and place, and identity in the Middle East. hakimhusban@hotmail.com

I Ketut Ardhana is head of Southeast Asian Studies Division, Research Center for Regional Resources, the Indonesian Institute of Sciences (PSDR-LIPI), Jakarta, Indonesia, and Professor of Asian History, Faculty of Letters at Udayana University, Bali, Indonesia.

Sencer Ayata is a professor in Middle East Technical University's Department of Sociology at Ankara, Turkey. He is also the dean of the Graduate School of Social Sciences and the director of the Confucius Institute in the same university. His areas of interests are urban sociology, urban communities, religious communities, political sociology, poverty, migration, and globalization. ayata@metu.edu.tr

Pradeep Chakkarath is a cultural psychologist and senior researcher at the Department of Social Anthropology and Social Psychology at the Ruhr University Bochum, Germany. His main research interests include culture and human development, the history and methodology of the social sciences, and indigenous psychologies. Pradeep.Chakkarath@rub.de

Aykan Erdemir is an assistant professor in the Department of Sociology and the deputy dean of the Graduate School of Social Sciences. Dr. Erdemir is the METU director of the German-Turkish Masters Program in Social Sciences. His research interests include social and political anthropology, the anthropology of Europe and the Middle East. aerdemir@metu.edu.tr

Sang-Jin Han is professor of sociology at Seoul National University and specializes in critical social theory, comparative study of democratic transformations, intercultural dialogue in terms of human rights and communitarian well-being, the middle class politics and civil society, and third way development. hansjin@snu.ac.kr

He Huang is assistant research fellow at the Institute of American Studies, Chinese Academy of Social Sciences. Political sociology, sociology of sciences, sociology of religion and civil society are among his interests. huanghe6@gmail.com

Jacques Kabbanji is professor of sociology and epistemology at the Lebanese University. He is also President of the Lebanese Association of Sociology. jacqueskabbanji@yahoo.ca

Michael Kuhn is director of KNOWWHY GLOBAL RESEARCH. His background is philosophy, political science and international economics. His research interests include epistemological and organisational implications of internationalizing social sciences communities. michaelkuhn@knowwhy.net

Olga Mamonova, M.Phil. in Sociology, PhD. in sociology, is a senior scientist in the Department of Sociology and executive secretary of the scientific and socio-cultural journal POISK at Moscow Institute of Socio-Cultural Programs, Department of Culture, Government of Moscow, Russia. Sociology and social ecology are among her specialties as well as socio-cultural issues and sociology of culture. foxie@inbox.ru

Yekti Maunati is a Researcher and Head of Research Center for Regional Resources at the Indonesian Institute of Sciences (PSDR-LIPI), Jakarta, Indonesia. yekti.maunati@lipi.go.id

Johann Mouton is professor of sociology and director of the Centre for Research on Science and Technology at Stellenbosch University, South Africa. His areas of specialisation and interest are the philosophy and sociology of science, research methodology and science policy studies. jm6@sun.ac.za

AUTHORS

Mahmoud Na'amneh is assistant professor in socio-cultural anthropology at Yarmouk University, Jordan. Identity, collective memory, and globalization in the Middle East are among his specialties.

Kazumi Okamoto is director of KNOWWHY GLOBAL RESEARCH. Her research interests focus on the involvement of social sciences communities and Higher Education in East Asian countries in the context of internationalization. okamoto@knowwhy.net

Renato Janine Ribeiro is professor of ethics and political philosophy at the University of Sao Paulo, Brazil. Political theory since Macchiavelli and especially Thomas Hobbes, democratic and republican conceptualization, representation at the crossroads of political theory and of the spectacle society are among his specialties, as well as evaluation of graduate studies programs and issues of higher education. rjanine@usp.br

Kwang-Yeong Shin is professor of sociology at Chung-Ang University Korea. His area of interests includes social stratification and inequality, comparative political economy, social theory, and political sociology. Recently he has been doing research on rising inequality in East Asia from a comparative perspective. kyshin@cau.ac.kr

Irina Sosunova, PhD. in sociology, Ph.D. in philosophy is professor and vice-rector at the International Independent University of Environmental and Political Sciences of Moscow, Russia. Her research interests include social science, sociology and social ecology. sosunova@mnepu.ru

Larissa G. Titarenko is professor of sociology at Belarusian State University in Minsk, Republic of Belarus. Social theory and history of sociology, sociology of culture and youth, gender and religion are among her research areas, as well as all issues related to post-communist transformation. larisa166@mail.ru

Hebe Vessuri is a social anthropologist at the Department of Science Studies of the Venezuelan Institute of Scientific Research (IVIC), Caracas. Dr. Vessuri's general interests are on the sociology and contemporary history of science in Latin America, science policy, and sociology of technology. She is also interested in the challenges and dilemmas of expertise and democracy in developing country contexts. hvessuri@ivic.ve

Doris Weidemann is a psychologist and professor of intercultural training with focus on the Greater China area at the University of Applied Sciences of Zwickau, Germany. Her research interests include cultural psychology, challenges of transnational research collaboration, and cross-cultural communication. doris.weidemann@fh-zwickau.de

Tomás Várnagy is professor of philosophy of law at Universidad Nacional de La Matanza, assistant professor of political theory at the University of Buenos Aires and director of the Masters Program at the National Defense School of Argentina. varnagy@hotmail.com

Igor Yegorov is a Department Head of the Centre for S&T Potential and Science History Studies (STEPS Centre), National Academy of Sciences of Ukraine and a principal researcher in the Institute of Statistics in Kiev, Ukraine. His main areas of interest are science and technology policy, economics and statistics of R&D and innovation, social problems of science. Igor.Yegorov@nas.gov.ua